Mastering Linux Device Driver Development

Write custom device drivers to support computer peripherals in Linux operating systems

John Madieu

BIRMINGHAM—MUMBAI

Mastering Linux Device Driver Development

Copyright © 2020 Packt Publishing

Commissioning Editor: Vijin Boricha

Senior Editor: Shazeen Iqbal

Content Development Editor: Ronn Kurien

Technical Editor: Sarvesh Jaywant

Copy Editor: Safis Editing

Project Coordinator: Neil Dmello

Proofreader: Safis Editing

Indexer: Rekha Nair

Production Designer: Alishon Mendonsa

First published: January 2021

Production reference: 1111220

Published by Packt Publishing Ltd.

Livery Place

35 Livery Street

Birmingham

B3 2PB, UK.

ISBN 978-1-78934-204-8

`www.packt.com`

`Packt.com`

Subscribe to our online digital library for full access to over 7,000 books and videos, as well as industry leading tools to help you plan your personal development and advance your career. For more information, please visit our website.

Why subscribe?

- Spend less time learning and more time coding with practical eBooks and Videos from over 4,000 industry professionals

- Improve your learning with Skill Plans built especially for you

- Get a free eBook or video every month

- Fully searchable for easy access to vital information

- Copy and paste, print, and bookmark content

Did you know that Packt offers eBook versions of every book published, with PDF and ePub files available? You can upgrade to the eBook version at `packt.com` and as a print book customer, you are entitled to a discount on the eBook copy. Get in touch with us at `customercare@packtpub.com` for more details.

At `www.packt.com`, you can also read a collection of free technical articles, sign up for a range of free newsletters, and receive exclusive discounts and offers on Packt books and eBooks.

Contributors

About the author

John Madieu is an embedded Linux and kernel engineer living in Paris, France. His main activities consist of developing device drivers and **Board Support Packages** (**BSPs**) for companies in domains such as IoT, automation, transport, healthcare, energy, and the military. John is the founder and chief consultant at LABCSMART, a company that provides training and services for embedded Linux and Linux kernel engineering. He is an open source and embedded systems enthusiast, convinced that it is only by sharing knowledge that we can learn more. He is passionate about boxing, which he practiced for 6 years professionally, and continues to channel this passion through training sessions that he provides voluntarily.

I would like to thank my other half, Claudia ATK, for her sleepless nights and the positive vibes she shared with me throughout the writing of this book, as well as Brigitte and François, my parents, for whom I have a lot of love and to whom I dedicate this book entirely.

I also thank Gebin George from Packt, without whom this book release would have been canceled, without forgetting Ronn Kurien and Suzanne Coutinho.

Finally, I would like to thank Cyprien Pacôme Nguefack, Stephane Capgras, and Gilberto Alberto, the mentors who inspire me in my daily life. I do not forget Sacker Ngoufack, my intern, Ramez Zagmo, for their services, and others who I could not name here.

About the reviewers

Salahaldeen Altous is an electrical engineer who holds a master's degree from TU Dortmund, Germany. He has spent almost 10 years working with embedded Linux through the complete software stack, from the user space down to the kernel space and firmware. Outside of work, he is an Arabic calligrapher and a cook.

I'd like to thank Almighty Allah first. Also, I'd like to thank my parents, my wife, Shatha, and our two children, Yahay and Shahd, for their continued support and patience.

Khem Raj holds a bachelor's degree (Hons) in electronics and communications engineering. In his career spanning 20 years in software systems, he has worked with organizations ranging from start-ups to Fortune 500 companies. During this time, he has worked on developing operating systems, compilers, computer programming languages, scalable build systems, and system software development and optimization. He is passionate about open source and is a prolific open source contributor, maintaining popular open source projects such as the Yocto Project. He is a frequent speaker at open source conferences. He is an avid reader and a lifelong learner.

I would like to thank Packt for giving me the opportunity to review this book. Above all, I to would like to thank my wife, Sweta, and our children, Himangi and Vihaan, for always being there for me and supporting me in all my endeavors.

Packt is searching for authors like you

If you're interested in becoming an author for Packt, please visit `authors.packtpub.com` and apply today. We have worked with thousands of developers and tech professionals, just like you, to help them share their insight with the global tech community. You can make a general application, apply for a specific hot topic that we are recruiting an author for, or submit your own idea.

Table of Contents

3
Delving into the MFD Subsystem and Syscon API

4
Storming the Common Clock Framework

Section 2: Multimedia and Power Saving in Embedded Linux Systems

5
ALSA SoC Framework – Leveraging Codec and Platform Class Drivers

6

ALSA SoC Framework – Delving into the Machine Class Drivers

7

Demystifying V4L2 and Video Capture Device Drivers

8

Integrating with V4L2 Async and Media Controller Frameworks

9
Leveraging the V4L2 API from the User Space

10
Linux Kernel Power Management

Section 3: Staying Up to Date with Other Linux Kernel Subsystems

11
Writing PCI Device Drivers

12
Leveraging the NVMEM Framework

13
Watchdog Device Drivers

14

Linux Kernel Debugging Tips and Best Practices

Other Books You May Enjoy

Index

Preface

Linux is one of the fastest-growing operating systems in the world, and in the last few years, the Linux kernel has evolved significantly to support a wide variety of embedded devices, with its improved subsystems and a lot of new features.

Mastering Linux Device Driver Development provides complete coverage of kernel topics such as video and audio frameworks that usually go unaddressed. You'll delve into some of the most complex and impactful Linux kernel frameworks, such as PCI, ALSA for SoC, and Video4Linux2, gaining expert tips and best practices along the way. In addition to this, you'll learn how to leverage frameworks such as NVMEM and Watchdog. Once the book has got you started with Linux kernel helpers, you'll gradually progress to working with special device types such as **Multi-Function Devices** (**MFDs**), followed by video and audio device drivers.

By the end of this book, you'll be able to write rock-solid device drivers and integrate them with some of the most complex Linux kernel frameworks, including V4L2 and ALSA SoC.

Who this book is for

This book is essentially intended for embedded enthusiasts and developers, Linux system administrators, and kernel hackers. Whether you are a software developer, a system architect, or a maker (electronics hobbyist), looking to dive into Linux driver development, this book is for you.

What this book covers

Chapter 1, *Linux Kernel Concepts for Embedded Developers*, walks through the Linux kernel helpers for locking, blocking I/O, deferring work, and interrupt management.

Chapter 2, *Leveraging the Regmap API and Simplifying the Code*, provides an overview of the Regmap framework and shows you how to leverage its APIs to ease interrupt management and abstract register access.

Chapter 3, Delving into the MFD Subsystem and Syscon API, focuses on MFD drivers in the Linux kernel, their APIs, and their structures, as well as introducing the syscon and simple-mfd helpers.

Chapter 4, Storming the Common Clock Framework, explains the Linux kernel clock framework and explores both producer and consumer device drivers, as well as their device tree bindings.

Chapter 5, ALSA SoC Framework – Leveraging Codec and Platform Class Drivers, discusses ALSA driver development for both codec and platform devices and introduces concepts such as kcontrol and **digital audio power management** (**DAPM**).

Chapter 6, ALSA SoC Framework – Delving into the Machine Class Drivers, dives into ALSA machine class driver development and shows you how to bind both codec and platform together and how to define audio routes.

Chapter 7, Demystifying V4L2 and Video Capture Device Drivers, describes V4L2's key concepts. It focuses on bridge video devices, introduces the concept of subdevices, and covers their respective device drivers.

Chapter 8, Integrating with V4L2 Async and Media Controller Frameworks, introduces the concept of asynchronous probing so that you don't have to care about bridge and subdevice probing order. Finally, this chapter introduces media controller frameworks in order to provide video routing and video pipe customizations.

Chapter 9, Leveraging V4L2 API from the User Space, closes our teaching series on V4L2 and deals with V4L2 from the user space. It first teaches you how to write C code in order to open, configure, and grab data from a video device. It then shows you how to write as little code as possible by leveraging user-space video-related tools such as v4l2-ctl and media-ctl.

Chapter 10, Linux Kernel Power Management, discusses power management on Linux-based systems and teaches you how to write power-aware device drivers.

Chapter 11, Writing PCI Device Drivers, deals with the PCI subsystem and introduces you to its Linux kernel implementation. This chapter also shows you how to write PCI device drivers.

Chapter 12, Leveraging the NVMEM Framework, describes the Linux **Non-Volatile Memory** (**NVEM**) subsystem. It first teaches you how to write both provider and consumer drivers as well as their device tree bindings. Then, it shows you how to take the most out of the device from user space.

Chapter 13, Watchdog Device Drivers, provides an accurate description of the Linux kernel Watchdog subsystem. It first introduces you to Watchdog device drivers and gradually takes you through the core of the subsystem, introducing some key concepts such as pre-timeout and governors. Toward the end, this chapter teaches you how to manage the subsystem from the user space.

Chapter 14, Linux Kernel Debugging Tips and Best Practices, highlights the most-used Linux kernel debugging and tracing techniques using kernel-embedded tools such as `ftrace` and oops message analysis.

To get the most out of this book

To get the most out of this book, some C and system programming knowledge is required. Moreover, the content of the book assumes that you are familiar with the Linux system and most of its basic commands.

Software/hardware covered in the book	OS requirements
A computer with good network bandwidth and enough disk space and RAM to download and build the Linux kernel	Preferably any Debian-based distribution
Any Cortex-A embedded board on the market (for example, Udoo, Raspberry Pi, and BeagleBone)	Either a Yocto/Buildroot or any vendor-specific OS

Any necessary packages not listed in the preceding table will be described in their respective chapters.

If you are using the digital version of this book, we advise you to type the code yourself. Doing so will help you avoid any potential errors related to the copying and pasting of code.

Download the color images

We also provide a PDF file that has color images of the screenshots/diagrams used in this book. You can download it here: `http://www.packtpub.com/sites/default/files/downloads/9781789342048_ColorImages.pdf`.

Conventions used

There are a number of text conventions used throughout this book.

Code in text: Indicates code words in text, database table names, folder names, filenames, file extensions, pathnames, dummy URLs, user input, and Twitter handles. Here is an example: "The parent IRQ is not requested here using any of the request_irq() family methods because gpiochip_set_chained_irqchip() will invoke irq_set_chained_handler_and_data() under the hood."

A block of code is set as follows:

```
static int fake_probe(struct i2c_client *client,
                      const struct i2c_device_id *id)
{
    [...]
    mutex_init(&data->mutex);
    [...]
}
```

When we wish to draw your attention to a particular part of a code block, the relevant lines or items are set in bold:

```
static int __init my_init(void)
{
    pr_info('Wait queue example\n');
    INIT_WORK(&wrk, work_handler);
    schedule_work(&wrk);
    pr_info('Going to sleep %s\n', __FUNCTION__);
    wait_event_interruptible(my_wq, condition != 0);
    pr_info('woken up by the work job\n');
    return 0;
}
```

Any command-line input or output is written as follows:

```
# echo 1 >/sys/module/printk/parameters/time
# cat /sys/module/printk/parameters/time
```

Bold: Indicates a new term, an important word, or words that you see onscreen. Here is an example: "The **simple-mfd** helper was introduced to handle zero conf/hacks subdevice registering, and **syscon** was introduced for sharing a device's memory region with other devices."

> **Tips or important notes**
> Appear like this.

Get in touch

Feedback from our readers is always welcome.

General feedback: If you have questions about any aspect of this book, mention the book title in the subject of your message and email us at `customercare@packtpub.com`.

Errata: Although we have taken every care to ensure the accuracy of our content, mistakes do happen. If you have found a mistake in this book, we would be grateful if you would report this to us. Please visit `www.packtpub.com/support/errata`, selecting your book, clicking on the Errata Submission Form link, and entering the details.

Piracy: If you come across any illegal copies of our works in any form on the Internet, we would be grateful if you would provide us with the location address or website name. Please contact us at `copyright@packt.com` with a link to the material.

If you are interested in becoming an author: If there is a topic that you have expertise in and you are interested in either writing or contributing to a book, please visit `authors.packtpub.com`.

Reviews

Please leave a review. Once you have read and used this book, why not leave a review on the site that you purchased it from? Potential readers can then see and use your unbiased opinion to make purchase decisions, we at Packt can understand what you think about our products, and our authors can see your feedback on their book. Thank you!

For more information about Packt, please visit `packt.com`.

Section 1: Kernel Core Frameworks for Embedded Device Driver Development

This section deals with the Linux kernel core, introducing the abstraction layers and facilities that the Linux kernel offers, in order to reduce the development efforts. Moreover, in this section we will learn about the Linux clock framework, thanks to which most peripherals on the system are driven.

This section contains the following chapters:

- *Chapter 1, Linux Kernel Concepts for Embedded Developers*
- *Chapter 2, Leveraging the Regmap API and Simplifying the Code*
- *Chapter 3, Delving into the MFD Subsystem and the Syscon API*
- *Chapter 4, Storming the Common Clock Framework*

1
Linux Kernel Concepts for Embedded Developers

As a standalone software, the Linux kernel implements a set of functions that help not to reinvent the wheel and ease device driver developments. The importance of these helpers is that it's not a requirement to use these for code to be accepted upstream. This is the kernel core that drivers rely on. We'll cover the most popular of these core functionalities in this book, though other ones also exist. We will begin by looking at the kernel locking API before discussing how to protect shared objects and avoid race conditions. Then, we will look at various work deferring mechanisms available, where you will learn what part of the code to defer in which execution context. Finally, you will learn how interrupts work and how to design interrupt handlers from within the Linux kernel.

This chapter will cover the following topics:

- The kernel locking API and shared objects
- Work deferring mechanisms
- Linux kernel interrupt management

Let's get started!

Technical requirements

To understand and follow this chapter's content, you will need the following:

- Advanced computer architecture knowledge and C programming skills
- Linux kernel 4.19 sources, available at `https://github.com/torvalds/linux`

The kernel locking API and shared objects

A resource is said to be shared when it can be accessed by several contenders, regardless of their exclusively. When they are exclusive, access must be synchronized so that only the allowed contender(s) may own the resource. Such resources might be memory locations or peripheral devices, while the contenders might be processors, processes, or threads. Operating systems perform mutual exclusion by atomically (that is, by means of an operation that can be interrupted) modifying a variable that holds the current state of the resource, making this visible to all contenders that might access the variable at the same time. This atomicity guarantees that the modification will either be successful, or not successful at all. Nowadays, modern operating systems rely on the hardware (which should allow atomic operations) used for implementing synchronization, though a simple system may ensure atomicity by disabling interrupts (and avoiding scheduling) around the critical code section.

In this section, we'll describe the following two synchronization mechanisms:

- **Locks**: Used for mutual exclusion. When one contender holds the lock, no other contender can hold it (others are excluded). The most well-known locking primitives in the kernel are spinlocks and mutexes.
- **Conditional variables**: Mostly used to sense or wait for a state change. These are implemented differently in the kernel, as we will see later, mainly in the *Waiting, sensing, and blocking in the Linux kernel* section.

When it comes to locking, it is up to the hardware to allow such synchronizations by means of atomic operations. The kernel then uses these to implement locking facilities. Synchronization primitives are data structures that are used for coordinating access to shared resources. Because only one contender can hold the lock (and thus access the shared resource), it might perform an arbitrary operation on the resource associated with the lock that would appear to be atomic to others.

Apart from dealing with the exclusive ownership of a given shared resource, there are situations where it is better to wait for the state of the resource to change; for example, waiting for a list to contain at least one object (its state then passes from empty to not empty) or for a task to complete (a DMA transaction, for example). The Linux kernel does not implement conditional variables. From our user space, we could think of using a conditional variable for both situations, but to achieve the same or even better, the kernel provides the following mechanisms:

- **Wait queue**: Mainly used to wait for a state change. It's designed to work in concert with locks.

- **Completion queue**: Used to wait for a given computation to complete.

Both mechanisms are supported by the Linux kernel and are exposed to drivers thanks to a reduced set of APIs (which significantly ease their use when used by a developer). We will discuss these in the upcoming sections.

Spinlocks

A spinlock is a hardware-based locking primitive. It depends on the capabilities of the hardware at hand to provide atomic operations (such as `test_and_set`, which, in a non-atomic implementation, would result in read, modify, and write operations). Spinlocks are essentially used in an atomic context where sleeping is not allowed or simply not needed (in interrupts, for example, or when you want to disable preemption), but also as an inter-CPU locking primitive.

It is the simplest locking primitive and also the base one. It works as follows:

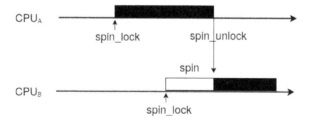

Figure 1.1 – Spinlock contention flow

Let's explore the diagram by looking at the following scenario:

When CPUB, which is running task B, wants to acquire the spinlock thanks to the spinlock's locking function and this spinlock is already held by another CPU (let's say CPUA, running task A, which has already called this spinlock's locking function), then CPUB will simply spin around a while loop, thus blocking task B until the other CPU releases the lock (task A calls the spinlock's release function). This spinning will only happen on multi-core machines, which is why the use case described previously, which involves more than one CPU since it's on a single core machine, cannot happen: the task either holds a spinlock and proceeds or doesn't run until the lock is released. I used to say that a spinlock is a lock held by a CPU, which is the opposite of a mutex (we will discuss this in the next section), which is a lock held by a task. A spinlock operates by disabling the scheduler on the local CPU (that is, the CPU running the task that called the spinlock's locking API). This also means that the task currently running on that CPU cannot be preempted by another task, except for IRQs if they are not disabled (more on this later). In other words, spinlocks protect resources that only one CPU can take/access at a time. This makes spinlocks suitable for SMP safety and for executing atomic tasks.

> **Important note**
>
> Spinlocks are not the only implementation that take advantage of hardware's atomic functions. In the Linux kernel, for example, the preemption status depends on a per-CPU variable that, if equal to 0, means preemption is enabled. However, if it's greater than 0, this means preemption is disabled (`schedule()` becomes inoperative). Thus, disabling preemption (`preempt_disable()`) consists of adding 1 to the current per-CPU variable (`preempt_count` actually), while `preempt_enable()` subtracts 1 (one) from the variable, checks whether the new value is 0, and calls `schedule()`. These addition/subtraction operations should then be atomic, and thus rely on the CPU being able to provide atomic addition/subtraction functions.

There are two ways to create and initialize a spinlock: either statically using the DEFINE_
SPINLOCK macro, which will declare and initialize the spinlock, or dynamically by calling
spin_lock_init() on an uninitialized spinlock.

First, we'll introduce how to use the DEFINE_SPINLOCK macro. To understand how this
works, we must look at the definition of this macro in include/linux/spinlock_
types.h, which is as follows:

```
#define DEFINE_SPINLOCK(x) spinlock_t x = __SPIN_LOCK_
UNLOCKED(x)
```

This can be used as follows:

```
static DEFINE_SPINLOCK(foo_lock)
```

After this, the spinlock will be accessible through its name, foo_lock. Note that its
address would be &foo_lock. However, for dynamic (runtime) allocation, you need to
embed the spinlock into a bigger structure, allocate memory for this structure, and then
call spin_lock_init() on the spinlock element:

```
struct bigger_struct {
    spinlock_t lock;
    unsigned int foo;
    [...]
};
```

```
static struct bigger_struct *fake_alloc_init_function()
{
    struct bigger_struct *bs;
    bs = kmalloc(sizeof(struct bigger_struct), GFP_KERNEL);
    if (!bs)
        return -ENOMEM;
    spin_lock_init(&bs->lock);
    return bs;
}
```

It is better to use DEFINE_SPINLOCK whenever possible. It offers compile-time
initialization and requires less lines of code with no real drawback. At this stage, we
can lock/unlock the spinlock using the spin_lock() and spin_unlock() inline
functions, both of which are defined in include/linux/spinlock.h:

```
void spin_unlock(spinlock_t *lock)
void spin_lock(spinlock_t *lock)
```

That being said, there are some limitations to using spinlocks this way. Though a spinlock prevents preemption on the local CPU, it does not prevent this CPU from being hogged by an interrupt (thus, executing this interrupt's handler). Imagine a situation where the CPU holds a "spinlock" in order to protect a given resource and an interrupt occurs. The CPU will stop its current task and branch out to this interrupt handler. So far, so good. Now, imagine that this IRQ handler needs to acquire this same spinlock (you've probably already guessed that the resource is shared with the interrupt handler). It will infinitely spin in place, trying to acquire a lock that's already been locked by a task that it has preempted. This situation is known as a deadlock.

To address this issue, the Linux kernel provides _irq variant functions for spinlocks, which, in addition to disabling/enabling the preemption, also disable/enable interrupts on the local CPU. These functions are spin_lock_irq() and spin_unlock_irq(), and they are defined as follows:

```
void spin_unlock_irq(spinlock_t *lock);
void spin_lock_irq(spinlock_t *lock);
```

You might think that this solution is sufficient, but it is not. The _irq variant partially solves this problem. Imagine that interrupts are already disabled on the processor before your code starts locking. So, when you call spin_unlock_irq(), you will not just release the lock, but also enable interrupts. However, this will probably happen in an erroneous manner since there is no way for spin_unlock_irq() to know which interrupts were enabled before locking and which weren't.

The following is a short example of this:

1. Let's say interrupts x and y were disabled before a spinlock was acquired, while z was not.

2. spin_lock_irq() will disable the interrupts (x, y, and z are now disabled) and take the lock.

3. spin_unlock_irq() will enable the interrupts. x, y, and z will all be enabled, which was not the case before the lock was acquired. This is where the problem arises.

This makes spin_lock_irq() unsafe when it's called from IRQs that are off-context as its counterpart, spin_unlock_irq(), will naively enable IRQs with the risk of enabling those that were not enabled while spin_lock_irq() was invoked. It only makes sense to use spin_lock_irq() when you know that interrupts are enabled; that is, you are sure nothing else might have disabled interrupts on the local CPU.

Now, imagine that you save the status of your interrupts in a variable before acquiring the lock and restoring them to how they were while they were releasing. In this situation, there would be no more issues. To achieve this, the kernel provides `_irqsave` variant functions. These behave just like the `_irq` ones, while also saving and restoring the interrupts status feature. These functions are `spin_lock_irqsave()` and `spin_lock_irqrestore()`, and they are defined as follows:

```
spin_lock_irqsave(spinlock_t *lock, unsigned long flags)
spin_unlock_irqrestore(spinlock_t *lock, unsigned long flags)
```

> **Important note**
>
> `spin_lock()` and all its variants automatically call `preempt_disable()`, which disables preemption on the local CPU. On the other hand, `spin_unlock()` and its variants call `preempt_enable()`, which try to enable (yes, try! – it depends on whether other spinlocks are locked, which would affect the value of the preemption counter) preemption, and which internally call `schedule()` if enabled (depending on the current value of the counter, which should be 0). `spin_unlock()` is then a preemption point and might reenable preemption.

Disabling interrupts versus only disabling preemption

Though disabling interrupts may prevent kernel preemption (a scheduler's timer interrupts would be disabled), nothing prevents the protected section from invoking the scheduler (the `schedule()` function). Lots of kernel functions indirectly invoke the scheduler, such as those that deal with spinlocks. As a result, even a simple `printk()` function may invoke the scheduler since it deals with the spinlock that protects the kernel message buffer. The kernel disables or enables the scheduler (performs preemption) by increasing or decreasing a kernel-global and per-CPU variable (that defaults to 0, meaning "enabled") called `preempt_count`. When this variable is greater than 0 (which is checked by the `schedule()` function), the scheduler simply returns and does nothing. Every time a spin_lock*-related helper gets invoked, this variable is increased by 1. On the other hand, releasing a spinlock (any `spin_unlock*` family function) decreases it by 1, and whenever it reaches 0, the scheduler is invoked, meaning that your critical section would not be very atomic.

Thus, if your code does not trigger preemption itself, it can only be protected from preemption by disabling interrupts. That being said, code that's locked a spinlock may not sleep as there would be no way to wake it up (remember, timer interrupts and schedulers are disabled on the local CPU).

Now that we are familiar with the spinlock and its subtilities, let's look at the mutex, which is our second locking primitive.

Mutexes

The mutex is the other locking primitive we will discuss in this chapter. It behaves just like the spinlock, with the only difference being that your code can sleep. If you try to lock a mutex that is already held by another task, your task will find itself suspended, and it will only be woken when the mutex is released. There's no spinning this time, which means that the CPU can process something else while your task is waiting. As I mentioned previously, *a spinlock is a lock held by a CPU, while a mutex is a lock held by a task.*

A mutex is a simple data structure that embeds a wait queue (to put contenders to sleep), while a spinlock protects access to this wait queue. The following is what `struct mutex` looks like:

```
struct mutex {
    atomic_long_t owner;
    spinlock_t wait_lock;
#ifdef CONFIG_MUTEX_SPIN_ON_OWNER
    struct optimistic_spin_queue osq; /* Spinner MCS lock */
#endif
    struct list_head wait_list;
[...]
};
```

In the preceding code, the elements that are only used in debugging mode have been removed for the sake of readability. However, as you can see, mutexes are built on top of spinlocks. `owner` represents the process that actually owns (hold) the lock. `wait_list` is the list in which the mutex's contenders are put to sleep. `wait_lock` is the spinlock that protects `wait_list` while contenders are inserted and are put to sleep. This helps keep `wait_list` coherent on SMP systems.

The mutex APIs can be found in the `include/linux/mutex.h` header file. Prior to acquiring and releasing a mutex, it must be initialized. As for other kernel core data structures, there may be a static initialization, as follows:

```
static DEFINE_MUTEX(my_mutex);
```

The following is the definition of the DEFINE_MUTEX() macro:

```
#define DEFINE_MUTEX(mutexname) \
        struct mutex mutexname = __MUTEX_INITIALIZER(mutexname)
```

The second approach the kernel offers is dynamic initialization. This can be done by making a call to the low-level __mutex_init() function, which is actually wrapped by a much more user-friendly macro known as mutex_init():

```
struct fake_data {
    struct i2c_client *client;
    u16 reg_conf;
    struct mutex mutex;
};

static int fake_probe(struct i2c_client *client,
                      const struct i2c_device_id *id)
{
    [...]
    mutex_init(&data->mutex);
    [...]
}
```

Acquiring (also known as locking) a mutex is as simple calling one of the following three functions:

```
void mutex_lock(struct mutex *lock);
int mutex_lock_interruptible(struct mutex *lock);
int mutex_lock_killable(struct mutex *lock);
```

If the mutex is free (unlocked), your task will immediately acquire it without going to sleep. Otherwise, your task will be put to sleep in a manner that depends on the locking function you use. With mutex_lock(), your task will be put in an uninterruptible sleep (TASK_UNINTERRUPTIBLE) while you wait for the mutex to be released (in case it is held by another task). mutex_lock_interruptible() will put your task in an interruptible sleep, in which the sleep can be interrupted by any signal. mutex_lock_killable() will allow your task's sleep to be interrupted, but only by signals that actually kill the task. Both functions return zero if the lock has been acquired successfully. Moreover, interruptible variants return -EINTR when the locking attempt is interrupted by a signal.

Whatever locking function is used, the mutex owner (and only the owner) should release the mutex using `mutex_unlock()`, which is defined as follows:

```
void mutex_unlock(struct mutex *lock);
```

If you wish to check the status of your mutex, you can use `mutex_is_locked()`:

```
static bool mutex_is_locked(struct mutex *lock)
```

This function simply checks whether the mutex owner is NULL and returns true if it is, or false otherwise.

> **Important note**   .
>
> It is only recommended to use `mutex_lock()` when you can guarantee that the mutex will not be held for a long time. Typically, you should use the interruptible variant instead.

There are specific rules when using mutexes. The most important are enumerated in the kernel's mutex API header file, `include/linux/mutex.h`. The following is an excerpt from it:

```
*  - only one task can hold the mutex at a time
*  - only the owner can unlock the mutex
*  - multiple unlocks are not permitted
*  - recursive locking is not permitted
*  - a mutex object must be initialized via the API
*  - a mutex object must not be initialized via memset or
*    copying
*  - task may not exit with mutex held
*  - memory areas where held locks reside must not be freed
*  - held mutexes must not be reinitialized
*  - mutexes may not be used in hardware or software interrupt
*    contexts such as tasklets and timers
```

The full version can be found in the same file.

Now, let's look at some cases where we can avoid putting the mutex to sleep while it is being held. This is known as the try-lock method.

The try-lock method

There are cases where we may need to acquire the lock if it is not already held by another elsewhere. Such methods try to acquire the lock and immediately (without spinning if we are using a spinlock, nor sleeping if we are using a mutex) return a status value. This tells us whether the lock has been successfully locked. They can be used if we do not need to access the data that's being protected by the lock when some other thread is holding the lock.

Both the spinlock and mutex APIs provide a try-lock method. They are called `spin_trylock()` and `mutex_trylock()`, respectively. Both methods return 0 on a failure (the lock is already locked) or 1 on a success (lock acquired). Thus, it makes sense to use these functions along with an statement:

```
int mutex_trylock(struct mutex *lock)
```

`spin_trylock()` actually targets spinlocks. It will lock the spinlock if it is not already locked in the same way that the `spin_lock()` method is. However, it immediately returns 0 without spinning if the spinlock is already locked:

```
static DEFINE_SPINLOCK(foo_lock);
[...]
static void foo(void)
{
[...]
    if (!spin_trylock(&foo_lock)) {
        /* Failure! the spinlock is already locked */
        [...]
        return;
    }
    /*
     * reaching this part of the code means
       that the
     * spinlock has been successfully locked
     */
[...]
    spin_unlock(&foo_lock);
[...]
}
```

On the other hand, `mutex_trylock()` targets mutexes. It will lock the mutex if it is not already locked in the same way that the `mutex_lock()` method is. However, it immediately returns 0 without sleeping if the mutex is already locked. The following is an example of this:

```
static DEFINE_MUTEX(bar_mutex);
[...]
static void bar (void)
{
[...]
    if (!mutex_trylock(&bar_mutex))
        /* Failure! the mutex is already locked */
        [...]
        return;
    }
    /*
     * reaching this part of the code means that the mutex has
     * been successfully locked
     */
[...]
    mutex_unlock(&bar_mutex);
[...]
}
```

In the preceding code, the try-lock is being used along with the `if` statement so that the driver can adapt its behavior.

Waiting, sensing, and blocking in the Linux kernel

This section could have been named *kernel sleeping mechanism* as the mechanisms we will deal with involve putting the processes involved to sleep. A device driver, during its life cycle, can initiate completely separate tasks, some of which depend on the completion of others. The Linux kernel addresses such dependencies with `struct completion` items. On the other hand, it may be necessary to wait for a particular condition to become true or the state of an object to change. This time, the kernel provides work queues to address this situation.

Waiting for completion or a state change

You may not necessarily be waiting exclusively for a resource, but for the state of a given object (shared or not) to change or for a task to complete. In kernel programming practices, it is common to initiate an activity outside the current thread, and then wait for that activity to complete. Completion is a good alternative to `sleep()` when you're waiting for a buffer to be used, for example. It is suitable for sensing data, as is the case with DMA transfers. Working with completions requires including the `<linux/completion.h>` header. Its structure looks as follows:

```
struct completion {
    unsigned int done;
    wait_queue_head_t wait;
};
```

You can create instances of the struct completion structure either statically using the static `DECLARE_COMPLETION(my_comp)` function or dynamically by wrapping the completion structure into a dynamic (allocated on the heap, which will be alive for the lifetime of the function/driver) data structure and invoking `init_completion(&dynamic_object->my_comp)`. When the device driver performs some work (a DMA transaction, for example) and others (threads, for example) need to be notified of their completion, the waiter has to call `wait_for_completion()` on the previously initialized struct completion object in order to be notified of this:

```
void wait_for_completion(struct completion *comp);
```

When the other part of the code has decided that the work has been completed (the transaction has been completed, in the case of DMA), it can wake up anyone (the code that needs to access the DMA buffer) who is waiting by either calling `complete()`, which will only wake one waiting process, or `complete_all()`, which will wake everyone waiting for this to complete:

```
void complete(struct completion *comp);
void complete_all(struct completion *comp);
```

A typical usage scenario is as follows (this excerpt has been taken from the kernel documentation):

```
CPU#1                                              CPU#2

struct completion setup_done;
init_completion(&setup_done);
initialize_work(...,&setup_done,...);

/* run non-dependent code */          /* do some setup */
[...]                                  [...]
wait_for_completion(&setup_done);      complete(setup_done);
```

The order in which `wait_for_completion()` and `complete()` are called does not matter. As semaphores, the completions API is designed so that they will work properly, even if `complete()` is called before `wait_for_completion()`. In such a case, the waiter will simply continue immediately once all the dependencies have been satisfied.

Note that `wait_for_completion()` will invoke `spin_lock_irq()` and `spin_unlock_irq()`, which, according to the *Spinlocks* section, are not recommended to be used from within an interrupt handler or with disabled IRQs. This is because it would result in spurious interrupts being enabled, which are hard to detect. Additionally, by default, `wait_for_completion()` marks the task as uninterruptible (`TASK_UNINTERRUPTIBLE`), making it unresponsive to any external signal (even kill). This may block for a long time, depending on the nature of the activity it's waiting for.

You may need the *wait* not to be done in an uninterruptible state, or at least you may need the *wait* being able to be interrupted either by any signal or only by signals that kill the process. The kernel provides the following APIs:

- `wait_for_completion_interruptible()`
- `wait_for_completion_interruptible_timeout()`
- `wait_for_completion_killable()`
- `wait_for_completion_killable_timeout()`

_killable variants will mark the task as TASK_KILLABLE, thus only making it responsive to signals that actually kill it, while _interruptible variants mark the task as TASK_INTERRUPTIBLE, allowing it to be interrupted by any signal. _timeout variants will, at most, wait for the specified timeout:

```
int wait_for_completion_interruptible(struct completion *done)
long wait_for_completion_interruptible_timeout(
         struct completion *done, unsigned long timeout)
long wait_for_completion_killable(struct completion *done)
long wait_for_completion_killable_timeout(
         struct completion *done, unsigned long timeout)
```

Since wait_for_completion*() may sleep, it can only be used in this process context. Because the interruptible, killable, or timeout variant may return before the underlying job has run until completion, their return values should be checked carefully so that you can adopt the right behavior. The killable and interruptible variants return -ERESTARTSYS if they're interrupted and 0 if they've been completed. On the other hand, the timeout variants will return -ERESTARTSYS if they're interrupted, 0 if they've timed out, and the number of jiffies (at least 1) left until timeout if they've completed before timeout. Please refer to kernel/sched/completion.c in the kernel source for more on this, as well as more functions that will not be covered in this book.

On the other hand, complete() and complete_all() never sleep and internally call spin_lock_irqsave()/spin_unlock_irqrestore(), making completion signaling, from an IRQ context, completely safe.

Linux kernel wait queues

Wait queues are high-level mechanisms that are used to process block I/O, wait for particular conditions to be true, wait for a given event to occur, or to sense data or resource availability. To understand how they work, let's have a look at the structure in include/linux/wait.h:

```
struct wait_queue_head {
    spinlock_t lock;
    struct list_head head;
};
```

A wait queue is nothing but a list (in which processes are put to sleep so that they can be awaken if some conditions are met) where there's a spinlock to protect access to this list. You can use a wait queue when more than one process wants to sleep and you're waiting for one or more events to occur so that it can be woke up. The head member is the list of processes waiting for the event(s). Each process that wants to sleep while waiting for the event to occur puts itself in this list before going to sleep. When a process is in the list, it is called a wait queue entry. When the event occurs, one or more processes on the list are woken up and moved off the list. We can declare and initialize a wait queue in two ways. First, we can declare and initialize it statically using DECLARE_WAIT_QUEUE_HEAD, as follows:

```
DECLARE_WAIT_QUEUE_HEAD(my_event);
```

We can also do this dynamically using init_waitqueue_head():

```
wait_queue_head_t my_event;
init_waitqueue_head(&my_event);
```

Any process that wants to sleep while waiting for my_event to occur can invoke either wait_event_interruptible() or wait_event(). Most of the time, the event is just the fact that a resource has become available. Thus, it only makes sense for a process to go to sleep after the availability of that resource has been checked. To make things easy for you, these functions both take an expression in place of the second argument so that the process is only put to sleep if the expression evaluates to false:

```
wait_event(&my_event, (event_occurred == 1) );
/* or */
wait_event_interruptible(&my_event, (event_occurred == 1) );
```

wait_event() and wait_event_interruptible() simply evaluate the condition when it's called. If the condition is false, the process is put into either a TASK_UNINTERRUPTIBLE or a TASK_INTERRUPTIBLE (for the _interruptible variant) state and removed from the running queue.

There may be cases where you need the condition to not only be true, but to time out after waiting a certain amount of time. You can address such cases using wait_event_timeout(), whose prototype is as follows:

```
wait_event_timeout(wq_head, condition, timeout)
```

This function has two behaviors, depending on the timeout having elapsed or not:

1. `timeout` has elapsed: The function returns 0 if the condition is evaluated to false or 1 if it is evaluated to true.

2. `timeout` has not elapsed yet: The function returns the remaining time (in jiffies – must at least be 1) if the condition is evaluated to true.

The time unit for the timeout is `jiffies`. So that you don't have to bother with seconds to `jiffies` conversion, you should use the `msecs_to_jiffies()` and `usecs_to_jiffies()` helpers, which convert milliseconds or microseconds into jiffies, respectively:

```
unsigned long msecs_to_jiffies(const unsigned int m)
unsigned long usecs_to_jiffies(const unsigned int u)
```

After a change has been made to any variable that could mangle the result of the wait condition, you must call the appropriate `wake_up*` family function. That being said, in order to wake up a process sleeping on a wait queue, you should call either `wake_up()`, `wake_up_all()`, `wake_up_interruptible()`, or `wake_up_interruptible_all()`. Whenever you call any of these functions, the condition is reevaluated. If the condition is true at this time, then a process (or all the processes for the `_all()` variant) in `wait queue` will be awakened, and its (their) state will be set to `TASK_RUNNING`; otherwise (the condition is false), nothing will happen:

```
/* wakes up only one process from the wait queue. */
wake_up(&my_event);


/* wakes up all the processes on the wait queue. */
wake_up_all(&my_event);:


/* wakes up only one process from the wait queue that is in
 * interruptible sleep.
 */
wake_up_interruptible(&my_event)


/* wakes up all the processes from the wait queue that
 * are in interruptible sleep.
 */
wake_up_interruptible_all(&my_event);
```

Since they can be interrupted by signals, you should check the return values of _
interruptible variants. A non-zero value means your sleep has been interrupted by
some sort of signal, so the driver should return ERESTARTSYS:

```
#include <linux/module.h>
#include <linux/init.h>
#include <linux/sched.h>
#include <linux/time.h>
#include <linux/delay.h>
#include<linux/workqueue.h>


static DECLARE_WAIT_QUEUE_HEAD(my_wq);
static int condition = 0;
/* declare a work queue*/
static struct work_struct wrk;


static void work_handler(struct work_struct *work)
{
    pr_info("Waitqueue module handler %s\n", __FUNCTION__);
    msleep(5000);
    pr_info("Wake up the sleeping module\n");
    condition = 1;
    wake_up_interruptible(&my_wq);
}


static int __init my_init(void)
{
    pr_info("Wait queue example\n");
    INIT_WORK(&wrk, work_handler);
    schedule_work(&wrk);
    pr_info("Going to sleep %s\n", __FUNCTION__);
    wait_event_interruptible(my_wq, condition != 0);
    pr_info("woken up by the work job\n");
    return 0;
}
void my_exit(void)
{
    pr_info("waitqueue example cleanup\n");
```

```
}
module_init(my_init);
module_exit(my_exit);
MODULE_AUTHOR("John Madieu <john.madieu@labcsmart.com>");
MODULE_LICENSE("GPL");
```

In the preceding example, the current process (actually, this is `insmod`) will be put to sleep in the wait queue for 5 seconds and woken up by the work handler. The output of `dmesg` is as follows:

```
[342081.385491]  Wait queue example
[342081.385505]  Going to sleep my_init
[342081.385515]  Waitqueue module handler work_handler
[342086.387017]  Wake up the sleeping module
[342086.387096]  woken up by the work job
[342092.912033]  waitqueue example cleanup
```

You may have noticed that I did not check the return value of `wait_event_interruptible()`. Sometimes (if not most of the time), this can lead to serious issues. The following is a true story: I've had to intervene in a company to fix a bug where killing (or sending a signal to) a user space task was making their kernel module crash the system (panic and reboot – of course, the system was configured so that it rebooted on panic). The reason this happened was because there was a thread in this user process that did an `ioctl()` on the `char` device exposed by their kernel module. This resulted in a call to `wait_event_interruptible()` in the kernel on a given flag, which meant there was some data that needed to be processed in the kernel (the `select()` system call could not be used).

So, what was their mistake? The signal that was sent to the process was making `wait_event_interruptible()` return without the flag being set (which meant data was still not available), and its code was not checking its return value, nor rechecking the flag or performing a sanity check on the data that was supposed to be available. The data was being accessed as if the flag had been set and it actually dereferenced an invalid pointer.

The solution could have been as simple as using the following code:

```
if (wait_event_interruptible(...)){
    pr_info("catching a signal supposed make us crashing\n");
    /* handle this case and do not access data */
    [....]
} else {
```

```
    /* accessing data and processing it */
    [...]
}
```

However, for some reason (historical to their design), we had to make it uninterruptible, which resulted in us using `wait_event()`. However, note that this function puts the process into an exclusive wait (an uninterruptible sleep), which means it can't be interrupted by signals. It should only be used for critical tasks. Interruptible functions are recommended in most situations.

Now that we are familiar with the kernel locking APIs, we will look at various work deferring mechanisms, all of which are heavily used when writing Linux device drivers.

Work deferring mechanisms

Work deferring is a mechanism the Linux kernel offers. It allows you to defer work/a task until the system's workload allows it to run smoothly or after a given time has lapsed. Depending on the type of work, the deferred task can run either in a process context or in an atomic context. It is common to using work deferring to complement the interrupt handler in order to compensate for some of its limitations, some of which are as follows:

- The interrupt handler must be as fast as possible, meaning that only a critical task should be performed in the handler so that the rest can be deferred later when the system is less busy.

- In the interrupt context, we cannot use blocking calls. The sleeping task should be scheduled in the process context.

The deferring work mechanism allows us to perform the minimum possible work in the interrupt handler (the so-called *top-half*, which runs in an interrupt context) and schedule an asynchronous action (the so-called *bottom-half*, which may – but not always – run in a user context) from the interrupt handler so that it can be run at a later time and execute the rest of the operations. Nowadays, the concept of bottom-half is mostly assimilated to deferred work running in a process context, since it was common to schedule work that might sleep (unlike rare work running in an interrupt context, which cannot happen). Linux now has three different implementations of this: **softIRQs**, **tasklets**, and **work queues**. Let's take a look at these:

- **SoftIRQs**: These are executed in an atomic context.

- **Tasklets**: These are also executed in an atomic context.

- **Work queues**: These run in a process context.

We will learn about each of them in detail in the next few sections.

SoftIRQs

As the name suggests, **softIRQ** stands for **software interrupt**. Such a handler can preempt all other tasks on the system except for hardware IRQ handlers, since they are executed with IRQs enabled. SoftIRQs are intended to be used for high frequency threaded job scheduling. Network and block devices are the only two subsystems in the kernel that make direct use of softIRQs. Even though softIRQ handlers run with interrupts enabled, they cannot sleep, and any shared data needs proper locking. The softIRQ API is defined as `kernel/softirq.c` in the kernel source tree, and any drivers that wish to use this API need to include `<linux/interrupt.h>`.

Note that you cannot dynamically register nor destroy softIRQs. They are statically allocated at compile time. Moreover, the usage of softIRQs is restricted to statically compiled kernel code; they cannot be used with dynamically loadable modules. SoftIRQs are represented by `struct softirq_action` structures defined in `<linux/interrupt.h>`, as follows:

```
struct softirq_action {
    void (*action)(struct softirq_action *);
};
```

This structure embeds a pointer to the function to be run when the `softirq` action is raised. Thus, the prototype of your softIRQ handler should look as follows:

```
void softirq_handler(struct softirq_action *h)
```

Running a softIRQ handler results in this action function being executed. It only has one parameter: a pointer to the corresponding `softirq_action` structure. You can register the softIRQ handler at runtime by means of the `open_softirq()` function:

```
void open_softirq(int nr,
                  void (*action)(struct softirq_action *))
```

nr represents the softIRQ's index, which is also considered as the softIRQ's priority (where 0 is the highest). action is a pointer to the softIRQ's handler. Any possible indexes are enumerated in the following enum:

```
enum
{
    HI_SOFTIRQ=0,    /* High-priority tasklets */
```

```
    TIMER_SOFTIRQ,   /* Timers */
    NET_TX_SOFTIRQ,  /* Send network packets */
    NET_RX_SOFTIRQ,  /* Receive network packets */
    BLOCK_SOFTIRQ,   /* Block devices */
    BLOCK_IOPOLL_SOFTIRQ, /* Block devices with I/O polling
                            blocked on other CPUs */
    TASKLET_SOFTIRQ, /* Normal Priority tasklets */
    SCHED_SOFTIRQ,   /* Scheduler */
    HRTIMER_SOFTIRQ, /* High-resolution timers */
    RCU_SOFTIRQ,     /* RCU locking */
    NR_SOFTIRQS      /* This only represent the number or
                      * softirqs type, 10 actually
                      */
};
```

SoftIRQs with lower indexes (highest priority) run before those with higher indexes (lowest priority). The names of all the available softIRQs in the kernel are listed in the following array:

```
const char * const softirq_to_name[NR_SOFTIRQS] = {
    "HI", "TIMER", "NET_TX", "NET_RX", "BLOCK", "BLOCK_IOPOLL",
        "TASKLET", "SCHED", "HRTIMER", "RCU"
};
```

It is easy to check the output of the /proc/softirqs virtual file, as follows:

```
~$ cat /proc/softirqs
                CPU0        CPU1        CPU2        CPU3
       HI:      14026         89         491         104
    TIMER:     862910     817640      816676      808172
   NET_TX:          0          2           1           3
   NET_RX:       1249        860         939        1184
    BLOCK:        130        100         138         145
 IRQ_POLL:          0          0           0           0
  TASKLET:      55947         23         108         188
    SCHED:    1192596     967411      882492      835607
  HRTIMER:          0          0           0           0
      RCU:     314100     302251      304380      298610
~$
```

A NR_SOFTIRQS entry array of `struct softirq_action` is declared in `kernel/softirq.c`:

```
static struct softirq_action softirq_vec[NR_SOFTIRQS] ;
```

Each entry in this array may contain one and only one softIRQ. As a consequence of this, there can be a maximum of NR_SOFTIRQS (10 in v4.19, which is the last version at the time of writing this) for registered softIRQs:

```
void open_softirq(int nr,
                    void (*action)(struct softirq_action *))
{
    softirq_vec[nr].action = action;
}
```

A concrete example of this is the network subsystem, which registers softIRQs that it needs (in `net/core/dev.c`) as follows:

```
open_softirq(NET_TX_SOFTIRQ, net_tx_action);
open_softirq(NET_RX_SOFTIRQ, net_rx_action);
```

Before a registered softIRQ gets a chance to run, it should be activated/scheduled. To do this, you must call `raise_softirq()` or `raise_softirq_irqoff()` (if interrupts are already off):

```
void __raise_softirq_irqoff(unsigned int nr)
void raise_softirq_irqoff(unsigned int nr)
void raise_softirq(unsigned int nr)
```

The first function simply sets the appropriate bit in the per-CPU softIRQ bitmap (the __softirq_pending field in the `struct irq_cpustat_t` data structure, which is allocated per-CPU in `kernel/softirq.c`), as follows:

```
irq_cpustat_t irq_stat[NR_CPUS] ____cacheline_aligned;
EXPORT_SYMBOL(irq_stat);
```

This allows it to run when the flag is checked. This function has been described here for study purposes and should not be used directly.

`raise_softirq_irqoff` needs be called with interrupts disabled. First, it internally calls `__raise_softirq_irqoff()`, as described previously, to activate the softIRQ. Then, it checks whether it has been called from within an interrupt (either hard or soft) context by means of the `in_interrupt()` macro (which simply returns the value of `current_thread_info( )->preempt_count`, where 0 means preemption is enabled. This states that we are not in an interrupt context. A value greater than 0 means we are in an interrupt context). If `in_interrupt() > 0`, this does nothing as we are in an interrupt context. This is because softIRQ flags are checked on the exit path of any I/O IRQ handler (`asm_do_IRQ()` for ARM or `do_IRQ()` for x86 platforms, which makes a call to `irq_exit()`). Here, softIRQs run in an interrupt context. However, if `in_interrupt() == 0`, then `wakeup_softirqd()` gets invoked. This is responsible for waking the local CPU `ksoftirqd` thread up (it schedules it) to ensure the softIRQ runs soon but in a process context this time.

`raise_softirq` first calls `local_irq_save()` (which disables interrupts on the local processor after saving its current interrupt flags). It then calls `raise_softirq_irqoff()`, as described previously, to schedule the softIRQ on the local CPU (remember, this function must be invoked with IRQs disabled on the local CPU). Finally, it calls `local_irq_restore()` to restore the previously saved interrupt flags.

There are a few things to remember about softIRQs:

- A softIRQ can never preempt another softIRQ. Only hardware interrupts can. SoftIRQs are executed at a high priority with scheduler preemption disabled, but with IRQs enabled. This makes softIRQs suitable for the most time-critical and important deferred processing on the system.

- While a handler runs on a CPU, other softIRQs on this CPU are disabled. SoftIRQs can run concurrently, however. While a softIRQ is running, another softIRQ (even the same one) can run on another processor. This is one of the main advantages of softIRQs over hardIRQs, and is the reason why they are used in the networking subsystem, which may require heavy CPU power.

- For locking between softIRQs (or even the same softIRQ as it may be running on a different CPU), you should use `spin_lock()` and `spin_unlock()`.

- SoftIRQs are mostly scheduled in the return paths of hardware interrupt handlers. **SoftIRQs that are scheduled outside of the interrupt context will run in a process context if they are still pending when the local** `ksoftirqd` **thread is given the CPU**. Their execution may be triggered in the following cases:

  --By the local per-CPU timer interrupt (on SMP systems only, with `CONFIG_SMP` enabled). See `timer_tick()`, `update_process_times()`, and `run_local_timers()` for more.

  --By making a call to the `local_bh_enable()` function (mostly invoked by the network subsystem for handling packet receiving/transmitting softIRQs).

  --On the exit path of any I/O IRQ handler (see `do_IRQ`, which makes a call to `irq_exit()`, which in turn invokes `invoke_softirq()`).

  --When the local `ksoftirqd` is given the CPU (that is, it's been awakened).

The actual kernel function responsible for walking through the softIRQ's pending bitmap and running them is `__do_softirq()`, which is defined in `kernel/softirq.c`. This function is always invoked with interrupts disabled on the local CPU. It performs the following tasks:

1. Once invoked, the function saves the current per-CPU pending softIRQ's bitmap in a so-called pending variable and locally disables softIRQs by means of `__local_bh_disable_ip`.

2. It then resets the current per-CPU pending bitmask (which has already been saved) and then reenables interrupts (softIRQs run with interrupts enabled).

3. After this, it enters a `while` loop, checking for pending softIRQs in the saved bitmap. If there is no softIRQ pending, nothing happens. Otherwise, it will execute the handlers of each pending softIRQ, taking care to increment their executions' statistics.

4. After all the pending handlers have been executed (we are out of the `while` loop), `__do_softirq()` once again reads the per-CPU pending bitmask (required to disable IRQs and save them into the same pending variable) in order to check if any softIRQs were scheduled when it was in the `while` loop. If there are any pending softIRQs, the whole process will restart (based on a `goto` loop), starting from *step 2*. This helps with handling, for example, softIRQs that have rescheduled themselves.

However, `__do_softirq()` will not repeat if one of the following conditions occurs:

- It has already repeated up to `MAX_SOFTIRQ_RESTART` times, which is set to `10` in `kernel/softirq.c`. This is actually the limit for the softIRQ processing loop, not the upper bound of the previously described `while` loop.

- It has hogged the CPU more than `MAX_SOFTIRQ_TIME`, which is set to 2 ms (`msecs_to_jiffies(2)`) in `kernel/softirq.c`, since this prevents the scheduler from being enabled.

If one of the two situations occurs, `__do_softirq()` will break its loop and call `wakeup_softirqd()` to wake the local `ksoftirqd` thread, which will later execute the pending softIRQs in the process context. Since `do_softirq` is called at many points in the kernel, it is likely that another invocation of `__do_softirqs` will handle pending softIRQs before `ksoftirqd` has the chance to run.

Note that softIRQs do not always run in an atomic context, but in this case, this is quite specific. The next section explains how and why softIRQs may be executed in a process context.

A word on ksoftirqd

A `ksoftirqd` is a per-CPU kernel thread that's raised in order to handle unserved software interrupts. It is spawned early on in the kernel boot process, as stated in `kernel/softirq.c`:

```
static __init int spawn_ksoftirqd(void)
{
    cpuhp_setup_state_nocalls(CPUHP_SOFTIRQ_DEAD,
                            "softirq:dead", NULL,
                            takeover_tasklets);
    BUG_ON(smpboot_register_percpu_thread(&softirq_threads));
    return 0;
}
early_initcall(spawn_ksoftirqd);
```

After running the `top` command, you will be able to see some `ksoftirqd/n` entries, where *n* is the logical CPU index of the CPU running the `ksoftirqd` thread. Since the `ksoftirqds` run in a process context, they are equal to classic processes/threads, and so are their competing claims for the CPU. `ksoftirqd` hogging CPUs for a long time may indicate a system under heavy load.

Now that we have finished looking at our first work deferring mechanism in the Linux kernel, we'll discuss tasklets, which are an alternative (from an atomic context point of view) to softIRQs, though the former are built using the latter.

Tasklets

Tasklets are *bottom halves* built on top of the `HI_SOFTIRQ` and `TASKLET_SOFTIRQ` softIRQs, with the only difference being that `HI_SOFTIRQ`-based tasklets run prior to the `TASKLET_SOFTIRQ`-based ones. This simply means tasklets are softIRQs, so they follow the same rules. *Unlike softIRQs however, two of the same tasklets never run concurrently.* The tasklet API is quite basic and intuitive.

Tasklets are represented by the `struct tasklet_struct` structure, which is defined in `<linux/interrupt.h>`. Each instance of this structure represents a unique tasklet:

```
struct tasklet_struct {
    struct tasklet_struct *next; /* next tasklet in the list */
    unsigned long state;            /* state of the tasklet,
                            * TASKLET_STATE_SCHED or
                            * TASKLET_STATE_RUN */
    atomic_t count;                 /* reference counter */
    void (*func)(unsigned long); /* tasklet handler function */
    unsigned long data; /* argument to the tasklet function */
};
```

The `func` member is the handler of the tasklet that will be executed by the underlying softIRQ. It is the equivalent of what `action` is to a softIRQ, with the same prototype and the same argument meaning. `data` will be passed as its sole argument.

You can use the `tasklet_init()` function to dynamically allocate and initialize tasklets at run-ime. For the static method, you can use the `DECLARE_TASKLET` macro. The option you choose will depend on your need (or requirement) to have a direct or indirect reference to the tasklet. Using `tasklet_init()` would require embedding the tasklet structure into a bigger and dynamically allocated object. An initialized tasklet can be scheduled by default – you could say it is enabled. `DECLARE_TASKLET_DISABLED` is an alternative to declaring default-disabled tasklets, and this will require the `tasklet_enable()` function to be invoked to make the tasklet schedulable. Tasklets are scheduled (similar to raising a softIRQ) via the `tasklet_schedule()` and `tasklet_hi_schedule()` functions. You can use `tasklet_disable()` to disable a tasklet. This function disables the tasklet and only returns when the tasklet has terminated its execution (assuming it was running). After this, the tasklet can still be scheduled, but it will not run on the CPU until it is enabled again. The asynchronous variant known as `tasklet_disable_nosync()` can be used too and returns immediately, even if termination has not occurred. Moreover, a tasklet that has been disabled several times should be enabled exactly the same number of times (this is allowed thanks to its `count` field):

```
DECLARE_TASKLET(name, func, data)
DECLARE_TASKLET_DISABLED(name, func, data);

tasklet_init(t, tasklet_handler, dev);
void tasklet_enable(struct tasklet_struct*);
void tasklet_disable(struct tasklet_struct *);
void tasklet_schedule(struct tasklet_struct *t);
void tasklet_hi_schedule(struct tasklet_struct *t);
```

The kernel maintains normal priority and high priority tasklets in two different queues. Queues are actually singly linked lists, and each CPU has its own queue pair (low and high priority). Each processor has its own pair. `tasklet_schedule()` adds the tasklet to the normal priority list, thereby scheduling the associated softIRQ with a `TASKLET_SOFTIRQ` flag. With `tasklet_hi_schedule()`, the tasklet is added to the high priority list, thereby scheduling the associated softIRQ with a `HI_SOFTIRQ` flag. Once the tasklet has been scheduled, its `TASKLET_STATE_SCHED` flag is set, and the tasklet is added to a queue. At the time of execution, the `TASKLET_STATE_RUN` flag is set and the `TASKLET_STATE_SCHED` state is removed, thus allowing the tasklet to be rescheduled during its execution, either by the tasklet itself or from within an interrupt handler.

High-priority tasklets are meant to be used for soft interrupt handlers with low latency requirements. Calling `tasklet_schedule()` on a tasklet that's already been scheduled, but whose execution has not started, will do nothing, resulting in the tasklet being executed only once. A tasklet can reschedule itself, which means you can safely call `tasklet_schedule()` in a tasklet. High-priority tasklets are always executed before normal ones and should be used carefully; otherwise, you may increase system latency. Stopping a tasklet is as simple as calling `tasklet_kill()`, which will prevent the tasklet from running again or waiting for it to complete before killing it if the tasklet is currently scheduled to run. If the tasklet reschedules itself, you should prevent the tasklet from rescheduling itself prior to calling this function:

```
void tasklet_kill(struct tasklet_struct *t);
```

That being said, let's take a look at the following example of tasklet code usage:

```
#include <linux/kernel.h>
#include <linux/module.h>
#include <linux/interrupt.h> /* for tasklets API */

char tasklet_data[] =
     "We use a string; but it could be pointer to a structure";

/* Tasklet handler, that just prints the data */
void tasklet_work(unsigned long data)
{
    printk("%s\n", (char *)data);
}
static DECLARE_TASKLET(my_tasklet, tasklet_function,
                       (unsigned long) tasklet_data);

static int __init my_init(void)
{
    tasklet_schedule(&my_tasklet);
    return 0;
}

void my_exit(void)
{
    tasklet_kill(&my_tasklet);
}

module_init(my_init);
```

```
module_exit(my_exit);
MODULE_AUTHOR("John Madieu <john.madieu@gmail.com>");
MODULE_LICENSE("GPL");
```

In the preceding code, we statically declared our `my_tasklet` tasklet and the function that's supposed to be invoked when this tasklet is scheduled, along with the data that will be given as an argument to this function.

Important note

Because the same tasklet never runs concurrently, the locking case between a tasklet and itself doesn't need to be addressed. However, any data that's shared between two tasklets should be protected with `spin_lock()` and `spin_unlock()`. Remember, tasklets are implemented on top of softIRQs.

Workqueues

In the previous section, we dealt with tasklets, which are atomically deferred mechanisms. Apart from atomic mechanisms, there are cases where we may want to sleep in the deferred task. Workqueues allow this.

A workqueue is an asynchronous work deferring mechanism that is widely used across kernels, allowing them to run a dedicated function asynchronously in a process execution context. This makes them suitable for long-running and lengthy tasks or work that needs to sleep, thus improving the user experience. At the core of the workqueue subsystem, there are two data structures that can explain the concept behind this mechanism:

- The work to be deferred (that is, the work item) is represented in the kernel by instances of `struct work_struct`, which indicates the handler function to be run. Typically, this structure is the first element of a user's structure of the work definition. If you need a delay before the work can be run after it has been submitted to the workqueue, the kernel provides `struct delayed_work` instead. A work item is a basic structure that holds a pointer to the function that is to be executed asynchronously. To summarize, we can enumerate two types of work item structures:

  --The `struct work_struct` structure, which schedules a task to be run at a later time (as soon as possible when the system allows it).

  --The `struct delayed_work` structure, which schedules a task to be run after at least a given time interval.

- The workqueue itself, which is represented by a `struct workqueue_struct`. This is the structure that work is placed on. It is a queue of work items.

Apart from these data structures, there are two generic terms you should be familiar with:

- **Worker threads**, which are dedicated threads that execute the functions off the queue, one by one, one after the other.
- **Workerpools** are a collection of worker threads (a thread pool) that are used to manage worker threads.

The first step in using work queues consists of creating a work item, represented by `struct work_struct` or `struct delayed_work` for the delayed variant, that's defined in `linux/workqueue.h`. The kernel provides either the `DECLARE_WORK` macro for statically declaring and initializing a work structure, or the `INIT_WORK` macro for doing the same by dynamically. If you need delayed work, you can use the `INIT_DELAYED_WORK` macro for dynamic allocation and initialization, or `DECLARE_DELAYED_WORK` for the static option:

```
DECLARE_WORK(name, function)
DECLARE_DELAYED_WORK(name, function)
INIT_WORK(work, func);
INIT_DELAYED_WORK(work, func);
```

The following code shows what our work item structure looks like:

```
struct work_struct {
    atomic_long_t data;
    struct list_head entry;
    work_func_t func;
#ifdef CONFIG_LOCKDEP
    struct lockdep_map lockdep_map;
#endif
};

struct delayed_work {
    struct work_struct work;
    struct timer_list timer;

    /* target workqueue and CPU ->timer uses to queue ->work */
    struct workqueue_struct *wq;
```

```
    int cpu;
};
```

The `func` field, which is of the `work_func_t` type, tells us a bit more about the header of a `work` function:

```
typedef void (*work_func_t)(struct work_struct *work);
```

`work` is an input parameter that corresponds to the work structure associated with your work. If you've submitted a delayed work, this would correspond to the `delayed_work.work` field. Here, it will be necessary to use the `to_delayed_work()` function to get the underlying delayed work structure:

```
struct delayed_work *to_delayed_work(struct work_struct *work)
```

Workqueues let your driver create a kernel thread, called a worker thread, to handle deferred work. A new workqueue can be created with these functions:

```
struct workqueue_struct *create_workqueue(const char *name
                                          name)


struct workqueue_struct
    *create_singlethread_workqueue(const char *name)
```

`create_workqueue()` creates a dedicated thread (a worker) per CPU on the system, which is probably not a good idea. On an 8-core system, this will result in 8 kernel threads being created to run work that's been submitted to your workqueue. In most cases, a single system-wide kernel thread should be enough. In this case, you should use `create_singlethread_workqueue()` instead, which creates, as its name suggests, a single threaded workqueue; that is, with one worker thread system-wide. Either normal or delayed work can be enqueued on the same queue. To schedule works on your created workqueue, you can use either `queue_work()` or `queue_delayed_work()`, depending on the nature of the work:

```
bool queue_work(struct workqueue_struct *wq,
                struct work_struct *work)
bool queue_delayed_work(struct workqueue_struct *wq,
                        struct delayed_work *dwork,
                        unsigned long delay)
```

These functions return false if the work was already on a queue and true otherwise. queue_dalayed_work() can be used to plan (delayed) work for execution with a given delay. The time unit for the delay is jiffies. If you don't want to bother with seconds-to-jiffies conversion, you can use the msecs_to_jiffies() and usecs_to_jiffies() helpers, which convert milliseconds or microseconds into jiffies, respectively:

```
unsigned long msecs_to_jiffies(const unsigned int m)
unsigned long usecs_to_jiffies(const unsigned int u)
```

The following example uses 200 ms as a delay:

```
schedule_delayed_work(&drvdata->tx_work, usecs_to_
                      jiffies(200));
```

Submitted work items can be canceled by calling either cancel_delayed_work(), cancel_delayed_work_sync(), or cancel_work_sync():

```
bool cancel_work_sync(struct work_struct *work)
bool cancel_delayed_work(struct delayed_work *dwork)
bool cancel_delayed_work_sync(struct delayed_work *dwork)
```

The following describes what these functions do:

- cancel_work_sync() synchronously cancels the given workqueue entry. In other words, it cancels work and waits for its execution to finish. The kernel guarantees that work won't be pending or executing on any CPU when it's return from this function, even if the work migrates to another workqueue or requeues itself. It returns true if work was pending, or false otherwise.

- cancel_delayed_work() asynchronously cancels a pending workqueue entry (a delayed one). It returns true (a non-zero value) if dwork was pending and canceled and false if it wasn't pending, probably because it is actually running, and thus might still be running after cancel_delayed_work(). To ensure the work really ran to its end, you may want to use flush_workqueue(), which flushes every work item in the given queue, or cancel_delayed_work_sync(), which is the synchronous version of cancel_delayed_work().

To wait for all the work items to finish, you can call `flush_workqueue()`. When you are done with a workqueue, you should destroy it with `destroy_workqueue()`. Both these options can be seen in the following code:

```
void flush_workqueue(struct worksqueue_struct * queue);
void destroy_workqueue(structure workqueque_struct *queue);
```

While you're waiting for any pending work to execute, the _sync variant functions sleep, which means they can only be called from a process context.

The kernel shared queue

In most situations, your code does not necessarily need to have the performance of its own dedicated set of threads, and because `create_workqueue()` creates one worker thread for each CPU, it may be a bad idea to use it on very large multi-CPU systems. In this situation, you may want to use the kernel shared queue, which has its own set of kernel threads preallocated (early during boot, via the `workqueue_init_early()` function) for running works.

This global kernel workqueue is the so-called `system_wq`, and is defined in `kernel/workqueue.c`. There is one instance per CPU, with each backed by a dedicated thread named `events/n`, where n is the processor number that the thread is bound to. You can queue work to the system's default workqueue using one of the following functions:

```
int schedule_work(struct work_struct *work);
int schedule_delayed_work(struct delayed_work *dwork,
                          unsigned long delay);
int schedule_work_on(int cpu, struct work_struct *work);
int schedule_delayed_work_on(int cpu,
                          struct delayed_work *dwork,
                          unsigned long delay);
```

`schedule_work()` immediately schedules the work that will be executed as soon as possible after the worker thread on the current processor wakes up. With `schedule_delayed_work()`, the work will be put in the queue in the future, after the delay timer has ticked. The _on variants are used to schedule the work on a specific CPU (this does not need to be the current one). Each of these function queues work on the system's shared workqueue, `system_wq`, which is defined in `kernel/workqueue.c`:

```
struct workqueue_struct *system_wq __read_mostly;
EXPORT_SYMBOL(system_wq);
```

To flush the kernel-global workqueue – that is, to ensure the given batch of work is completed – you can use `flush_scheduled_work()`:

```
void flush_scheduled_work(void);
```

`flush_scheduled_work()` is a wrapper that calls `flush_workqueue()` on `system_wq`. Note that there may be work in `system_wq` that you have not submitted and have no control over. Due to this, flushing this workqueue entirely is overkill. It is recommended to use `cancel_delayed_work_sync()` or `cancel_work_sync()` instead.

> **Tip**
> Unless you have a strong reason to create a dedicated thread, the default (kernel-global) thread is preferred.

Workqueues – a new generation

The original (now legacy) workqueue implementation used two kinds of workqueues: those with a **single thread system-wide**, and those with a **thread per-CPU**. However, due to the increasing number of CPUs, this led to some limitations:

- On very large systems, the kernel could run out of process IDs (defaulted to 32k) just at boot, before the init was started.

- Multi-threaded workqueues provided poor concurrency management as their threads competed for the CPU with other threads on the system. Since there were more CPU contenders, this introduced some overhead; that is, more context switches than necessary.

- The consumption of much more resources than what was really needed.

Moreover, subsystems that needed a dynamic or fine-grained level of concurrency had to implement their own thread pools. As a result of this, a new workqueue API has been designed and the legacy workqueue API (`create_workqueue()`, `create_singlethread_workqueue()`, and `create_freezable_workqueue()`) has been scheduled to be removed. However, these are actually wrappers around the new ones – the so-called concurrency-managed workqueues. This is done using per-CPU worker pools that are shared by all the workqueues in order to automatically provide a dynamic and flexible level of concurrency, thus abstracting such details for API users.

Concurrency-managed workqueues

The concurrency-managed workqueue is an upgrade of the workqueue API. Using this new API implies that you must choose between two macros to create the workqueue: `alloc_workqueue()` and `alloc_ordered_workqueue()`. These macros both allocate a workqueue and return a pointer to it on success, and NULL on failure. The returned workqueue can be freed using the `destroy_workqueue()` function:

```
#define alloc_workqueue(fmt, flags, max_active, args...)
#define alloc_ordered_workqueue(fmt, flags, args...)
void destroy_workqueue(struct workqueue_struct *wq)
```

`fmt` is the `printf` format for the name of the workqueue, while `args...` are arguments for `fmt`. `destroy_workqueue()` is to be called on the workqueue once you are done with it. All work that's currently pending will be completed first, before the kernel destroys the workqueue. `alloc_workqueue()` creates a workqueue based on `max_active`, which defines the concurrency level by limiting the number of work (tasks) that can be executing (workers in a runnable sate) simultaneously from this workqueue on any given CPU. For example, a `max_active` of 5 would mean that, at most, five work items on this workqueue can be executing at the same time per CPU. On the other hand, `alloc_ordered_workqueue()` creates a workqueue that processes each work item one by one in the queued order (that is, FIFO order).

`flags` controls how and when work items are queued, assigned execution resources, scheduled, and executed. Various flags are used in this new API. Let's take a look at some of them:

- `WQ_UNBOUND`: Legacy workqueues had a worker thread per CPU and were designed to run tasks on the CPU where they were submitted. The kernel scheduler had no choice but to always schedule a worker on the CPU that it was defined on. With this approach, even a single workqueue could prevent a CPU from idling and being turned off, which leads to increased power consumption or poor scheduling policies. `WQ_UNBOUND` turns off this behavior. Work is not bound to a CPU anymore, hence the name unbound workqueues. There is no more locality, and the scheduler can reschedule the worker on any CPU as it sees fit. The scheduler has the last word now and can balance CPU load, especially for long and sometimes CPU-intensive work.

- WQ_MEM_RECLAIM: This flag is to be set for workqueues that need to guarantee forward progress during a memory reclaim path (when free memory is running dangerously low; here, the system is under memory pressure. In this case, GFP_KERNEL allocations may block and deadlock the entire workqueue). The workqueue is then guaranteed to have a ready-to-use worker thread, a so-called rescuer thread reserved for it, regardless of memory pressure, so that it can progress. One rescuer thread is allocated for each workqueue that has this flag set.

Let's consider a situation where we have three work items (*w1*, *w2*, and *w3*) in our workqueue, *W*. *w1* does some work and then waits for *w3* to complete (let's say it depends on the computation result of *w3*). Afterward, *w2* (which is independent of the others) does some kmalloc() allocation (GFP_KERNEL). Now, it seems like there is not enough memory. While *w2* is blocked, it still occupies the workqueue of *W*. This results in *w3* not being able to run, despite the fact that there is no dependency between *w2* and *w3*. Since there is not enough memory available, there is no way to allocate a new thread to run *w3*. A pre-allocated thread would definitely solve this problem, not by magically allocating the memory for *w2*, but by running *w3* so that *w1* can continue its job, and so on. *w2* will continue its progression as soon as possible, when there is enough available memory to allocate. This pre-allocated thread is the so-called rescuer thread. You must set this WQ_MEM_RECLAIM flag if you think the workqueue might be used in the memory reclaim path. This flag replaces the old WQ_RESCUER flag as of the following commit: https://git.kernel.org/pub/scm/linux/kernel/git/torvalds/linux.git/commit/?id=493008a8e475771a2126e0ce95a73e35b371d277.

- WQ_FREEZABLE: This flag is used for power management purposes. A workqueue with this flag set will be frozen when the system is suspended or hibernates. On the freezing path, all current work(s) of the worker(s) will be processed. When the freeze is complete, no new work items will be executed until the system is unfrozen. Filesystem-related workqueue(s) may use this flag to ensure that modifications that are made to files are pushed to disk or create the hibernation image on the freezing path and that no modifications are made on-disk after the hibernation image has been created. In this situation, non-freezable items may do things differently that could lead to filesystem corruption. As an example, all of the XFS internal workqueues have this flag set (see `fs/xfs/xfs_super.c`) to ensure no further changes are made on disk once the freezer infrastructure freezes the kernel threads and creates the hibernation image. You should not set this flag if your workqueue can run tasks as part of the hibernation/suspend/resume process of the system. More information on this topic can be found in `Documentation/power/freezing-of-tasks.txt`, as well as by taking a look at the kernel's internal `freeze_workqueues_begin()` and `thaw_workqueues()` functions.

- WQ_HIGHPRI: Tasks that have this flag set run immediately and do not wait for the CPU to become available. This flag is used for workqueues that queue work items that require high priority for execution. Such workqueues have worker threads with a high priority level (a lower `nice` value).

  In the early days of the CMWQ, high-priority work items were just queued at the head of a global normal priority worklist so that they could immediately run. Nowadays, there is no interaction between normal priority and high-priority workqueues as each has its own worklist and its own worker pool. The work items of a high-priority workqueue are queued to the high-priority worker pool of the target CPU. Tasks in this workqueue should not block much. Use this flag if you do not want your work item competing for CPU with normal or lower-priority tasks. Crypto and Block subsystems use this, for example.

- WQ_CPU_INTENSIVE: Work items that are part of a CPU-intensive workqueue may burn a lot of CPU cycles and will not participate in the workqueue's concurrency management. Instead, their execution is regulated by the system scheduler, just like any other task. This makes this flag useful for bound work items that may hog CPU cycles. Though their execution is regulated by the system scheduler, the start of their execution is still regulated by concurrency management, and runnable non-CPU-intensive work items can delay the execution of CPU-intensive work items. Actually, the crypto and dm-crypt subsystems use such workqueues. To prevent such tasks from delaying the execution of other non-CPU-intensive work items, they will not be taken into account when the workqueue code determines whether the CPU is available.

In order to be compliant with the old workqueue API, the following mappings are made to keep this API compatible with the original one:

- `create_workqueue(name)` is mapped to `alloc_workqueue(name, WQ_MEM_RECLAIM, 1)`.

- `create_singlethread_workqueue(name)` is mapped to `alloc_ordered_workqueue(name, WQ_MEM_RECLAIM)`.

- `create_freezable_workqueue(name)` is mapped to `alloc_workqueue(name, WQ_FREEZABLE | WQ_UNBOUND|WQ_MEM_RECLAIM, 1)`.

To summarize, `alloc_ordered_workqueue()` actually replaces `create_freezable_workqueue()` and `create_singlethread_workqueue()` (as per the following commit: `https://git.kernel.org/pub/scm/linux/kernel/git/next/linux-next.git/commit/?id=81dcaf6516d8`). Workqueues allocated with `alloc_ordered_workqueue()` are unbound and have `max_active` set to 1.

When it comes to scheduled items in a workqueue, the work items that have been queued to a specific CPU using `queue_work_on()` will execute on that CPU. Work items that have been queued via `queue_work()` will prefer the queueing CPU, though this locality is not guaranteed.

> **Important Note**
>
> Note that `schedule_work()` is a wrapper that calls `queue_work()` on the system workqueue (`system_wq`), while `schedule_work_on()` is a wrapper around `queue_work_on()`. Also, keep in mind that `system_wq = alloc_workqueue("events", 0, 0);`. Take a look at the `workqueue_init_early()` function in `kernel/workqueue.c` in the kernel sources to see how other system-wide workqueues are created.
>
> Memory reclaim is a Linux kernel mechanism on the memory allocation path. This consists of allocating memory after throwing the current content of that memory somewhere else.

With that, we have finished looking at workqueues and the concurrency-managed ones in particular. Next, we'll introduce Linux kernel interrupt management, which is where most of the previous mechanisms will be solicited.

Linux kernel interrupt management

Apart from servicing processes and user requests, another job of the Linux kernel is managing and speaking with hardware. This is either from the CPU to the device or from the device to the CPU. This is achieved by means of interrupts. An interrupt is a signal that's sent to the processor by an external hardware device requesting immediate attention. Prior to an interrupt being visible to the CPU, this interrupt should be enabled by the interrupt controller, which is a device on its own, and whose main job consists of routing interrupts to CPUs.

An interrupt may have five states:

- **Active**: An interrupt that has been acknowledged by a **processing element** (**PE**) and is being handled. While being handled, another assertion of the same interrupt is not presented as an interrupt to a processing element, until the initial interrupt is no longer active.

- **Pending (asserted)**: An interrupt that is recognized as asserted in hardware, or generated by software, and is waiting to be handled by the target PE. It is a common behavior for most hardware devices not to generate other interrupts until their "interrupt pending" bit has been cleared. A disabled interrupt can't be pending as it is never asserted, and it is immediately dropped by the interrupt controller.

- **Active and pending**: An interrupt that is active from one assertion of the interrupt and is pending from a subsequent assertion.

- **Inactive**: An interrupt that is not active or pending. Deactivation clears the active state of the interrupt, and thereby allows the interrupt, when it is pending, to be taken again.

- **Disabled/Deactivated**: This is unknown to the CPU and not even seen by the interrupt controller. This will never be asserted. Disabled interrupts are lost.

> Important note
>
> There are interrupt controllers where disabling an interrupt means masking that interrupt, or vice versa. In the remainder of this book, we will consider disabling to be the same as masking, though this is not always true.

Upon reset, the processor disables all the interrupts until they are enabled again by the initialization code (this is the job of the Linux kernel in our case). The interrupts are enabled/disabled by setting/clearing the bits in the processor status/control registers. Upon an interrupt assertion (an interrupt occurred), the processor will check whether the interrupts are masked or not and will do nothing if they are masked. Once unmasked, the processor will pick one pending interrupt, if any (the order does not matter since it will do this for each pending interrupt until they are all serviced), and will execute a specially purposed function called the **Interrupt Service Routine** (**ISR**) that is associated with this interrupt. This ISR must be registered by the code (that is, our device driver, which relies on the kernel irq core code) at a special location called the vector table. Right before the processor starts executing this ISR, it does some context saving (including the unmasked status of interrupts) and then masks the interrupts on the local CPU (interrupts can be asserted and will be serviced once unmasked). Once the ISR is running, we can say that the interrupt is being serviced.

The following is the complete IRQ handling flow on ARM Linux. This happens when an interrupt occurs and the interrupts are enabled in the PSR:

1. The ARM core will disable further interrupts occurring on the local CPU.

2. The ARM core will then put the **Current Program Status Register** (**CPSR**) in the **Saved Program Status Register** (**SPSR**), put the current **Program Counter** (**PC**) in the **Link Register** (**LR**), and then switch to IRQ mode.

3. Finally, the ARM processor will refer to the vector table and jumps to the exception handler. In our case, it jumps to the exception handler of IRQ, which in the Linux kernel corresponds to the vector_stub macro defined in arch/arm/kernel/entry-armv.S.

    These three steps are done by the ARM processor itself. Now, the kernel jumps into action:

4. The vector_stub macro checks from what processor mode we used to get here – either kernel mode or user mode – and determines the macro to call accordingly; either __irq_user or __irq_svc.

5. __irq_svc() will save the registers (from r0 to r12) on the kernel stack and then call the irq_handler() macro, which either calls handle_arch_irq() (present in arch/arm/include/asm/entry-macro-multi.S) if CONFIG_MULTI_IRQ_HANDLER is defined, or arch_irq_handler_default() otherwise, with handle_arch_irq being a global pointer to the function that's set in arch/arm/kernel/setup.c (from within the setup_arch() function).

6. Now, we need to identify the hardware-IRQ number, which is what `asm_do_IRQ()` does. It then calls `handle_IRQ()` on that hardware-IRQ, which in turn calls `__handle_domain_irq()`, which will translate the hardware-irq into its corresponding Linux IRQ number (`irq = irq_find_mapping(domain, hwirq)`) and call `generic_handle_irq()` on the decoded Linux IRQ (`generic_handle_irq(irq)`).

7. `generic_handle_irq()` will look for the IRQ descriptor structure (Linux's view of an interrupt) that corresponds to the decoded Linux IRQ (`struct irq_desc *desc = irq_to_desc(irq)`) and calling `generic_handle_irq_desc()` on this descriptor), which will result in `desc->handle_irq(desc)`. `desc->handle_irq` corresponding to the high-level IRQ handler that was set using `irq_set_chip_and_handler()` during the mapping of this IRQ.

8. `desc->handle_irq()` may result in a call to `handle_level_irq()`, `handle_simple_irq()`, `handle_edge_irq()`, and so on.

9. The high-level IRQ handler calls our ISR.

10. Once the ISR has been completed, `irq_svc` will return and restore the processor state by restoring registers (r0-r12), the PC, and the CSPR.

> **Important note**
>
> Going back to *step 1*, during an interrupt, the ARM core disables further IRQs on the local CPU. It is worth mentioning that in the earlier Linux kernel days, there were two families of interrupt handlers: those running with interrupts disabled (that is, with the old `IRQF_DISABLED` flag set) and those running with interrupts enabled: they were then interruptible. The former were called **fast handlers**, while the latter were called **slow handlers**. For the latter, interrupts were actually reenabled by the kernel prior to invoking the handler. Since the interrupt context has a really small stack size compared to the process stack, it makes no sense that we may run into a stack overflow if we are in an interrupt context (running a given IRQ handler) while other interrupts keep occurring, even the one being serviced. This is confirmed by the commit at `https://git.kernel.org/pub/scm/linux/kernel/git/next/linux-next.git/commit/?id=e58aa3d2d0cc`, which deprecated the fact of running interrupt handlers with IRQs enabled. As of this patch, IRQs remain disabled (left untouched after ARM core disabled them on the local CPU) during the execution of an IRQ handler. Additionally, the aforementioned flags have been entirely removed by the commit at `https://git.kernel.org/pub/scm/linux/kernel/git/next/linux-next.git/commit/?id=d8bf368d0631`, since Linux v4.1.

Designing an interrupt handler

Now that we're familiar with the concept of bottom halves and deferring mechanisms, the time for us to implement interrupt handlers has come. In this section, we'll take care of some specifics. Nowadays, the fact that interrupt handlers run with interrupts disabled (on the local CPU) means that we need to respect certain constraints in the ISR design:

- **Execution time:** Since IRQ handlers run with interrupts disabled on the local CPU, the code must be as short and as small as possible, as well as fast enough to ensure the previously disabled CPU-local interrupts are reenabled quickly in so that other IRQs are not missed. Time-consuming IRQ handlers may considerably alter the real-time properties of the system and slow it down.

- **Execution context**: Since interrupt handlers are executed in an atomic context, sleeping (or any other mechanism that may sleep, such as mutexes, copying data from kernel to user space or vice versa, and so on) is forbidden. Any part of the code that requires or involves sleeping must be deferred into another, safer context (that is, a process context).

An IRQ handler needs to be given two arguments: the interrupt line to install the handler for, and a unique device identifier of the peripheral (mostly used as a context data structure; that is, the pointer to the per-device or private structure of the associated hardware device):

```
typedef irqreturn_t (*irq_handler_t)(int, void *);
```

The device driver that wants to enable a given interrupt and register an ISR for it should call `request_irq()`, which is declared in `<linux/interrupt.h>`. This must be included in the driver code:

```
int request_irq(unsigned int irq,
                irq_handler_t handler,
                unsigned long flags,
                const char *name,
                void *dev)
```

While the aforementioned API would require the caller to free the IRQ when it is no longer needed (that is, on driver detach), you can use the device managed variant, `devm_request_irq()`, which contains internal logic that allows it to take care of releasing the IRQ line automatically. It has the following prototype:

```
int devm_request_irq(struct device *dev, unsigned int irq,
                     irq_handler_t handler,
                     unsigned long flags,
                     const char *name, void *dev)
```

Except for the extra `dev` parameter (which is the device that requires the interrupt), both `devm_request_irq()` and `request_irq()` expect the following arguments:

- `irq`, which is the interrupt line (that is, the interrupt number of the issuing device). Prior to validating the request, the kernel will make sure the requested interrupt is valid and that it is not already assigned to another device, unless both devices request that this IRQ line needs to be shared (with the help of flags).

- `handler`, which is a function pointer to the interrupt handler.

- `flags`, which represents the interrupt flags.

- `name`, an ASCII string representing the name of the device generating or claiming this interrupt.

- `dev` should be unique to each registered handler. This cannot be NULL for shared IRQs since it is used to identify the device via the kernel IRQ core. The most common way of using it is to provide a pointer to the device structure or a pointer to any per-device (that's potentially useful to the handler) data structure. This is because when an interrupt occurs, both the interrupt line (`irq`) and this parameter will be passed to the registered handler, which can use this data as context data for further processing.

`flags` mangle the state or behavior of the IRQ line or its handler by means of the following masks, which can be ORed to form the final desired bit mask according to your needs:

```
#define IRQF_TRIGGER_RISING    0x00000001
#define IRQF_TRIGGER_FALLING   0x00000002
#define IRQF_TRIGGER_HIGH      0x00000004
#define IRQF_TRIGGER_LOW       0x00000008


#define IRQF_SHARED            0x00000080
```

```
#define IRQF_PROBE_SHARED        0x00000100
#define IRQF_NOBALANCING         0x00000800
#define IRQF_IRQPOLL             0x00001000
#define IRQF_ONESHOT             0x00002000
#define IRQF_NO_SUSPEND          0x00004000
#define IRQF_FORCE_RESUME        0x00008000
#define IRQF_NO_THREAD           0x00010000
#define IRQF_EARLY_RESUME        0x00020000
#define IRQF_COND_SUSPEND        0x00040000
```

Note that flags can also be zero. Let's take a look at some important flags. I'll leave the rest for you to explore in `include/linux/interrupt.h`:

- `IRQF_TRIGGER_HIGH` and `IRQF_TRIGGER_LOW` flags are to be used for level-sensitive interrupts. The former is for interrupts triggered at high level and the latter is for the low-level triggered interrupts. Level-sensitive interrupts are triggered as long as the physical interrupt signal is high. If the interrupt source is not cleared by the end of its interrupt handler in the kernel, the operating system will repeatedly call that kernel interrupt handler, which may lead platform to hang. In other words, when the handler services the interrupt and returns, if the IRQ line is still asserted, the CPU will signal the interrupt again immediately. To prevent such a situation, the interrupt must be acknowledged (that is, cleared or de-asserted) by the kernel interrupt handler immediately when it is received.

  However, those flags are safe with regard to interrupt sharing because if several devices pull the line active, an interrupt will be signaled (assuming the IRQ is enabled or as soon as it becomes so) until all drivers have serviced their devices. The only drawback is that it may lead to lockup if a driver fails to clear its interrupt source.

- `IRQF_TRIGGER_RISING` and `IRQF_TRIGGER_FALLING` concern edge-triggered interrupts, rising and falling edges respectively. Such interrupts are signaled when the line changes from inactive to active state, but only once. To get a new request the line must go back to inactive and then to active again. Most of the time, no special action is required in software in order to acknowledge this type of interrupt.

When using edge-triggered interrupts however, interrupts may be lost, especially in the context of a shared interrupt line: if one device pulls the line active for too long a time, when another device pulls the line active, no edge will be generated, the second request will not be seen by the processor and then will be ignored. With a shared edge-triggered interrupts, if a hardware does not de-assert the IRQ line, no other interrupt will be notified for either shared device.

> **Important note**
>
> As a quick reminder, you can just remember that level triggered interrupts signal a state, while edge triggered ones signal an event.
>
> Moreover, when requesting an interrupt without specifying an IRQF_ TRIGGER flag, the setting should be assumed to be *as already configured*, which may be as per machine or firmware initialization. In such cases, you can refer to the device tree (if specified in there) for example to see what this *assumed configuration* is.

- IRQF_SHARED: This allows the interrupt line to be shared among several devices. However, each device driver that needs to share the given interrupt line must set this flag; otherwise, the registration will fail.

- IRQF_NOBALANCING: This excludes the interrupt from *IRQ balancing*, which is a mechanism that consists of distributing/relocating interrupts across CPUs, with the goal of increasing performance. This prevents the CPU affinity of this IRQ from being changed. This flag can be used to provide a flexible setup for *clocksources* in order to prevent the event from being misattributed to the wrong core. This misattribution may result in the IRQ being disabled because if the CPU handling the interrupt is not the one that triggered it, the handler will return IRQ_NONE. This flag is only meaningful on multicore systems.

- IRQF_IRQPOLL: This flag allows the *irqpoll* mechanism to be used, which fixes interrupt problems. This means that this handler should be added to the list of known interrupt handlers that can be looked for when a given interrupt is not handled.

- IRQF_ONESHOT: Normally, the actual interrupt line being serviced is enabled after its hard-IRQ handler completes, whether it awakes a threaded handler or not. This flag keeps the interrupt line disabled after the hard-IRQ handler completes. This flag must be set on threaded interrupts (we will discuss this later) for which the interrupt line must remain disabled until the threaded handler has completed. After this, it will be enabled.

- `IRQF_NO_SUSPEND`: This does not disable the IRQ during system hibernation/ suspension. This means that the interrupt is able to save the system from a suspended state. Such IRQs may be timer interrupts, which may trigger and need to be handled while the system is suspended. The whole IRQ line is affected by this flag in that if the IRQ is shared, every registered handler for this shared line will be executed, not just the one who installed this flag. You should avoid using `IRQF_NO_SUSPEND` and `IRQF_SHARED` at the same time as much as possible.

- `IRQF_FORCE_RESUME`: This enables the IRQ in the system resume path, even if `IRQF_NO_SUSPEND` is set.

- `IRQF_NO_THREAD`: This prevents the interrupt handler from being threaded. This flag overrides the `threadirqs` kernel (used on RT kernels, such as when applying the `PREEMPT_RT` patch) command-line option, which forces every interrupt to be threaded. This flag was introduced to address the non-threadability of some interrupts (for example, timers, which cannot be threaded even when all the interrupt handlers are forced to be threaded).

- `IRQF_TIMER`: This marks this handler as being specific to the system timer interrupts. It helps not to disable the timer IRQ during system suspend to ensure that it resumes normally and does not thread them when full preemption (see `PREEMPT_RT`) is enabled. It is just an alias for `IRQF_NO_SUSPEND | IRQF_NO_THREAD`.

- `IRQF_EARLY_RESUME`: This resumes IRQ early at the resume time of system core (syscore) operations instead of at device resume time. Go to `https://lkml.org/lkml/2013/11/20/89` to see the commit introducing its support.

We must also consider the return type, `irqreturn_t`, of interrupt handlers since they may involve further actions once the handler is returned:

- `IRQ_NONE`: On a shared interrupt line, once the interrupt occurs, the kernel irqcore successively walks through the handlers that have been registered for this line and executes them in the order they have been registered. The driver then has the responsibility of checking whether it is their device that issued the interrupt. If the interrupt does not come from its device, it must return `IRQ_NONE` in order to instruct the kernel to call the next registered interrupt handler. This return value is mostly used on shared interrupt lines since it informs the kernel that the interrupt does not come from our device. However, if **99,900** of the previous **100,000** interrupts of a given IRQ line have not been handled, the kernel assumes that this IRQ is stuck in some manner, drops a diagnostic, and tries to turn the IRQ off. For more information on this, have a look at the `__report_bad_irq()` function in the kernel source tree.

- `IRQ_HANDLED`: This value should be returned if the interrupt has been handled successfully. On a threaded IRQ, this value acknowledges the interrupt without waking the thread handler up.

- `IRQ_WAKE_THREAD`: On a thread IRQ handler, this value must be returned the by hard-IRQ handler in order to wake the handler thread. In this case, `IRQ_HANDLED` must only be returned by the threaded handler that was previously registered with `request_threaded_irq()`. We will discuss this later in the *Threaded IRQ handlers* section of this chapter.

> **Important note**
>
> You must be very careful when reenabling interrupts in the handler. Actually, you must never reenable IRQs from within your IRQ handler as this would involve allowing "interrupts reentrancy". In this case, it is your responsibility to address this.

In the unloading path of your driver (or once you think you do not need the IRQ line anymore during your driver runtime life cycle, which is quite rare), you must release your IRQ resource by unregistering your interrupt handler and potentially disabling the interrupt line. The `free_irq()` interface does this for you:

```
void free_irq(unsigned int irq, void *dev_id)
```

That being said, if an IRQ allocated with `devm_request_irq()` needs to be freed separately, `devm_free_irq()` must be used. It has the following prototype:

```
void devm_free_irq(struct device *dev,
                   unsigned int irq,
                   void *dev_id)
```

This function has an extra `dev` argument, which is the device to free the IRQ for. This is usually the same as the one that the IRQ has been registered for. Except for `dev`, this function takes the same arguments and performs the same function as `free_irq()`. However, instead of `free_irq()`, it should be used to manually free IRQs that have been allocated with `devm_request_irq()`.

Both `devm_request_irq()` and `free_irq()` remove the handler (identified by `dev_id` when it comes to shared interrupts) and disable the line. If the interrupt line is shared, the handler is simply removed from the list of handlers for this IRQ, and the interrupt line is disabled in the future when the last handler is removed. Moreover, if possible, your code must ensure the interrupt is really disabled on the card it drives before calling this function, since omitting this may leads to spurious IRQs.

There are few things that are worth mentioning here about interrupts that you should never forget:

- Since interrupt handlers in Linux run with IRQs disabled on the local CPU and the current line is masked in all other cores, they don't need to be reentrant, since the same interrupt will never be received until the current handler has completed. However, all other interrupts (on other cores) remain enabled (or should we say untouched), so other interrupts keep being serviced, even though the current line is always disabled, as well as further interrupts on the local CPU. Consequently, the same interrupt handler is never invoked concurrently to service a nested interrupt. This greatly simplifies writing your interrupt handler.

- Critical regions that need to run with interrupts disabled should be limited as much as possible. To remember this, tell yourselves that your interrupt handler has interrupted other code and needs to give CPU back.

- Interrupt handlers cannot block as they do not run in a process context.

- They may not transfer data to/from user space since this may block.

- They may not sleep or rely on code that may lead to sleep, such as invoking `wait_event()`, memory allocation with anything other than `GFP_ATOMIC`, or using a mutex/semaphore. The threaded handler can handle this.

- They may not trigger nor call `schedule()`.

- Only one interrupt on a given line can be pending (its interrupt flag bits get set when its interrupt condition occurs, regardless of the state of its corresponding enabled bit or the global enabled bit). Any further interrupt of this line is lost. For example, if you are processing an RX interrupt while five more packets are received at the same time, you should not expect five times more interrupts to appear sequentially. You'll only be notified once. If the processor doesn't service the ISR first, there's no way to check how many RX interrupts will occur later. This means that if the device generates another interrupt before the handler function returns `IRQ_HANDLED`, the interrupt controller will be notified of the pending interrupt flag and the handler will get called again (only once), so you may miss some interrupts if you are not fast enough. Multiple interrupts will happen while you are still handling the first one.

> **Important note**
>
> If an interrupt occurs while it is disabled (or masked), it will not be processed at all (masked in the flow handler), but will be recognized as asserted and will remain pending so that it will be processed when enabled (or unmasked).
>
> The interrupt context has its own (fixed and quite low) stack size. Therefore, it totally makes sense to disable IRQs while running an ISR as reentrancy could cause stack overflow if too many preemptions happen.
>
> The concept of non-reentrancy for an interrupt means that if an interrupt is already in an active state, it cannot enter it again until the active status is cleared.

The concept of top and bottom halves

External devices send interrupt requests to the CPU either to signal a particular event or to request a service. As stated in the previous section, bad interrupt management may considerably increase system latency and decrease its real-time quality. We also stated that interrupt processing – that is, the hard-IRQ handler – must be very fast, not only to keep the system responsive, but also so that it doesn't miss other interrupt events.

Take a look at the following diagram:

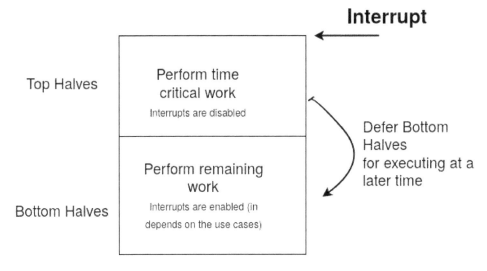

Figure 1.2 – Interrupt splitting flow

The basic idea is that you split the interrupt handler into two parts. The first part is a function) that will run in a so-called hard-IRQ context, with interrupts disabled, and perform the minimum required work (such as doing some quick sanity checks, time-sensitive tasks, read/write hardware registers, and processing this data and acknowledging the interrupt on the device that raised it). This first part is the so-called top-half on Linux systems. The top-half then schedules a (sometimes threaded) handler, which then runs a so-called bottom-half function, with interrupts re-enabled. This is the second part of the interrupt. The bottom-half may then perform time-consuming tasks (such as buffer processing) – tasks that may sleep, depending on the deferring mechanism.

This splitting would considerably increase the system's responsiveness as the time spent with IRQs disabled is reduced to its minimum. When the bottom halves are run in kernel threads, they compete for the CPU with other processes on the runqueue. Moreover, they may have their real-time properties set. The top half is actually the handler that's registered using `request_irq()`. When using `request_threaded_irq()`, as we will see in the next section, the top half is the first handler that's given to the function.

As we described previously, a bottom half represents any task (or work) that's scheduled from within an interrupt handler. Bottom halves are designed using a work-deferring mechanism, which we have seen previously. Depending on which one you choose, it may run in a (software) interrupt context or in a process context. This includes *SoftIRQs*, *tasklets, workqueues*, and *threaded IRQs*.

> **Important note**
> Tasklets and SoftIRQs do not actually fit into the so-called "thread interrupts" mechanism since they run in their own special contexts.

Since softIRQ handlers run at a high priority with scheduler preemption disabled, they do not relinquish the CPU to processes/threads until they complete, so care must be taken while using them for bottom-half delegation. Nowadays, since the quantum that's allocated for a particular process may vary, there is no strict rule regarding how long the softIRQ handler should take to complete so that it doesn't slow the system down as the kernel would not be able to give CPU time to other processes. I would say that this should be no longer than a half of jiffy.

The hard-IRQ handler (the top half) has to be as fast as possible, and most of time, it should just be reading and writing in I/O memory. Any other computation should be deferred to the bottom half, whose main goal is to perform any time-consuming and minimal interrupt-related work that's not performed by the top half. There are no clear guidelines on repartitioning work between the top and bottom halves. The following is some advice:

- Hardware-related work and time-sensitive work should be performed in the top half.

- If the work doesn't need to be interrupted, perform it in the top half.

- From my point of view, everything else can be deferred – that is, performed in the bottom half – so that it runs with interrupts enabled and when the system is less busy.

- If the hard-IRQ handler is fast enough to process and acknowledge interrupts consistently within a few microseconds, then there is absolutely no need to use bottom-half delegations at all.

Next, we will look at threaded IRQ handlers.

Threaded IRQ handlers

Threaded interrupt handlers were introduced to reduce the time spent in the interrupt handler and deferring the rest of the work (that is, processing) out to kernel threads. So, the top half (hard-IRQ handler) would consist of quick sanity checks such as ensuring that the interrupt comes from its device and waking the bottom half accordingly. A threaded interrupt handler runs in its own thread, either in the thread of their parent (if they have one) or in a separate kernel thread. Moreover, the dedicated kernel thread can have its real-time priority set, though it runs at normal real-time priority (that is, MAX_USER_RT_PRIO/2 as shown in the setup_irq_thread() function in kernel/irq/manage.c).

The general rule behind threaded interrupts is simple: keep the hard-IRQ handler as minimal as possible and defer as much work to the kernel thread as possible (preferably all work). You should use request_threaded_irq() (defined in kernel/irq/manage.c) if you want to request a threaded interrupt handler:

```
int
request_threaded_irq(unsigned int irq,
                     irq_handler_t handler,
                     irq_handler_t thread_fn,
                     unsigned long irqflags,
```

```
                    const char *devname,
                    void *dev_id)
```

This function accepts two special parameters `handler` and `thread_fn`. The other parameters are the same as they are for `request_irq()`:

- `handler` immediately runs when the interrupt occurs in the interrupt context, and acts as a hard-IRQ handler. Its job usually consists of reading the interrupt cause (in the device's status register) to determine whether or how to handle the interrupt (this is frequent on MMIO devices). If the interrupt does not come from its device, this function should return `IRQ_NONE`. This return value usually only makes sense on shared interrupt lines. In the other case, if this hard-IRQ handler can finish interrupt processing fast enough (this is not a universal rule, but let's say no longer than half a jiffy – that is, no longer than 500 µs if `CONFIG_HZ`, which defines the value of a jiffy, is set to 1,000) for a set of interrupt causes, it should return `IRQ_HANDLED` after processing in order to acknowledge the interrupts. Interrupt processing that does not fall into this time lapse should be deferred to the threaded IRQ handler. In this case, the hard-IRQ handler should return `IRQ_WAKE_T HREAD` in order to awake the threaded handler. Returning `IRQ_WAKE_THREAD` only makes sense when the `thread_fn` handler is also registered.

- `thread_fn` is the threaded handler that's added to the scheduler runqueue when the hard-IRQ handler function returns `IRQ_WAKE_THREAD`. If `thread_fn` is `NULL` while `handler` is set and it returns `IRQ_WAKE_THREAD`, nothing happens at the return path of the hard-IRQ handler except for a simple warning message being shown. Have a look at the `__irq_wake_thread()` function in the kernel sources for more information. As `thread_fn` competes for the CPU with other processes on the runqueue, it may be executed immediately or later in the future when the system has less load. This function should return `IRQ_HANDLED` when it has completed the interrupt handling process successfully. At this stage, the associated kthread will be taken off the runqueue and put in a blocked state until it's woken up again by the hard-IRQ function.

A default hard-IRQ handler will be installed by the kernel if `handler` is `NULL` and `thread_fn != NULL`. This is the default primary handler. It is an almost empty handler that simply returns `IRQ_WAKE_THREAD` in order to wake up the associated kernel thread that will execute the `thread_fn` handler. This makes it possible to move the execution of interrupt handlers entirely to the process context, thus preventing buggy drivers (buggy IRQ handlers) from breaking the whole system and reducing interrupt latency. A dedicated handler's kthreads will be visible in `ps ax`:

```
/*
 * Default primary interrupt handler for threaded interrupts is
 * assigned as primary handler when request_threaded_irq is
 * called with handler == NULL. Useful for one-shot interrupts.
 */
static irqreturn_t irq_default_primary_handler(int irq,
                                                void *dev_id)
{
    return IRQ_WAKE_THREAD;
}


int
request_threaded_irq(unsigned int irq,
                     irq_handler_t handler,
                     irq_handler_t thread_fn,
                     unsigned long irqflags,
                     const char *devname,
                     void *dev_id)
{
    [...]
    if (!handler) {
        if (!thread_fn)
            return -EINVAL;
        handler = irq_default_primary_handler;
    }
    [...]
}
EXPORT_SYMBOL(request_threaded_irq);
```

Important note

Nowadays, `request_irq()` is just a wrapper around `request_threaded_irq()`, with the `thread_fn` parameter set to NULL.

Note that the interrupt is acknowledged at the interrupt controller level when you return from the hard-IRQ handler (whatever the return value is), thus allowing you to take other interrupts into account. In such a situation, if the interrupt hasn't been acknowledged at the device level, the interrupt will fire again and again, resulting in stack overflows (or being stuck in the hard-IRQ handler forever) for level-triggered interrupts since the issuing device still has the interrupt line asserted. Before threaded IRQs were a thing, when you needed to run the bottom-half in a thread, you would instruct the top half to disable the IRQ at the device level, prior to waking the thread up. This way, even if the controller is ready to accept another interrupt, it is not raised again by the device.

The `IRQF_ONESHOT` flag resolves this problem. It must be set when it comes to use a threaded interrupt (at the `request_threaded_irq()` call); otherwise, the request will fail with the following error:

```
pr_err(
  "Threaded irq requested with handler=NULL and !ONESHOT for irq
  %d\n",
  irq);
```

For more information on this, please have a look at the `__setup_irq()` function in the kernel source tree.

The following is an excerpt from the message that introduced the `IRQF_ONESHOT` flag and explains what it does (the entire message can be found at `http://lkml.iu.edu/hypermail/linux/kernel/0908.1/02114.html`):

> *"It allows drivers to request that the interrupt is not unmasked (at the controller level) after the hard interrupt context handler has been executed and the thread has been woken. The interrupt line is unmasked after the thread handler function has been executed."*

Important note

If you omit the `IRQF_ONESHOT` flag, you'll have to provide a hard-IRQ handler (in which you should disable the interrupt line); otherwise, the request will fail.

An example of a thread-only IRQ is as follows:

```
static irqreturn_t data_event_handler(int irq, void *dev_id)
{
    struct big_structure *bs = dev_id;
    process_data(bs->buffer);
    return IRQ_HANDLED;
}
static int my_probe(struct i2c_client *client,
                    const struct i2c_device_id *id)
{
    [...]
    if (client->irq > 0) {
        ret = request_threaded_irq(client->irq,
                        NULL,
                        &data_event_handler,
                        IRQF_TRIGGER_LOW | IRQF_ONESHOT,
                        id->name,
                        private);
        if (ret)
            goto error_irq;
    }
    [...]
    return 0;
error_irq:
    do_cleanup();
    return ret;
}
```

In the preceding example, our device sits on an I2C bus. Thus, accessing the available data may cause it to sleep, so this should not be performed in the hard-IRQ handler. This is why our handler parameter is NULL.

> **Tip**
>
> If the IRQ line where you need threaded ISR handling to be shared among
> several devices (for example, some SoCs share the same interrupt among their
> internal ADCs and the touchscreen module), you must implement the hard-
> IRQ handler, which should check whether the interrupt has been raised by
> your device or not. If the interrupt does come from your device, you should
> disable the interrupt at the device level and return `IRQ_WAKE_THREAD` to
> wake the threaded handler. The interrupt should be enabled back at the device
> level in the return path of the threaded handler. If the interrupt does not come
> from your device, you should return `IRQ_NONE` directly from the hard-IRQ
> handler.

Moreover, if one driver has set either the `IRQF_SHARED` or `IRQF_ONESHOT` flag
on the line, every other driver sharing the line must set the same flags. The `/proc/
interrupts` file lists the IRQs and their processing per CPU, the IRQ name that was
given during the requesting step, and a comma-separated list of drivers that registered an
ISR for that interrupt.

Threaded IRQs are the best choice for interrupt processing as they can hog too many CPU
cycles (exceeding a jiffy in most cases), such as bulk data processing. Threading IRQs
allow the priority and CPU affinity of their associated thread to be managed individually.
Since this concept comes from the real-time kernel tree (from *Thomas Gleixner*), it fulfills
many requirements of a real-time system, such as allowing a fine-grained priority model
to be used and reducing interrupt latency in the kernel.

Take a look at `/proc/irq/IRQ_NUMBER/smp_affinity`, which can be used to get
or set the corresponding `IRQ_NUMBER` affinity. This file returns and accepts a bitmask
that represents which processors can handle ISRs that have been registered for this IRQ.
This way, you can, for example, decide to set the affinity of a hard-IRQ to one CPU while
setting the affinity of the threaded handler to another CPU.

Requesting a context IRQ

A driver requesting an IRQ must know the nature of the interrupt in advance and decide
whether its handler can run in the hard-IRQ context in order to call `request_irq()` or
`request_threaded_irq()` accordingly.

There is a problem when it comes to request IRQ lines provided by discrete and non-MMIO-based interrupt controllers, such as I2C/SPI gpio-expanders. Since accessing those buses may cause them to sleep, it would be a disaster to run the handler of such slow controllers in a hard-IRQ context. Since the driver does not contain any information about the nature of the interrupt line/controller, the IRQ core provides the `request_any_context_irq()` API. This function determines whether the interrupt controller/line can sleep and calls the appropriate requesting function:

```
int request_any_context_irq(unsigned int irq,
                            irq_handler_t handler,
                            unsigned long flags,
                            const char *name,
                            void *dev_id)
```

`request_any_context_irq()` and `request_irq()` have the same interface but different semantics. Depending on the underlying context (the hardware platform), `request_any_context_irq()` selects either a hardIRQ handling method using `request_irq()` or a threaded handling method using `request_threaded_irq()`. It returns a negative error value on failure, while on success, it returns either `IRQC_IS_HARDIRQ` (meaning hardI-RQ handling is used) or `IRQC_IS_NESTED` (meaning the threaded version is used). With this function, the behavior of the interrupt handler is decided at runtime. For more information, take a look at the comment introducing it in the kernel by following this link: `https://git.kernel.org/pub/scm/linux/kernel/git/next/linux-next.git/commit/?id=ae731f8d0785`.

The advantage of using `request_any_context_irq()` is that you don't need to care about what can be done in the IRQ handler. This is because the context in which the handler will run depends on the interrupt controller that provides the IRQ line. For example, for a gpio-IRQ-based device driver, if the gpio belongs to a controller that seats on an I2C or SPI bus (in which case gpio access may sleep), the handler will be threaded. Otherwise (the gpio access may not sleep and is memory mapped as it belongs to the SoC), the handler will run in the hardIRQ context.

In the following example, the device expects an IRQ line mapped to a gpio. The driver cannot assume that the given gpio line will be memory mapped since it's coming from the SoC. It may come from a discrete I2C or SPI gpio controller as well. A good practice would be to use `request_any_context_irq()` here:

```
static irqreturn_t packt_btn_interrupt(int irq, void *dev_id)
{
    struct btn_data *priv = dev_id;
```

```
        input_report_key(priv->i_dev,
                         BTN_0,
                         gpiod_get_value(priv->btn_gpiod) & 1);
        input_sync(priv->i_dev);
        return IRQ_HANDLED;
}

static int btn_probe(struct platform_device *pdev)
{
        struct gpio_desc *gpiod;
        int ret, irq;

        gpiod = gpiod_get(&pdev->dev, "button", GPIOD_IN);
        if (IS_ERR(gpiod))
            return -ENODEV;

        priv->irq = gpiod_to_irq(priv->btn_gpiod);
        priv->btn_gpiod = gpiod;
        [...]
        ret = request_any_context_irq(
                      priv->irq,
                      packt_btn_interrupt,
                      (IRQF_TRIGGER_FALLING | IRQF_TRIGGER_RISING),
                      "packt-input-button",
                      priv);
        if (ret < 0)
            goto err_btn;
    return 0;

err_btn:
        do_cleanup();
        return ret;
}
```

The preceding code is simple enough but is quite safe thanks to `request_any_context_irq()`, which prevents us from mistaking the type of the underlying gpio.

Using a workqueue to defer a bottom-half

Since we have already discussed the workqueue API, we will provide an example of how to use it here. This example is not error-free and has not been tested. It is just a demonstration that highlights the concept of bottom-half deferring by means of a workqueue.

Let's start by defining the data structure that will hold the elements we need for further development:

```
struct private_struct {
    int counter;
    struct work_struct my_work;
    void __iomem *reg_base;
    spinlock_t lock;
    int irq;
    /* Other fields */
    [...]
};
```

In the preceding data structure, our work structure is represented by the my_work element. We aren't using the pointer here because we will need to use the container_ of() macro to grab a pointer to the initial data structure. Next, we can define the method that will be invoked in the worker thread:

```
static void work_handler(struct work_struct *work)
{
    int i;
    unsigned long flags;
    struct private_data *my_data =
            container_of(work, struct private_data, my_work);
    /*
    * let's proccessing at least half of MIN_REQUIRED_FIFO_SIZE
    * prior to re-enabling the irq at device level, and so that
    * buffer further data
    */
    for (i = 0, i < MIN_REQUIRED_FIFO_SIZE, i++) {
        device_pop_and_process_data_buffer();
        if (i == MIN_REQUIRED_FIFO_SIZE / 2)
            enable_irq_at_device_level();
```

```
    }

    spin_lock_irqsave(&my_data->lock, flags);
    my_data->buf_counter -= MIN_REQUIRED_FIFO_SIZE;
    spin_unlock_irqrestore(&my_data->lock, flags);
}
```

In the preceding code, we start data processing when enough data has been buffered. Now, we can provide our IRQ handler, which is responsible for scheduling our work, as follows:

```
/* This is our hard-IRQ handler.*/
static irqreturn_t my_interrupt_handler(int irq, void *dev_id)
{
    u32 status;
    unsigned long flags;
    struct private_struct *my_data = dev_id;

    /* Let's read the status register in order to determine how
     * and what to do
     */
    status = readl(my_data->reg_base + REG_STATUS_OFFSET);

    /*
     * Let's ack this irq at device level. Even if it raises
     * another irq, we are safe since this irq remain disabled
     * at controller level while we are in this handler
     */
    writel(my_data->reg_base + REG_STATUS_OFFSET,
        status | MASK_IRQ_ACK);
    /*
     * Protecting the shared resource, since the worker also
     * accesses this counter
     */
    spin_lock_irqsave(&my_data->lock, flags);
    my_data->buf_counter++;
    spin_unlock_irqrestore(&my_data->lock, flags);

    /*
```

```
        * Ok. Our device raised an interrupt in order to inform it
        * has some new data in its fifo. But is it enough for us
        * to be processed
        */
    if (my_data->buf_counter != MIN_REQUIRED_FIFO_SIZE)) {
        /* ack and re-enable this irq at controller level */
        return IRQ_HANDLED;
    } else {
        /*
         * Right. prior to schedule the worker and returning
         * from this handler, we need to disable the irq at
         * device level
         */
        writel(my_data->reg_base + REG_STATUS_OFFSET,
                MASK_IRQ_DISABLE);
                schedule_work(&my_work);
    }

        /* This will re-enable the irq at controller level */
        return IRQ_HANDLED;
};
```

The comments in the IRQ handler code are meaningful enough. `schedule_work()` is the function that schedules our work. Finally, we can write our `probe` method, which will request our IRQ and register the previous handler:

```
static int foo_probe(struct platform_device *pdev)
{
    struct resource *mem;
    struct private_struct *my_data;

    my_data = alloc_some_memory(sizeof(struct private_struct));
    mem = platform_get_resource(pdev, IORESOURCE_MEM, 0);
    my_data->reg_base =
        ioremap(ioremap(mem->start, resource_size(mem)));

    if (IS_ERR(my_data->reg_base))
        return PTR_ERR(my_data->reg_base);
```

```
    /*
     * work queue initialization. "work_handler" is the
     * callback that will be executed when our work is
     * scheduled.
     */
    INIT_WORK(&my_data->my_work, work_handler);
    spin_lock_init(&my_data->lock);
    my_data->irq = platform_get_irq(pdev, 0);

    if (request_irq(my_data->irq, my_interrupt_handler,
                    0, pdev->name, my_data))
        handler_this_error()
    return 0;
}
```

The structure of the preceding probe method shows without a doubt that we are facing a platform device driver. Generic IRQ and workqueue APIs have been used here to initialize our workqueue and register our handler.

Locking from within an interrupt handler

If a resource is shared between two or more use contexts (kthread, work, threaded IRQ, and so on) and only with a threaded bottom-half (that is, they're never accessed by the hard-IRQ), then mutex locking is the way to go, as shown in the following example:

```
static int my_probe(struct platform_device *pdev)
{
    int irq;
    int ret;

    irq = platform_get_irq(pdev, i);
    ret = devm_request_threaded_irq(dev, irq, NULL,
                                        my_threaded_irq,
                                        IRQF_ONESHOT, dev_
                                        name(dev),
                                        my_data);
    [...]
    return ret;
```

```
}

static irqreturn_t my_threaded_irq(int irq, void *dev_id)
{
    struct priv_struct *my_data = dev_id;

    /* Save FIFO Underrun & Transfer Error status */
    mutex_lock(&my_data->fifo_lock);

    /* accessing the device's buffer through i2c */
    [...]

    mutex_unlock(&ldev->fifo_lock);
    return IRQ_HANDLED;
}
```

In the preceding code, both the user task (kthread, work, and so on) and the threaded bottom half must hold the mutex before accessing the resource.

The preceding case is the simplest one to exemplify. The following are some rules that will help you lock between hard-IRQ contexts and others:

- *If a resource is shared between a user context and a hard interrupt handler*, you will want to use the spinlock variant, which disables interrupts; that is, the simple `_irq` or `_irqsave/_irq_restore` variants. This ensures that the user context is never preempted by this IRQ when it's accessing the resource. This can be seen in the following example:

```
static int my_probe(struct platform_device *pdev)
{
    int irq;
    int ret;
    [...]
    irq = platform_get_irq(pdev, 0);
    if (irq < 0)
        goto handle_get_irq_error;

    ret = devm_request_threaded_irq(&pdev->dev,
                                    irq,
```

```
                                        my_hardirq,
                                        my_threaded_irq,
                                        IRQF_ONESHOT,
                                        dev_name(dev),
                                        my_data);
    if (ret < 0)
        goto err_cleanup_irq;
    [...]
    return 0;
}
static irqreturn_t my_hardirq(int irq, void *dev_id)
{
    struct priv_struct *my_data = dev_id;
    unsigned long flags;

    /* No need to protect the shared resource */
    my_data->status = __raw_readl(
            my_data->mmio_base + my_data->foo.reg_offset);
    /* Let us schedule the bottom-half */
    return IRQ_WAKE_THREAD;
}


static irqreturn_t my_threaded_irq(int irq, void *dev_id)
{
    struct priv_struct *my_data = dev_id;
    spin_lock_irqsave(&my_data->lock, flags);
    /* Processing the status status */
    process_status(my_data->status);
    spin_unlock_irqrestore(&my_data->lock, flags);
    [...]
    return IRQ_HANDLED;
}
```

In the preceding code, the hard-IRQ handler doesn't need to hold the spinlock as it can never be preempted. Only the user context must be held. There is a case where protection may not be necessary between the hard-IRQ and its threaded counterpart; that is, when the `IRQF_ONESHOT` flag is set while requesting the IRQ line. This flag keeps the interrupt disabled after the hard-IRQ handler has finished. With this flag set, the IRQ line remains disabled until the threaded handler has been run until its completion. This way, the hard-IRQ handler and its threaded counterpart will never compete and a lock for a resource shared between the two might not be necessary.

- When the resource is shared between user context and softIRQ, there are two things you need to guard against: the fact the user context can be interrupted by the softIRQ (remember, softIRQs run on the return path of hard-IRQ handlers) and the fact that the critical region can be entered from another CPU (remember, the same softIRQ may run concurrently on another CPU). In this case, you should use spinlock API variants that will disable softIRQs; that is, `spin_lock_bh()` and `spin_unlock_bh()`. The _bh prefix means the bottom half. Because those APIs have not been discussed in detail in this chapter, you can use the _irq or even _irqsave variants, which disable hardware interrupts as well.

- The same applies to tasklets (because tasklets are built on top of softIRQs), with the only difference that a tasklet never runs concurrently (it never runs on more than one CPU at once); a tasklet is exclusive by design.

- There are two things to guard against when it comes to locking between hard IRQ and softIRQ: the softIRQ can be interrupted by the hard-IRQ and the critical region can be entered (1 for either by another hard-IRQ if designed in this way, 2 by the same softIRQ, or 3 by another softIRQ) from another CPU. Because the softIRQ can never run when the hard-IRQ handler is running, hard-IRQ handlers only need to use the `spin_lock()` and `spin_unlock()` APIs, which prevent concurrent access by other hard handlers on another CPU. However, softIRQ needs to use the locking API that actually disables interrupts – that is, the _irq() or irqsave() variants – with a preference for the latter.

- Because softIRQs may run concurrently, *locking may be necessary between two different softIRQs, or even between a softIRQ and itself* (running on another CPU). In this case, `spinlock()`/`spin_unlock()` should be used. There's no need to disable hardware interrupts.

At this point, we are done looking at interrupt locking, which means we have come to the end of this chapter.

Summary

This chapter introduced some core kernel functionalities that will be used in the next few chapters of this book. The concepts we covered concerned bit manipulation to Linux kernel interrupt design and implementation, through locking helpers and work deferring mechanisms. By now, you should be able to decide whether you should split your interrupt handler into two parts or not, as well as know what locking primitive suits your needs.

In the next chapter, we'll cover Linux kernel managed resources, which is an interface that's used to offload allocated resource management to the kernel core.

2

Leveraging the Regmap API and Simplifying the Code

This chapter introduces the Linux kernel register mapping abstraction layer and shows how to simplify and delegate I/O operations to the regmap subsystem. Dealing with devices, whether they are built-in in the SoC (**memory mapped I/O**, also known as **MMIO**) or seated on I2C/SPI buses, consists of accessing (reading/modifying/updating) registers. Regmap became necessary because a lot of device drivers open-coded their register access routines. **Regmap** stands for **Register Map**. It was primarily developed for **ALSA SoC (ASoC)** in order to get rid of redundant open-coded SPI/I2C register access routines in codec drivers. At its origin, regmap provided a set of APIs for reading/writing non-memory-map I/O (for example, I2C and SPI read/write). Since then, MMIO regmap has been upgraded so that we can use regmap to access MMIO.

Nowadays, this framework abstracts I2C, SPI, and MMIO register access, and not only handles locking when necessary, but also manages the register cache, as well as register readability and writability. It also handles IRQ chips and IRQs. This chapter will discuss regmap and explain the way to use it to abstract register access with I2C, SPI, and MMIO devices. We will also describe how to use regmap to manage IRQ and IRQ controllers.

This chapter will cover the following topics:

- Introduction to regmap and its data structures: I2C, SPI, and MMIO
- Regmap and IRQ management
- Regmap IRQ API and data structures

Technical requirements

In order to be comfortable when going through this chapter, you'll need the following:

- Good C programming skills
- Familiarity with the concept of the device tree
- Linux kernel v4.19.X sources, available at `https://git.kernel.org/pub/scm/linux/kernel/git/stable/linux.git/refs/tags`

Introduction to regmap and its data structures – I2C, SPI, and MMIO

Regmap is an abstraction register access mechanism provided by the Linux kernel that mainly targets SPI, I2C, and memory-mapped registers.

APIs in this framework are bus agnostic and handle the underlying configuration under the hood. That being said, the main data structure in this framework is `struct regmap_config`, defined in `include/linux/regmap.h` in the kernel source tree as follows:

```
struct regmap_config {
    const char *name;
    int reg_bits;
    int reg_stride;
    int pad_bits;
    int val_bits;
    bool (*writeable_reg)(struct device *dev, unsigned int reg);
    bool (*readable_reg)(struct device *dev, unsigned int reg);
    bool (*volatile_reg)(struct device *dev, unsigned int reg);
    bool (*precious_reg)(struct device *dev, unsigned int reg);

    int (*reg_read)(void *context, unsigned int reg,
```

```
                       unsigned int *val);
    int (*reg_write)(void *context, unsigned int reg,
                     unsigned int val);

    bool disable_locking;
    regmap_lock lock;
    regmap_unlock unlock;
    void *lock_arg;
    bool fast_io;

    unsigned int max_register;
    const struct regmap_access_table *wr_table;
    const struct regmap_access_table *rd_table;
    const struct regmap_access_table *volatile_table;
    const struct regmap_access_table *precious_table;
    const struct reg_default *reg_defaults;
    unsigned int num_reg_defaults;

    unsigned long read_flag_mask;
    unsigned long write_flag_mask;
    enum regcache_type cache_type;
    bool use_single_rw;
    bool can_multi_write;
};
```

For simplicity, some of the fields in this structure have been removed and are not
discussed in this chapter. As long as `struct regmap_config` is properly completed,
users may ignore underlying bus mechanisms. Let's introduce the fields in this data
structure:

- `reg_bits` indicates the size of a register in terms of bits. In other words, it is the
  number of bits in a register's address.

- `reg_stride` is the stride of the register address. A register address is valid if it is
  a multiple of this value. If set to `0`, a value of `1` will be used, meaning any address
  is valid. Any read/write to an address that is not a multiple of this value will return
  `-EINVAL`.

- `pad_bits` is the number of bits of padding between the register and the value. This is the number of bits to shift the register's value left when formatting.

- `val_bits`: This represents the number of bits used to store a register's value. It is a mandatory field.

- `writeable_reg`: If provided, this optional callback will be called on each regmap write operation to check whether the given address is writable or not. If this function returns `false` on an address given to a regmap write transaction, the transaction will return `-EIO`. The following excerpt shows how this callback can be implemented:

```
static bool foo_writeable_register(struct device *dev,
                                    unsigned int reg)
{
    switch (reg) {
    case 0x30 ... 0x38:
    case 0x40 ... 0x45:
    case 0x50 ... 0x57:
    case 0x60 ... 0x6e:
    case 0xb0 ... 0xb2:
        return true;
    default:
        return false;
    }
}
```

- `readable_reg`: This is the same as `writeable_reg` but for register read operations.

- `volatile_reg`: This is an optional callback that, if provided, will be called every time a register needs to be read or written through the regmap cache. If the register is volatile (the register value can't be cached), the function should return `true`. A direct read/write is then performed on the register. If `false` is returned, it means the register is cacheable. In this case, the cache will be used for a read operation, and the cache will be written to in the case of a write operation. The following is an example, with fake register addresses chosen randomly:

```
static bool volatile_reg(struct device *dev,
                          unsigned int reg)
{
    switch (reg) {
```

```
    case 0x30:
    case 0x31:
    [...]
    case 0xb3:
        return false;
    case 0xb4:
        return true;
    default:
        if ((reg >= 0xb5) && (reg <= 0xcc))
            return false;
    [...]
        break;
    }
    return true;
}
```

- `reg_read`: If your device needs *special hacks* for reading operations, you can provide a custom read callback and make this field point to it so that instead of using standard regmap read functions, this callback is used. That said, most devices do not need this.

- `reg_write`: This is the same as `reg_read` but for write operations.

- `disable_locking`: This shows whether the `lock/unlock` callbacks should be used or not. If `false`, no locking mechanisms will be used. It means this regmap is either protected by external means or is guaranteed not to be accessed from multiple threads.

- `lock/unlock`: These are optional lock/unlock callbacks that override the regmap's default lock/unlock functions. These are based on spinlock or mutex, depending on whether accessing the underlying device may sleep or not.

- `lock_arg`: This is the only argument of the `lock/unlock` functions (it will be ignored if the regular lock/unlock functions are not overridden).

- `fast_io`: This indicates that the register's I/O is fast. If set, the regmap will use a spinlock instead of a mutex to perform locking. This field is ignored if custom lock/unlock (not discussed here) functions are used (see the `lock/unlock` fields of `struct regmap_config` in the kernel sources). It should be used only for "no bus" cases (MMIO devices), not for slow buses such as I2C, SPI, or similar buses whose accesses may sleep.

- `wr_table`: This is an alternative to the `writeable_reg()` callback, of type `regmap_access_table`, which is a structure holding a `yes_range` and a `no_range` field, both of which are pointers to `struct regmap_range`. Any register that belongs to a `yes_range` entry is considered writable, and is considered not writable if it belongs to `no_range` or is not specified in `yes_range`.

- `rd_table`: This is the same as `wr_table`, but for any read operation.

- `volatile_table`: Instead of `volatile_reg`, you could provide `volatile_table`. The principle is the same as `wr_table` and `rd_table`, but for the caching mechanism.

- `max_register`: This is optional; it specifies the maximum valid register address upon which no operation is permitted.

- `reg_defaults` is an array of elements of type `reg_default`, where each element is a `{reg, value}` pair that represents the power-on reset values for a given register. This is used along with the cache so that reading an address that exists in this array and that has not been written since a power-on reset will return the default register value in this array without performing any read transactions on the device. An example of this is the IIO device driver, which you can find out more about at `https://elixir.bootlin.com/linux/v4.19/source/drivers/iio/light/apds9960.c`.

- `use_single_rw`: This is a Boolean that, if set, will instruct the regmap to convert any bulk write or read operations on the device into a series of single write or read operations. This is useful for devices that do not support bulk read and/or write operations.

- `can_multi_write`: This only targets write operations. If set, it indicates that this device supports the multi-write mode of bulk write operations. If it's empty, multi-write requests will be split into individual write operations.

- `num_reg_defaults`: This is the number of elements in `reg_defaults`.

- `read_flag_mask`: This is a mask to be set in the highest bytes of the register when doing a read. Normally, in SPI or I2C, a write or a read will have the highest bit set in the top byte to differentiate write and read operations.

- `write_flag_mask`: This is a mask to be set in the highest bytes of the register when doing a write.

- `cache_type`: This is the actual cache type, which can be either `REGCACHE_NONE`, `REGCACHE_RBTREE`, `REGCACHE_COMPRESSED`, or `REGCACHE_FLAT`.

Initializing a regmap is as simple as calling one of the following functions depending on the bus behind which our device sits:

```
struct regmap * devm_regmap_init_i2c(
                struct i2c_client *client,
                struct regmap_config *config)

struct regmap * devm_regmap_init_spi(
                struct spi_device *spi,
                const struct regmap_config);

struct regmap * devm_regmap_init_mmio(
                struct device *dev,
                void __iomem *regs,
                const struct regmap_config *config)

#define devm_regmap_init_spmi_base(dev, config) \
    __regmap_lockdep_wrapper(__devm_regmap_init_spmi_base, \
                        #config, dev, config)

#define devm_regmap_init_w1(w1_dev, config) \
    __regmap_lockdep_wrapper(__devm_regmap_init_w1, #config, \
                        w1_dev, config)
```

In the preceding prototypes, the return value will be a valid pointer to struct regmap or ERR_PTR() if there is an error. The regmap will be automatically freed by the device management code. regs is a pointer to the memory-mapped IO region (returned by devm_ioremap_resource() or any ioremap* family function). dev is the device (of type struct device) that will be interacted with. The following example is an excerpt of drivers/mfd/sun4i-gpadc.c in the kernel source code:

```
struct sun4i_gpadc_dev {
    struct device *dev;
    struct regmap *regmap;
    struct regmap_irq_chip_data *regmap_irqc;
    void __iomem *base;
};
```

```
static const struct regmap_config sun4i_gpadc_regmap_config = {
    .reg_bits = 32,
    .val_bits = 32,
    .reg_stride = 4,
    .fast_io = true,
};

static int sun4i_gpadc_probe(struct platform_device *pdev)
{
    struct sun4i_gpadc_dev *dev;
    struct resource *mem;
    [...]
    mem = platform_get_resource(pdev, IORESOURCE_MEM, 0);
    dev->base = devm_ioremap_resource(&pdev->dev, mem);
    if (IS_ERR(dev->base))
        return PTR_ERR(dev->base);
    dev->dev = &pdev->dev;
    dev_set_drvdata(dev->dev, dev);
    dev->regmap = devm_regmap_init_mmio(dev->dev, dev->base,
                                &sun4i_gpadc_regmap_config);

    if (IS_ERR(dev->regmap)) {
        ret = PTR_ERR(dev->regmap);
        dev_err(&pdev->dev, "failed to init regmap: %d\n", ret);
        return ret;
    }
    [...]
```

This excerpt shows how to create a regmap. Though this excerpt is MMIO-oriented,
the concept remains the same for other types. Instead of using devm_regmap_init_
MMIO(), we would use devm_regmap_init_spi() or devm_regmap_init_i2c()
respectively for an SPI- or I2C-based regmap.

Accessing device registers

There are two main functions for accessing device registers. These are `regmap_write()` and `regmap_read()`, which take care of locking and abstracting the underlying bus:

```
int regmap_write(struct regmap *map,
                 unsigned int reg,
                 unsigned int val);

int regmap_read(struct regmap *map,
                unsigned int reg,
                unsigned int *val);
```

In the preceding two functions, the first argument, `map`, is the regmap structure returned during initialization. `reg` is the register address to write/read data to/from. `val` is the data to be written in a write operation, or the read value in a read operation. The following is a detailed description of these APIs:

- `regmap_write` is used to write data to the device. The following are the steps performed by this function:

    1) First, it checks whether `reg` is aligned with the `regmap_config.reg_stride`. If not, it returns `-EINVAL` and the function fails.

    2) It then takes the lock depending on the `fast_io`, `lock`, and `unlock` fields. If a `lock` callback is provided, it will be used to take the lock. Otherwise, the regmap core will use its internal default lock function, using a spinlock or a mutex depending on whether `fast_io` has been set or not. Next, the regmap core performs some sanity checks on the register address passed as follows:

    --If `max_register` is set, it will check whether this register's address is less than `max_register`. If the address is not less than `max_register`, then `regmap_write()` fails, returning an `-EIO` (invalid I/O) error code

    --Then, if the `writeable_reg` callback is set, this callback is called with the register as a parameter. If this callback returns `false`, then `regmap_write()` fails, returning `-EIO`. If `writeable_reg` is not set but `wr_table` is set, the regmap core will check whether the register address lies within `no_range`. If it does, then `regmap_write()` fails and returns `-EIO`. If it doesn't, the regmap core will check whether the register address lies in `yes_range`. If it is not present there, then `regmap_write()` fails and returns `-EIO`.

3) If the `cache_type` field is set, then caching will be used. The value to be written will be cached for future reference instead of being written to the hardware.

4) If `cache_type` is not set, then the write routine is invoked immediately to write the value into the hardware register. This routine will first apply `write_flag_mask` to the first byte of the register address before writing the value into this register.

5) Finally, the lock is released using the appropriate unlocking function.

- `regmap_read` is used to read data from the device. This function performs the same security and sanity checks as `regmap_write()`, but replaces `writable_reg` and `wr_table` with `readable_reg` and `rd_table`. When it comes to caching, if it is enabled, the register value is read from the cache. If caching is not enabled, the read routine is called to read the value from the hardware register instead. That routine will apply `read_flag_mask` to the highest byte of the register address prior to the read operation, and `*val` is updated with the new value read. After this, the lock is released using the appropriate unlocking function.

While the preceding accessors target a single register at a time, others can perform bulk accesses, as we will see in the next section.

Reading/writing multiple registers in a single shot

Sometimes you may want to perform bulk read/write operations of data from/to a register range at the same time. Even if you use `regmap_read()` or `regmap_write()` in a loop, the best solution would be to use the regmap APIs provided for such situations. These functions are `regmap_bulk_read()` and `regmap_bulk_write()`:

```
int regmap_bulk_read(struct regmap *map, unsigned int reg,
                      void *val, size_tval_count);
int regmap_bulk_write(struct regmap *map, unsigned int reg,
                       const void *val, size_t val_count)
```

These functions read/write multiple registers from/to the device. map is the regmap used to perform operations. For a read operation, reg is the first register from where reading should start, val is a pointer to the buffer where read values should be stored in *native register size* of the device (it means if the device register size is 4 bytes, the read value will be stored in 4 bytes units), and val_count is the number of registers to read. For a write operation, reg is the first register to be written from, val is a pointer to the block of data to be written in *native register size* of the device, and val_count is the number of registers to write. For both of these functions, a value of 0 will be returned on success and a negative errno will be returned if there is an error.

> **Tip**
>
> There are other interesting read/write functions provided by this framework. Take a look at the kernel header file for more information. An interesting one is regmap_multi_reg_write(), which writes multiple registers in a set of {register, value} pairs supplied in any order, possibly not all in a single range, to the device given as a parameter.

Now that we are familiar with register access, we can go further by managing register content at a bit level.

Updating bits in registers

To update a bit in a given register, we have regmap_update_bits(), a three-in-one function. Its prototype is as follows:

```
int regmap_update_bits(struct regmap *map, unsigned int reg,
                       unsigned int mask, unsigned int val)
```

It performs a read/modify/write cycle on the register map. It is a wrapper of _regmap_update_bits(), which looks as follows:

```
static int _regmap_update_bits(
               struct regmap *map, unsigned int reg,
               unsigned int mask, unsigned int val,
               bool *change, bool force_write)
{
    int ret;
    unsigned int tmp, orig;

    if (change)
```

```c
        *change = false;

    if (regmap_volatile(map, reg) && map->reg_update_bits) {
        ret = map->reg_update_bits(map->bus_context,
                                        reg, mask, val);
        if (ret == 0 && change)
            *change = true;
    } else {
        ret = _regmap_read(map, reg, &orig);
        if (ret != 0)
            return ret;

        tmp = orig & ~mask;
        tmp |= val & mask;
        if (force_write || (tmp != orig)) {
            ret = _regmap_write(map, reg, tmp);
            if (ret == 0 && change)
                *change = true;
        }
    }
    return ret;
}
```

Bits that need to be updated should be set to 1 in mask, and the corresponding bits will be given the value of the bit of the same position in val. As an example, to set the first (BIT(0)) and third (BIT(2)) bits to 1, mask should be 0b00000101 and the value should be 0bxxxxxx1x1. To clear the seventh bit (BIT(6)), mask must be 0b01000000 and the value should be 0bx0xxxxxx, and so on.

Tip

For debugging purpose, you can use the debugfs filesystem to dump the content of the regmap managed registers, as the following excerpt shows:

mount -t debugfs none /sys/kernel/debug

cat /sys/kernel/debug/regmap/1-0008/registers

This will dump the register addresses along with their values in <addr:value> format.

In this section, we have seen how easy it is to access hardware registers. Moreover, we have learned some fancy tricks for playing with registers at the bit level, which is often used in status and configuration registers. Next, we will have a look at IRQ management.

Regmap and IRQ management

Regmap does not only abstract access to registers. Here, we will see how this framework abstracts IRQ management at a lower level, such as IRQ chip handling, thus hiding boilerplate operations.

Quick recap on Linux kernel IRQ management

IRQs are exposed to devices by means of special devices called interrupt controllers. From a software point of view, an interrupt controller device driver manages and exposes these lines using the virtual IRQ concept, known as the IRQ domain in the Linux kernel. Interrupt management is built on top of the following structures:

- `struct irq_chip`: This structure is the Linux representation of an IRQ controller and implements a set of methods to drive the interrupt controller that are directly called by the core IRQ code. If necessary, this structure should be filled by the driver, providing a set of callbacks allowing us to manage IRQs on the IRQ chip, such as `irq_startup`, `irq_shutdown`, `irq_enable`, `irq_disable`, `irq_ack`, `irq_mask`, `irq_unmask`, `irq_eoi`, and `irq_set_affinity`. Dumb IRQ chip devices (chip that does not allow IRQ management, for example) should use the kernel-provided `dummy_irq_chip`.

- `struct irq_domain`: Each interrupt controller is given a domain, which is for the controller what the address space is for a process. The `struct irq_domain` structure stores mappings between hardware IRQs and Linux IRQs (that is, virtual IRQs, or virq). It is the hardware interrupt number translation object. This structure provides the following:

  --A pointer to the firmware node for a given interrupt controller (`fwnode`).

  --A method to convert a firmware (device tree) description of an IRQ into an ID local to the interrupt controller (the **hardware IRQ** number, known as the **hwirq**). For gpio chips that also act as IRQ controllers, the hardware IRQ number (hwirq) for a given gpio line corresponds to the local index of this line in the chip most of times.

  --A way to retrieve the Linux view of an IRQ from the hwirq.

- `struct irq_desc`: This structure is the Linux kernel view of an interrupt, containing all of the core stuff and one-to-one mapping to the Linux interrupt number.

- `struct irq_action`: This is the structure Linux uses to describe an IRQ handler.

- `struct irq_data`: This structure is embedded in the `struct irq_desc` structure, and contains the following:

  --The data that is relevant to the `irq_chip` managing this interrupt

  --Both the Linux IRQ number and the hwirq

  --A pointer to the `irq_chip`

  --A pointer to the interrupt translation domain (`irq_domain`)

Always keep in mind that **the irq_domain is for the interrupt controller what an address space is for a process, as it stores mappings between virqs and hwirqs**.

An interrupt controller driver creates and registers `irq_domain` by calling one of the `irq_domain_add_<mapping_method>()` functions. These functions are actually `irq_domain_add_linear()`, `irq_domain_add_tree()`, and `irq_domain_add_nomap()`. In fact, `<mapping_method>` is the method by which `hwirqs` should be mapped to `virqs`.

`irq_domain_add_linear()` creates an empty and fixed-size table, indexed by the hwirq number. `struct irq_desc` is allocated for each hwirq that gets mapped. The allocated IRQ descriptor is then stored in the table, at the index that equals the hwirq to which it has been allocated. This linear mapping is suitable for fixed and small numbers of hwirqs (lower than 256).

While the main advantages of this mapping are the fact that the IRQ number lookup time is fixed and that `irq_desc` is allocated for in-use IRQs only, the major drawback comes from the size of the table, which can be as large as the largest possible `hwirq` number. The majority of drivers should use the linear map. This function has the following prototype:

```
struct irq_domain *irq_domain_add_linear(
                    struct device_node *of_node,
                    unsigned int size,
                    const struct irq_domain_ops *ops,
                    void *host_data)
```

`irq_domain_add_tree()` creates an empty `irq_domain` that maintains the mapping between Linux IRQs and `hwirq` numbers in a radix tree. When an hwirq is mapped, a `struct irq_desc` is allocated, and the hwirq is used as the lookup key for the radix tree. A tree map is a good choice if the hwirq number is very large, since it does not need to allocate a table as large as the largest hwirq number. The disadvantage is that the `hwirq-to-IRQ` number lookup is dependent on how many entries are in the table. Very few drivers should need this mapping. It has the following prototype:

```
struct irq_domain *irq_domain_add_tree(
                      struct device_node *of_node,
                      const struct irq_domain_ops *ops,
                      void *host_data)
```

`irq_domain_add_nomap()` is something you will probably never use; however, its entire description is available in `Documentation/IRQ-domain.txt`, in the kernel source tree. Its prototype is as follows:

```
struct irq_domain *irq_domain_add_nomap(
                      struct device_node *of_node,
                      unsigned int max_irq,
                      const struct irq_domain_ops *ops,
                      void *host_data)
```

In all of those prototypes, `of_node` is a pointer to the interrupt controller's DT node. `size` represents the number of interrupts in the domain in case of linear mapping. `ops` represents map/unmap domain callbacks, and `host_data` is the controller's private data pointer. As these three functions all create empty `irq` domains, you should use the `irq_create_mapping()` function with the hwirq and a pointer to the `irq` domain passed to it in order to create a mapping, and insert this mapping into the domain:

```
unsigned int irq_create_mapping(struct irq_domain *domain,
                           irq_hw_number_t hwirq)
```

In the preceding prototype, `domain` is the domain to which this hardware interrupt belongs. A NULL value means the default domain. `hwirq` is the hardware IRQ number you need to create a mapping for. This function maps a hardware interrupt into the Linux IRQ space and returns a Linux IRQ number. Also, keep in mind that only one mapping per hardware interrupt is permitted. The following is an example of creating a mapping:

```
unsigned int virq = 0;
virq = irq_create_mapping(irq_domain, hwirq);
```

```
if (!virq) {
    ret = -EINVAL;
    goto err_irq;
}
```

In the preceding code, `virq` is the Linux kernel IRQ (the **virtual IRQ number**, **virq**) corresponding to the mapping.

> **Important note**
>
> When writing drivers for GPIO controllers that are also interrupt controllers, `irq_create_mapping()` is called from within the `gpio_chip.to_irq()` callback, and the virq is returned as `return irq_create_mapping(gpiochip->irq_domain, hwirq)`, where `hwirq` is the GPIO offset from the GPIO chip.

Some drivers prefer creating the mappings and populating the domain for each hwirq in advance inside the `probe()` function, as shown here:

```
for (j = 0; j < gpiochip->chip.ngpio; j++) {
    irq = irq_create_mapping(gpiochip ->irq_domain, j);
}
```

After this, such drivers just call `irq_find_mapping()` (given the hwirq) into the `to_irq()` callback function. `irq_create_mapping()` will allocate a new `struct irq_desc` structure if no mapping already exists for the given `hwirq`, associate it with the hwirq, and call the `irq_domain_ops.map()` callback (by using the `irq_domain_associate()` function) so that the driver can perform any required hardware setup.

The struct irq_domain_ops

This structure exposes some callbacks that are specific to the irq domain. As mappings are created in a given irq domain, each mapping (actually each `irq_desc`) should be given an irq configuration, some private data, and a translation function (given a device tree node and an interrupt specifier, the translation function decodes the hardware irq number and Linux irq type value). This is what callbacks in this structure do:

```
struct irq_domain_ops {
    int (*map)(struct irq_domain *d, unsigned int virq,
                  irq_hw_number_t hw);
    void (*unmap)(struct irq_domain *d, unsigned int virq);
```

```
    int (*xlate)(struct irq_domain *d, struct device_node *node,
                 const u32 *intspec, unsigned int intsize,
                 unsigned long *out_hwirq,
                 unsigned int *out_type);
};
```

Each Linux kernel IRQ management of the elements in the preceding data structure deserves a section on its own to describe it.

irq_domain_ops.map()

The following is the prototype of this callback:

```
int (*map)(struct irq_domain *d, unsigned int virq,
           irq_hw_number_t hw);
```

Before describing what this function does, let's describe its arguments:

- d: The IRQ domain used by this IRQ chip
- virq: The global IRQ number used by this GPIO-based IRQ chip
- hw: The local IRQ/GPIO line offset on this GPIO chip

.map() creates or updates a mapping between a virq and an hwirq. This callback sets up the IRQ configuration. It is called (internally by the irq core) only once for a given mapping. This is where we set the irq chip data for the given irq, which could be done using irq_set_chip_data(), which has this prototype:

```
int irq_set_chip_data(unsigned int irq, void *data);
```

Depending on the type of the IRQ chip (nested or chained), additional actions can be performed.

irq_domain_ops.xlate()

Given a DT node and an interrupt specifier, this callback decodes the hardware IRQ number along with its Linux IRQ type value. Depending on the #interrupt-cells property specified in your DT controller node, the kernel provides a generic translation function:

- irq_domain_xlate_twocell(): This generic translation function is for direct two-cell binding. The DT IRQ specifier works with two-cell bindings, where the cell values map directly to the hwirq number and Linux IRQ flags.

- irq_domain_xlate_onecell(): This is a generic xlate function for direct one-cell bindings.

- irq_domain_xlate_onetwocell(): This is a generic xlate function for one- or two-cell bindings.

An example of the domain operation is as follows:

```
static struct irq_domain_ops mcp23016_irq_domain_ops = {
    .map = my_irq_domain_map,
    .xlate = irq_domain_xlate_twocell,
};
```

The distinctive feature of the preceding data structure is the value assigned to the .xlate element, that is, irq_domain_xlate_twocell. This means we are expecting a two-cell irq specifier in the device tree in which the first cell would specify the irq, and the second would specify its flags.

Chaining IRQs

When an interrupt occurs, the irq_find_mapping() helper function can be used to find the Linux IRQ number from the hwirq number. This hwirq number could be, for example, the GPIO offset in a bank of GPIO controllers. Once a valid virq has been found and returned, you should call either handle_nested_irq() or generic_handle_irq() on this virq. The magic comes from the previous two functions, which manage the irq-flow handlers.

This means that there are two ways to play with interrupt handlers. Hard interrupt handlers, or **chained interrupts**, are atomic and run with irqs disabled and may schedule the threaded handler; there are also the simply threaded interrupt handlers, known as **nested interrupts**, which may be interrupted by other interrupts.

Chained interrupts

This approach is used for a controller that may not sleep, such as the SoC's internal GPIO controller, which is memory-mapped and whose accesses do not sleep. *Chained* means that those interrupts are just chains of function calls (for example, the SoC's GPIO controller interrupt handler is being called from within the GIC interrupt handler, just like a function call). With this approach, child IRQ handlers are being called inside the parent hwirq handler. generic_handle_irq() must be used here for chaining child IRQ handlers inside the parent hwirq handler. Even from within the child interrupt handlers, we are still in an atomic context (hardware interrupt). You cannot call functions that may sleep.

For chained (and only chained) IRQ chips, irq_domain_ops.map() is also the right place to assign a high-level irq-type flow handler to the given irq using irq_set_chip_and_handler(), so that this high-level code, depending on what it is, will do some hacks before calling the corresponding irq handler. The magic operates here thanks to the irq_set_chip_and_handler() function:

```
void irq_set_chip_and_handler(unsigned int irq,
                              struct irq_chip *chip,
                              irq_flow_handler_t handle)
```

In the preceding prototype, irq represents the Linux IRQ (the virq), given as a parameter to the irq_domain_ops.map() function; chip is your irq_chip structure; and handle is your high-level interrupt flow handler.

> **Important note**
>
> Some controllers are quite dumb and need almost nothing in their irq_chip structure. In this case, you should pass dummy_irq_chip to irq_set_chip_and_handler(). dummy_irq_chip is defined in kernel/irq/dummychip.c.

The following code flow summarizes what `irq_set_chip_and_handler()` does:

```
void irq_set_chip_and_handler(unsigned int irq,
                              struct irq_chip *chip,
                              irq_flow_handler_t handle)
{
    struct irq_desc *desc = irq_get_desc(irq);
    desc->irq_data.chip = chip;
    desc->handle_irq = handle;
}
```

These are some possible high-level IRQ flow handlers provided by the generic layer:

```
/*
 * Built-in IRQ handlers for various IRQ types,
 * callable via desc->handle_irq()
 */
void handle_level_irq(struct irq_desc *desc);
void handle_fasteoi_irq(struct irq_desc *desc);
void handle_edge_irq(struct irq_desc *desc);
void handle_edge_eoi_irq(struct irq_desc *desc);
void handle_simple_irq(struct irq_desc *desc);
void handle_untracked_irq(struct irq_desc *desc);
void handle_percpu_irq(struct irq_desc *desc);
void handle_percpu_devid_irq(struct irq_desc *desc);
void handle_bad_irq(struct irq_desc *desc);
```

Each function name describes quite well the type of IRQ it handles. This is what `irq_domain_ops.map()` may look like for a chained IRQ chip:

```
static int my_chained_irq_domain_map(struct irq_domain *d,
                                     unsigned int virq,
                                     irq_hw_number_t hw)
{
    irq_set_chip_data(virq, d->host_data);
    irq_set_chip_and_handler(virq, &dummy_irq_chip,
                             handle_ edge_irq);
    return 0;
}
```

While writing the parent irq handler for a chained IRQ chip, the code should call `generic_handle_irq()` on each child `irq`. This function simply calls `irq_desc->handle_irq()`, which points to the high-level interrupt handler assigned to the given child IRQ using `irq_set_chip_and_handler()`. The underlying high-level `irq` event handler (let's say `handle_level_irq()`) will first do some hacks, then will run the hard `irq-handler` (`irq_desc->action->handler`) and, depending on the return value, will run the threaded handler (`irq_desc->action->thread_fn`) if provided.

Here is an example of the parent IRQ handler for a chained IRQ chip, whose original code is located in `drivers/pinctrl/pinctrl-at91.c` in the kernel source:

```
static void parent_hwirq_handler(struct irq_desc *desc)
{
    struct irq_chip *chip = irq_desc_get_chip(desc);
    struct gpio_chip *gpio_chip =
    irq_desc_get_handler_ data(desc);
    struct at91_gpio_chip *at91_gpio = gpiochip_get_data
                                        (gpio_ chip);
    void __iomem *pio = at91_gpio->regbase;
    unsigned long isr;
    int n;

    chained_irq_enter(chip, desc);
    for (;;) {
        /* Reading ISR acks pending (edge triggered) GPIO
         * interrupts. When there are none pending, we're
         * finished unless we need to process multiple banks
         * (like ID_PIOCDE on sam9263).
         */
        isr = readl_relaxed(pio + PIO_ISR) &
                        readl_relaxed(pio + PIO_IMR);
        if (!isr) {
            if (!at91_gpio->next)
                break;
            at91_gpio = at91_gpio->next;
            pio = at91_gpio->regbase;
            gpio_chip = &at91_gpio->chip;
```

```
            continue;
        }
        for_each_set_bit(n, &isr, BITS_PER_LONG) {
            generic_handle_irq(
                    irq_find_mapping(gpio_chip->irq.domain, n));
        }
    }
    chained_irq_exit(chip, desc);
    /* now it may re-trigger */
    [...]
}
```

Chained IRQ chip drivers do not need to register the parent `irq` handler using `devm_request_threaded_irq()` or `devm_request_irq()`. This handler is automatically registered when the driver calls `irq_set_chained_handler_and_data()` on this parent irq, given the associated handler as parameter, along with some private data:

```
void irq_set_chained_handler_and_data(unsigned int irq,
                                    irq_flow_handler_t
                                    handle,
                                    void *data)
```

The parameters of this function are quite self-explanatory. You should call this function in the `probe` function as follows:

```
static int my_probe(struct platform_device *pdev)
{
    int parent_irq, i;
    struct irq_domain *my_domain;

    parent_irq = platform_get_irq(pdev, 0);
    if (!parent_irq) {
     pr_err("failed to map parent interrupt %d\n", parent_irq);
        return -EINVAL;
    }

    my_domain =
        irq_domain_add_linear(np, nr_irq, &my_irq_domain_ops,
```

```
                                    my_private_data);
    if (WARN_ON(!my_domain)) {
        pr_warn("%s: irq domain init failed\n", __func__);
        return;
    }


    /* This may be done elsewhere */
    for(i = 0; i < nr_irq; i++) {
        int virqno = irq_create_mapping(my_domain, i);

        /*
         * May need to mask and clear all IRQs before
         * registering a handler
         */
          [...]


        irq_set_chained_handler_and_data(parent_irq,
                                         parent_hwirq_handler,
                                         my_private_data);
        /*
         * May need to call irq_set_chip_data() on
         * the virqno too
         */
        [...]
    }
    [...]
}
```

In the preceding fake `probe` method, a linear domain is created using `irq_domain_add_linear()`, and an irq mapping (virtual irq) is created in this domain with `irq_create_mapping()`. Finally, we set a high-level chained flow handler and its data for the main (or parent) IRQ.

Important note

Note that `irq_set_chained_handler_and_data()` automatically enables the interrupt (specified in the first parameter), assigns its handler (also given as a parameter), and marks this interrupt as `IRQ_NOREQUEST`, `IRQ_NOPROBE`, or `IRQ_NOTHREAD`, which mean this interrupt cannot be requested via `request_irq()` anymore, cannot be probed by auto probing, and cannot be threaded at all (it is chained), respectively.

Nested interrupts

The nested flow method is used by IRQ chips that may sleep, such as those that are on slow buses, such as I2C (for example, an I2C GPIO expander). "Nested" refers to those interrupt handlers that do not run in the hardware context (they are not really hwirq, and are not in an atomic context), but are threaded instead and can be preempted. Here, the handler function is called inside the calling threads context. For nested (and only nested) IRQ chips, the `irq_domain_ops.map()` callback is also the right place to set up `irq` configuration flags. The most important configuration flags are as follows:

- `IRQ_NESTED_THREAD`: This is a flag that indicates that on `devm_request_threaded_irq()`, no dedicated interrupt thread should be created for the irq handler, as it is called nested in the context of a demultiplexing interrupt handler thread (there's more information about this in the `__setup_irq()` function, implemented in `kernel/irq/manage.c` in the kernel source). You can use `void irq_set_nested_thread(unsigned int irq, int nest)` to act on this flag, where `irq` corresponds to the global interrupt number and `nest` should be 0 to clear or 1 to set the `IRQ_NESTED_THREAD` flag.

- `IRQ_NOTHREAD`: This flag can be set using `void irq_set_nothread(unsigned int irq)`. It is used to mark the given IRQ as non-threadable.

This is what `irq_domain_ops.map()` may look like for a nested IRQ chip:

```
static int my_nested_irq_domain_map(struct irq_domain *d,
                                    unsigned int virq,
                                    irq_hw_number_t hw)
{
```

```
    irq_set_chip_data(virq, d->host_data);
    irq_set_nested_thread(virq, 1);
    irq_set_noprobe(virq);
    return 0;
}
```

While writing the parent irq handler for a nested IRQ chip, the code should call `handle_nested_irq()` in order to handle child irq handlers so that they run from the parent irq thread. `handle_nested_irq()` does not care about `irq_desc->action->handler`, which is the hard irq handler. It simply runs `irq_desc->action->thread_fn`:

```
static irqreturn_t mcp23016_irq(int irq, void *data)
{
    struct mcp23016 *mcp = data;
    unsigned int child_irq, i;
    /* Do some stuff */
    [...]
    for (i = 0; i < mcp->chip.ngpio; i++) {
        if (gpio_value_changed_and_raised_irq(i)) {
            child_irq = irq_find_mapping(mcp->chip.irqdomain,
                                         i);
            handle_nested_irq(child_irq);
        }
    }
    [...]
}
```

Nested IRQ chip drivers **must** register the parent irq handler using `devm_request_threaded_irq()`, as there is no function like `irq_set_chained_handler_and_data()` for this kind of IRQ chip. It does not make sense to use this API for nested IRQ chips. Nested IRQ chips, most of the time, are GPIO chip-based. Thus, we would be better off using the GPIO chip-based IRQ chip API, or using the regmap-based IRQ chip API, as shown in the next section. However, let's see what such an example looks like:

```
static int my_probe(struct i2c_client *client,
                    const struct i2c_device_id *id)
{
    int parent_irq, i;
```

```
    struct irq_domain *my_domain;
    [...]

    int irq_nr = get_number_of_needed_irqs();

    /* Do we have an interrupt line ? Enable the IRQ chip */
    if (client->irq) {
        domain = irq_domain_add_linear(
                        client->dev.of_node, irq_nr,
                        &my_irq_domain_ops, my_private_data);
        if (!domain) {
            dev_err(&client->dev,
                    "could not create irq domain\n");
            return -ENODEV;
        }
        /*
         * May be creating irq mapping in this domain using
         * irq_create_mapping() or let the mfd core doing
         * this if it is an MFD chip device
         */
        [...]

        ret =
            devm_request_threaded_irq(
                &client->dev, client->irq,
                NULL, my_parent_irq_thread,
                IRQF_TRIGGER_FALLING | IRQF_ONESHOT,
                "my-parent-irq", my_private_data);
        [...]
    }
    [...]
}
```

In the preceding `probe` method, there are two main differences with the chained flow:

- First, the way the main IRQ is registered: While chained IRQ chips used `irq_set_chained_handler_and_data()`, which automatically registered the handler, the nested flow method has to register its handler explicitly using the `request_threaded_irq()` family method.

- Second, the way the main IRQ handler invokes underlying irq handlers: In the chained flow, `handle_nested_irq()` is called in the main IRQ handler, which invokes the handlers of each underlying irq as a chain of function calls, which are executed in the same context as the main handler, that is, atomically (the atomicity is also known as `hard-irq`). However, the nested flow handler had to call `handle_nested_irq()`, which executes the handler (`thread_fn`) of the underlying irq in the thread context of the parent.

These are the main differences between chained and nested flows.

irqchip and gpiolib API – new generation

Since each `irq-gpiochip` driver open-coded its own `irqdomain` handling, this led to a lot of redundant code. Kernel developers decided to move that code to the gpiolib framework, thus providing the `GPIOLIB_IRQCHIP` Kconfig symbol, enabling us to use a unified irq domain management API for GPIO chips. That portion of code helps with handling the management of GPIO IRQ chips and the associated `irq_domain` and resource allocation callbacks, as well as their setup, using the reduced set of helper functions. These are `gpiochip_irqchip_add()` or `gpiochip_irqchip_add_nested()`, and `gpiochip_set_chained_irqchip()` or `gpiochip_set_nested_irqchip()`. `gpiochip_irqchip_add()` or `gpiochip_irqchip_add_nested()` both add an IRQ chip to a GPIO chip. Here are their respective prototypes:

```
static inline int gpiochip_irqchip_add(
                                  struct gpio_chip *gpiochip,
                                  struct irq_chip *irqchip,
                                  unsigned int first_irq,
                                  irq_flow_handler_t handler,
                                  unsigned int type)


static inline int gpiochip_irqchip_add_nested(
                           struct gpio_chip *gpiochip,
                           struct irq_chip *irqchip,
                           unsigned int first_irq,
```

```
                              irq_flow_handler_t handler,
                              unsigned int type)
```

In the preceding prototypes, the `gpiochip` parameter is the GPIO chip to add the `irqchip` to. `irqchip` is the IRQ chip to be added to the GPIO chip in order to extend its capabilities so that it can act as an IRQ controller as well. This IRQ chip has to be configured properly, either by the driver or by the IRQ core code (if `dummy_irq_chip` is given as a parameter). If it's not dynamically assigned, `first_irq` will be the base (first) IRQ to allocate GPIO chip IRQs from. `handler` is the primary IRQ handler to use (often one of the predefined high-level IRQ core functions). `type` is the default type for IRQs on this `IRQ chip`; pass `IRQ_TYPE_NONE` here and let the drivers configure this upon request.

A summary of each of these function actions is as follows:

- The first one allocates a `struct irq_domain` to the GPIO chip using the `irq_domain_add_simple()` function. This IRQ domain's ops is set with the kernel IRQ core domain ops variable called `gpiochip_domain_ops`. This domain ops is defined in `drivers/gpio/gpiolib.c`, with the `irq_domain_ops.xlate` field set to `irq_domain_xlate_twocell`, meaning that this gpio chip will handle two-celled IRQs.

- Sets the `gpiochip.to_irq` field to `gpiochip_to_irq`, which is a callback that returns `irq_create_mapping(chip->irq.domain, offset)`, creating an IRQ mapping that corresponds to the GPIO offset. This is performed when we invoke `gpiod_to_irq()` on that GPIO. This function assumes that each of the pins on the `gpiochip` can generate a unique IRQ. The following is how the `gpiochip_domain_ops` IRQ domain is defined:

```
static const struct irq_domain_ops gpiochip_domain_ops =
{
    .map = gpiochip_irq_map,
    .unmap = gpiochip_irq_unmap,
    /* Virtually all GPIO-based IRQ chips are two-celled */
    .xlate = irq_domain_xlate_twocell,
};
```

The only difference between `gpiochip_irqchip_add_nested()` and `gpiochip_irqchip_add()` is that the former adds a nested IRQ chip to the GPIO chip (it sets the `gpio_chip->irq.threaded` field to `true`), while the later adds a chained IRQ chip to a GPIO chip and sets this field to `false`. On the other hand, `gpiochip_set_chained_irqchip()` and `gpiochip_set_nested_irqchip()` respectively assign/connect a chained or a nested IRQ chip to the GPIO chip. The following are the prototypes of those two functions:

```
void gpiochip_set_chained_irqchip(
                        struct gpio_chip *gpiochip,
                        struct irq_chip *irqchip,
                        unsigned int parent_irq,
                        irq_flow_handler_t parent_handler)

void gpiochip_set_nested_irqchip(struct gpio_chip *gpiochip,
                        struct irq_chip *irqchip,
                        unsigned int parent_irq)
```

In the preceding prototypes, `gpiochip` is the GPIO chip to set the `irqchip` chain to. `irqchip` represents the IRQ chip to chain to the GPIO chip. `parent_irq` is the irq number corresponding to the parent IRQ for this chained IRQ chip. In other words, it is the IRQ number to which this chip is connected. `parent_handler` is the parent interrupt handler for the accumulated IRQ coming out of the GPIO chip. It is actually the hwirq handler. This is not used for nested IRQ chips, as the parent handler is threaded. The chained variant will internally call `irq_set_chained_handler_and_data()` on `parent_handler`.

Chained gpiochip-based IRQ chips

`gpiochip_irqchip_add()` and `gpiochip_set_chained_irqchip()` are to be used on chained GPIO chip-based IRQ chips, while `gpiochip_irqchip_add_nested()` and `gpiochip_set_nested_irqchip()` are used on nested GPIO chip-based IRQ chips only. With chained GPIO chip-based IRQ chips, `gpiochip_set_chained_irqchip()` will configure the parent hwirq's handler. There's no need to call any `devm_request_*` irq family function. However, the parent hwirq's handler has to call `generic_handle_irq()` on the raised child `irqs`, as in the following example (from `drivers/pinctrl/pinctrl-at91.c` in the kernel sources), somewhat similar to a standard chained IRQ chip:

```
static void gpio_irq_handler(struct irq_desc *desc)
{
```

```c
    unsigned long isr;
    int n;

    struct irq_chip *chip = irq_desc_get_chip(desc);
    struct gpio_chip *gpio_chip =
    irq_desc_get_handler_data(desc);
    struct at91_gpio_chip *at91_gpio =
                        gpiochip_get_data(gpio_chip);
    void __iomem *pio = at91_gpio->regbase;

    chained_irq_enter(chip, desc);
    for (;;) {
        isr = readl_relaxed(pio + PIO_ISR) &
                readl_relaxed(pio + PIO_IMR);
        [...]
        for_each_set_bit(n, &isr, BITS_PER_LONG) {
            generic_handle_irq(irq_find_mapping(
                        gpio_chip->irq.domain, n));
        }
    }
    chained_irq_exit(chip, desc);
    [...]
}
```

In the preceding code, the interrupt handler is introduced first. Upon an interrupt issued by the GPIO chip, its whole gpio status bank is read in order to detect each bit that is set there, which would mean a potential IRQ triggered by the device behind the corresponding gpio line.

generic_handle_irq() is then invoked on each irq descriptor whose index in the domain corresponds to the index of a bit set in the gpio status bank. This method in turn will invoke each handler registered for each descriptor found in the previous step in an atomic context (the hard-irq context), except if the underlying driver for the device for which the gpio is used as an irq line requested the handler to be threaded.

Now we can introduce the probe method, an example of which is as follows:

```c
static int at91_gpio_probe(struct platform_device *pdev)
{
```

```
    [...]
    ret = gpiochip_irqchip_add(&at91_gpio->chip,
                               &gpio_irqchip,
                               0,
                               handle_edge_irq,
                               IRQ_TYPE_NONE);
    if (ret) {
        dev_err(
            &pdev->dev,
            "at91_gpio.%d: Couldn't add irqchip to gpiochip.\n",
            at91_gpio->pioc_idx);
        return ret;
    }
    [...]
    /* Then register the chain on the parent IRQ */
    gpiochip_set_chained_irqchip(&at91_gpio->chip,
                                 &gpio_irqchip,
                                 at91_gpio->pioc_virq,
                                 gpio_irq_handler);
    return 0;
}
```

There's nothing special there. The mechanism here somehow follows what we have seen in the generic IRQ chips. The parent IRQ is not requested here using any of the `request_irq()` family methods because `gpiochip_set_chained_irqchip()` will invoke `irq_set_chained_handler_and_data()` under the hood.

Nested gpiochip-based irqchips

The following excerpt shows how nested GPIO chip-based IRQ chips are registered by their drivers. This is somewhat similar to standalone nested IRQ chips:

```
static irqreturn_t pcf857x_irq(int irq, void *data)
{
    struct pcf857x *gpio = data;
    unsigned long change, i, status;
    status = gpio->read(gpio->client);

    /*
```

```
     * call the interrupt handler if gpio is used as
     * interrupt source, just to avoid bad irqs
     */
    mutex_lock(&gpio->lock);
    change = (gpio->status ^ status) & gpio->irq_enabled;
    gpio->status = status;
    mutex_unlock(&gpio->lock);

    for_each_set_bit(i, &change, gpio->chip.ngpio)
        handle_nested_irq(
            irq_find_mapping(gpio->chip.irq.domain, i));
    return IRQ_HANDLED;
}
```

The preceding code is the IRQ handler. As we can see, it uses `handle_nested_irq()`, which is nothing new for us. Let's now inspect the `probe` method:

```
static int pcf857x_probe(struct i2c_client *client,
                         const struct i2c_device_id *id)
{
    struct pcf857x *gpio;
    [...]
    /* Enable irqchip only if we have an interrupt line */
    if (client->irq) {
        status = gpiochip_irqchip_add_nested(&gpio->chip,
                                             &gpio->irqchip,
                                             0,
                                             handle_level_irq,
                                             IRQ_TYPE_NONE);
        if (status) {
            dev_err(&client->dev, "cannot add irqchip\n");
            goto fail;
        }
        status = devm_request_threaded_irq(
                &client->dev, client->irq,
                NULL, pcf857x_irq,
                IRQF_ONESHOT |IRQF_TRIGGER_FALLING |
                IRQF_SHARED,
```

```
                dev_name(&client->dev), gpio);
        if (status)
            goto fail;

        gpiochip_set_nested_irqchip(&gpio->chip,
                                    &gpio->irqchip,
                                    client->irq);
    }
[...]
}
```

Here, the parent irq handler is threaded and has to be registered using `devm_request_threaded_irq()`. This explains why its IRQ handler has to call `handle_nested_irq()` on child irqs in order to invoke their handlers. Once more, this looks like generic nested `irqchips`, except for the fact that gpiolib has wrapped some of the underlying nested `irqchip` APIs. To confirm this, you can have a look into the body of the `gpiochip_set_nested_irqchip()` and `gpiochip_irqchip_add_nested()` methods.

Regmap IRQ API and data structures

The regmap IRQ API is implemented in `drivers/base/regmap/regmap-irq.c`. It is mainly built on top of two essential functions, `devm_regmap_add_irq_chip()` and `regmap_irq_get_virq()`, and three data structures, `struct regmap_irq_chip`, `struct regmap_irq_chip_data`, and `struct regmap_irq`.

> **Important note**
>
> Regmap's `irqchip` API entirely uses threaded IRQs. Thus, only what we have seen in the *Nested interrupts* section will apply here.

Regmap IRQ data structures

As mentioned earlier, we need to introduce the three data structures of the `regmap irq api` in order to understand how it abstracts IRQ management.

struct regmap_irq_chip and struct regmap_irq

The `struct regmap_irq_chip` structure describes a generic `regmap irq_chip`. Prior to discussing this structure, let's first introduce `struct regmap_irq`, which stores the register and the mask description of an IRQ for `regmap irq_chip`:

```
struct regmap_irq {
    unsigned int reg_offset;
    unsigned int mask;
    unsigned int type_reg_offset;
    unsigned int type_rising_mask;
    unsigned int type_falling_mask;
};
```

The following are descriptions of the fields in the preceding structure:

- `reg_offset` is the offset of the status/mask register within the bank. This bank may actually be the `{status/mask/unmask/ack/wake}_base` register of the IRQ chip.

- `mask` is the mask used to flag/control this IRQ status register. When disabling the IRQ, the mask value will be *ORed* with the actual content of `reg_offset` from the regmap's `irq_chip.status_base` register. For `irq` enabling, `~mask` will be ANDed.

- `type_reg_offset` is the offset register (from the `irqchip` status base register) for the IRQ type setting.

- `type_rising_mask` is the mask bit to configure *rising* type IRQs. This value will be ORed with the actual content of `type_reg_offset` when setting the type of the IRQ to `IRQ_TYPE_EDGE_RISING`.

- `type_falling_mask` is the mask bit to configure *falling* type IRQs. This value will be ORed with the actual content of `type_reg_offset` when setting the type of the IRQ to `IRQ_TYPE_EDGE_FALLING`. For the `IRQ_TYPE_EDGE_BOTH` type, (`type_falling_mask | irq_data->type_rising_mask`) will be used as a mask.

Now that we are familiar with `struct regmap_irq`, let's describe `struct regmap_irq_chip`, the structure of which looks as follows:

```
struct regmap_irq_chip {
    const char *name;
```

```
        unsigned int status_base;
        unsigned int mask_base;
        unsigned int unmask_base;
        unsigned int ack_base;
        unsigned int wake_base;
        unsigned int type_base;
        unsigned int irq_reg_stride;
        bool mask_writeonly:1;
        bool init_ack_masked:1;
        bool mask_invert:1;
        bool use_ack:1;
        bool ack_invert:1;
        bool wake_invert:1;
        bool type_invert:1;

        int num_regs;

        const struct regmap_irq *irqs;
        int num_irqs;

        int num_type_reg;
        unsigned int type_reg_stride;

        int (*handle_pre_irq)(void *irq_drv_data);
        int (*handle_post_irq)(void *irq_drv_data);
        void *irq_drv_data;
};
```

This structure describes a generic `regmap_irq_chip`, which can handle most interrupt controllers (not all of them, as we will see later). The following list describes the fields in this data structure:

- `name` is a descriptive name for the IRQ controller.

- `status_base` is the base status register address to which the regmap IRQ core adds `regmap_irq.reg_offset` prior to obtaining the final status register for the given `regmap_irq`.

- `mask_writeonly` states whether the base mask register is write-only or not. If yes, `regmap_write_bits()` is used to write into the register, otherwise `regmap_update_bits()` is used.

- `unmask_base` is the base unmask register address, which has to be specified for chips that have separate mask and unmask registers.

- `ack_base` is the acknowledgement base register address. Using a value of 0 is possible with the `use_ack` bit.

- `wake_base` is the base address for `wake enable`, used to control the irq power management wakeups. If the value is 0, it means this is unsupported.

- `type_base` is the base address for the IRQ type to which the regmap IRQ core adds `regmap_irq.type_reg_offset` prior to obtaining the final type register for the given `regmap_irq`. If it is 0, it means this is unsupported.

- `irq_reg_stride` is the stride to use for chips where registers are not contiguous.

- `init_ack_masked` states whether the regmap IRQ core should acknowledge all masked interrupts once during initialization.

- `mask_invert`, if `true`, means the mask register is inverted. It means cleared bit indexes correspond to masked out interrupts.

- `use_ack`, if `true`, means the acknowledgement register should be used even if it is 0.

- `ack_invert`, if `true`, means the acknowledgement register is inverted: corresponding bit is cleared for a acknowledge.

- `wake_invert`, if `true`, means the wake register is inverted: cleared bits correspond to wake enabled.

- `type_invert`, if `true`, means inverted type flags are used.

- `num_regs` is the number of registers in each control bank. The number of registers to read when using `regmap_bulk_read()` will be given. Have a look at the definition of `regmap_irq_thread()` for more information.

- `irqs` is an array of descriptors for individual IRQs, and `num_irqs` is the total number of descriptors in the array. Interrupt numbers are assigned based on the index in this array.

- num_type_reg is the number of type registers, while type_reg_stride is the stride to use for chips where type registers are not contiguous. Regmap IRQ implements the generic interrupt service routine, which is common for most devices.

- Some devices, such as MAX77620 or MAX20024, need special handling before and after servicing the interrupt. This is where handle_pre_irq and handle_post_irq come in. These are driver-specific callbacks to handle interrupts from devices before regmap_irq_handler processes the interrupts. irq_drv_data is then the data that is passed as a parameter to those pre-/post-interrupt handlers. For example, the MAX77620 programming guidelines for interrupt servicing says the following:

--When interrupt occurs from PMIC, mask the PMIC interrupt by setting GLBLM.

--Read IRQTOP and service the interrupt accordingly.

--Once all interrupts have been checked and serviced, the interrupt service routine un-masks the hardware interrupt line by clearing GLBLM.

Back to the regmap_irq_chip.irqs field, this field is of the regmap_irq type, introduced earlier.

struct regmap_irq_chip_data

This structure is the runtime data structure for the regmap IRQ controller, allocated on the successful return path of devm_regmap_add_irq_chip(). It has to be stored in a large and private data structure for later use. Its definition is as follows:

```
struct regmap_irq_chip_data {
    struct mutex lock;
    struct irq_chip irq_chip;
    struct regmap *map;
    const struct regmap_irq_chip *chip;
    int irq_base;
    struct irq_domain *domain;
    int irq;
    [...]
};
```

For simplicity, some fields in the structure have been removed. Here is a description of the fields in this structure:

- `lock` is the lock used to protect accesses to the `irq_chip` to which `regmap_irq_chip_data` belongs. As regmap IRQs are totally threaded, it is safe to use a mutex.

- `irq_chip` is the underlying interrupt chip descriptor structure (providing IRQ-related operations) for this regmap-enabled `irqchip`, set with `regmap_irq_chip`, defined as follows in `drivers/base/regmap/regmap-irq.c`:

```c
static const struct irq_chip regmap_irq_chip = {
    .irq_bus_lock = regmap_irq_lock,
    .irq_bus_sync_unlock = regmap_irq_sync_unlock,
    .irq_disable = regmap_irq_disable,
    .irq_enable = regmap_irq_enable,
    .irq_set_type = regmap_irq_set_type,
    .irq_set_wake = regmap_irq_set_wake,
};
```

- `map` is the regmap structure for the aforementioned `irq_chip`.

- `chip` is a pointer to the generic regmap `irq_chip`, which should have been set up in the driver. It is given as a parameter to `devm_regmap_add_irq_chip()`.

- `base`, if more than zero, is the base from which it allocates a specific IRQ number. In other words, the numbering of IRQ starts at `base`.

- domain is the IRQ domain for the underlying IRQ chip, with ops set to regmap_domain_ops, defined as follows:

```
static const struct irq_domain_ops regmap_domain_ops = {
    .map = regmap_irq_map,
    .xlate = irq_domain_xlate_onetwocell,
};
```

- irq is the parent (base) IRQ for irq_chip. It corresponds to the irq parameter given to devm_regmap_add_irq_chip().

Regmap IRQ API

Earlier in the chapter, we introduced both devm_regmap_add_irq_chip() and regmap_irq_get_virq() as two essential functions the regmap IRQ API is made of. These are actually the most important functions for regmap IRQ management and the following are their respective prototypes:

```
int
devm_regmap_add_irq_chip(struct device *dev,
                         struct regmap *map,
                         int irq, int irq_flags,
                         int irq_base,
                         const struct regmap_irq_chip *chip,
                         struct regmap_irq_chip_data **data)


int regmap_irq_get_virq(struct regmap_irq_chip_data *data,
                        int irq)
```

In the preceding code, dev is the device pointer to which irq_chip belongs. map is a valid and initialized regmap for the device. irq_base, if more than zero, will be the number of the first allocated IRQ. chip is the configuration for the interrupt controller. In the prototype of regmap_irq_get_virq(), *data is an initialized input parameter that must have been returned by devm_regmap_add_irq_chip() through **data.

`devm_regmap_add_irq_chip()` is the function you should use to add regmap-based irqchip support in the code. Its `data` parameter is an output argument that represents the runtime data structure for the controller, allocated at the success of this function call. Its `irq` argument is the parent and primary IRQ for the irqchip. It is the IRQ the device uses to signal interrupts, while `irq_flags` is a mask of `IRQF_` flags to use for this primary interrupt. If this function succeeds (that is, returns 0), then output data will be set with a fresh allocated and well-configured structure of type `regmap_irq_chip_data`. This function returns `errno` on failure. `devm_regmap_add_irq_chip()` is a combination of the following:

- Allocating and initializing `struct regmap_irq_chip_data`.

- `irq_domain_add_linear()` (if `irq_base == 0`), which allocates an IRQ domain given the number of IRQs needed in the domain. On success, the IRQ domain will be assigned to the `.domain` field of the previously allocated IRQ chip's data. This domain's `ops.map` function will configure each IRQ child as nested into the parent thread, and `ops.xlate` will be set to `irq_domain_xlate_onetwocell`. If `irq_base > 0`, `irq_domain_add_legacy()` is used instead of `irq_domain_add_linear()`.

- `request_threaded_irq()`, in order to register the parent IRQ thread handler. Regmap uses its own defined threaded handler, `regmap_irq_thread()`, which does some hacks prior to calling `handle_nested_irq()` on the child `irqs`.

The following is an excerpt that summarizes the preceding actions:

```
static int regmap_irq_map(struct irq_domain *h,
                          unsigned int virq,
                          irq_hw_number_t hw)
{
    struct regmap_irq_chip_data *data = h->host_data;
    irq_set_chip_data(virq, data);
    irq_set_chip(virq, &data->irq_chip);
    irq_set_nested_thread(virq, 1);
    irq_set_parent(virq, data->irq);
    irq_set_noprobe(virq);
    return 0;
}

static const struct irq_domain_ops regmap_domain_ops = {
    .map    = regmap_irq_map,
```

```
        .xlate = irq_domain_xlate_onetwocell,
};

static irqreturn_t regmap_irq_thread(int irq, void *d)
{
    [...]
    for (i = 0; i < chip->num_irqs; i++) {
        if (data->status_buf[chip->irqs[i].reg_offset /
            map->reg_stride] & chip->irqs[i].mask) {
            handle_nested_irq(irq_find_mapping(data->domain,
            i));
            handled = true;
        }
    }
    [...]

    if (handled)
        return IRQ_HANDLED;
    else
        return IRQ_NONE;
}

int regmap_add_irq_chip(struct regmap *map, int irq,
                        int irq_ flags,
                        int irq_base,
                        const struct regmap_irq_chip *chip,
                        struct regmap_irq_chip_data **data)
{
    struct regmap_irq_chip_data *d;

    [...]
    d = kzalloc(sizeof(*d), GFP_KERNEL);
    if (!d)
        return -ENOMEM;

    /* The below is just for simplicity */
    initialize_irq_chip_data(d);
```

```
    if (irq_base)
        d->domain = irq_domain_add_legacy(map->dev->of_node,
                                          chip->num_irqs,
                                          irq_base, 0,
                                          &regmap_domain_ops,
                                          d);
    else
        d->domain = irq_domain_add_linear(map->dev->of_node,
                                          chip->num_irqs,
                                          &regmap_domain_ops,
                                          d);

    ret = request_threaded_irq(irq, NULL, regmap_irq_thread,
                               irq_flags | IRQF_ONESHOT,
                               chip->name, d);
    [...]
    *data = d;

    return 0;
}
```

`regmap_irq_get_virq()` maps an interrupt on a chip to a virtual IRQ. It simply returns `irq_create_mapping(data->domain, irq)` on the given `irq` and domain, as we saw earlier. Its `irq` parameter is the index of the interrupt requested in the chip IRQs.

Regmap IRQ API example

Let's use the `max7760` GPIO controller's driver to see how the concepts behind the regmap IRQ API are applied. This driver is located at `drivers/gpio/gpio-max77620.c` in the kernel source, and the following is a simplified excerpt of the way this driver uses regmap to handle IRQ management.

Let's start by defining the data structure that will be used throughout the writing of the code:

```
struct max77620_gpio {
    struct gpio_chip gpio_chip;
    struct regmap *rmap;
```

```
        struct device *dev;
};

struct max77620_chip {
    struct device *dev;
    struct regmap *rmap;
    int chip_irq;
    int irq_base;
    [...]
    struct regmap_irq_chip_data *top_irq_data;
    struct regmap_irq_chip_data *gpio_irq_data;
};
```

The meaning of the preceding data structure will become clear when you go through the code. Next, let's define our regmap IRQs array, as follows:

```
static const struct regmap_irq max77620_gpio_irqs[] = {
    [0] = {
        .mask = MAX77620_IRQ_LVL2_GPIO_EDGE0,
        .type_rising_mask = MAX77620_CNFG_GPIO_INT_RISING,
        .type_falling_mask = MAX77620_CNFG_GPIO_INT_FALLING,
        .reg_offset = 0,
        .type_reg_offset = 0,
    },
    [1] = {
        .mask = MAX77620_IRQ_LVL2_GPIO_EDGE1,
        .type_rising_mask = MAX77620_CNFG_GPIO_INT_RISING,
        .type_falling_mask = MAX77620_CNFG_GPIO_INT_FALLING,
        .reg_offset = 0,
        .type_reg_offset = 1,
    },
    [2] = {
        .mask = MAX77620_IRQ_LVL2_GPIO_EDGE2,
        .type_rising_mask = MAX77620_CNFG_GPIO_INT_RISING,
        .type_falling_mask = MAX77620_CNFG_GPIO_INT_FALLING,
        .reg_offset = 0,
        .type_reg_offset = 2,
```

```
    },
    [...]
    [7] = {
        .mask = MAX77620_IRQ_LVL2_GPIO_EDGE7,
        .type_rising_mask = MAX77620_CNFG_GPIO_INT_RISING,
        .type_falling_mask = MAX77620_CNFG_GPIO_INT_FALLING,
        .reg_offset = 0,
        .type_reg_offset = 7,
    },
};
```

You may have noticed the array has been truncated for the sake of readability. This array can then be assigned to the `regmap_irq_chip` data structure, as follows:

```
static const struct regmap_irq_chip max77620_gpio_irq_chip = {
    .name = "max77620-gpio",
    .irqs = max77620_gpio_irqs,
    .num_irqs = ARRAY_SIZE(max77620_gpio_irqs),
    .num_regs = 1,
    .num_type_reg = 8,
    .irq_reg_stride = 1,
    .type_reg_stride = 1,
    .status_base = MAX77620_REG_IRQ_LVL2_GPIO,
    .type_base = MAX77620_REG_GPIO0,
};
```

To summarize the preceding excerpts, the driver fills an array (`max77620_gpio_irqs[]`) of `regmap_irq` and uses it to build a `regmap_irq_chip` structure (`max77620_gpio_irq_chip`). Once the `regmap_irq_chip` data structure is ready, we start writing an `irqchip` callback, as required by the kernel `gpiochip` core:

```
static int max77620_gpio_to_irq(struct gpio_chip *gc,
                                    unsigned int offset)
{
    struct max77620_gpio *mgpio = gpiochip_get_data(gc);
    struct max77620_chip *chip =
                        dev_get_drvdata(mgpio->dev- >parent);
    return regmap_irq_get_virq(chip->gpio_irq_data, offset);
}
```

In the preceding snippet, we have only defined the callback that will be assigned to the `.to_irq` field of the GPIO chip. Other callbacks can be found in the original driver. Again, the code has been truncated here. At this stage, we can talk about the `probe` method, which will use all of the previously defined functions:

```
static int max77620_gpio_probe(struct platform_device *pdev)
{
    struct max77620_chip *chip =
    dev_get_drvdata(pdev->dev.parent);
    struct max77620_gpio *mgpio;
    int gpio_irq;
    int ret;

    gpio_irq = platform_get_irq(pdev, 0);
    [...]
    mgpio = devm_kzalloc(&pdev->dev, sizeof(*mgpio),
                         GFP_KERNEL);
    if (!mgpio)
        return -ENOMEM;

    mgpio->rmap = chip->rmap;
    mgpio->dev = &pdev->dev;

    /* setting gpiochip stuffs*/
    mgpio->gpio_chip.direction_input =
                                max77620_gpio_dir_input;
    mgpio->gpio_chip.get = max77620_gpio_get;
    mgpio->gpio_chip.direction_output =
                                max77620_gpio_dir_output;
    mgpio->gpio_chip.set = max77620_gpio_set;
    mgpio->gpio_chip.set_config = max77620_gpio_set_config;
    mgpio->gpio_chip.to_irq = max77620_gpio_to_irq;
    mgpio->gpio_chip.ngpio = MAX77620_GPIO_NR;
    mgpio->gpio_chip.can_sleep = 1;
    mgpio->gpio_chip.base = -1;
#ifdef CONFIG_OF_GPIO
    mgpio->gpio_chip.of_node = pdev->dev.parent->of_node;
#endif
```

```
    ret = devm_gpiochip_add_data(&pdev->dev,
                            &mgpio->gpio_chip, mgpio);
    [...]
    ret = devm_regmap_add_irq_chip(&pdev->dev,
                            chip->rmap, gpio_irq,
                            IRQF_ONESHOT, -1,
                            &max77620_gpio_irq_chip,
                            &chip->gpio_irq_data);
    [...]
    return 0;
}
```

In this `probe` method excerpt (which has no error checks), `max77620_gpio_irq_chip` is finally given to `devm_regmap_add_irq_chip` in order to populate the irqchip with IRQs and then add the IRQ chip to the regmap core. This function also sets `chip->gpio_irq_data` with a valid `regmap_irq_chip_data` structure, and `chip` is the private data structure allowing us to store this IRQ chip data for later use. Since this IRQ controller is built on top of a GPIO controller (`gpiochip`), the `gpio_chip.to_irq` field had to be set, and here it is the `max77620_gpio_to_irq` callback. This callback simply returns the value returned by `regmap_irq_get_virq()`, which creates and returns a valid `irq` mapping in `regmap_irq_chip_data.domain` according to the offset given as a parameter. The other functions have already been introduced and are not new for us.

In this section, we introduced the entirety of IRQ management using regmap. You are ready to move your MMIO-based IRQ management to regmap.

Summary

This chapter essentially dealt with regmap core. We introduced the framework, walked through its APIs, and described some use cases. Apart from register access, we have also learned how to use regmap for MMIO-based IRQ management. The next chapter, which deals with MFD devices and the syscon framework, will make intense use of the concepts learned in this chapter. By the end of this chapter, you should be able to develop regmap-enabled IRQ controllers, and you won't find yourself reinventing the wheel and leveraging this framework for register access.

3
Delving into the MFD Subsystem and Syscon API

The increasingly dense integration of devices has led to a kind of device that is made up of several other devices or IPs that can achieve a dedicated function. With the advent of this device, a new subsystem appeared in the Linux kernel. These are **MFDs**, which stands for **multi-function devices**. These devices are physically seen as standalone devices, but from a software point of view, these are represented in a parent-child relationship, where the children are subdevices.

While some I2C- and SPI-based devices/subdevices might need either some hacks or configurations prior to being added to the system, there are also MMIO-based devices/subdevices where zero conf/hacks are required as they just need to share the main device's register region between subdevices. The simple-mfd helper has then been introduced to handle zero conf/hacks subdevice registering, and syscon has been introduced for sharing a device's memory region with other devices. Since regmap was handling MMIO registers and managed locking (aka synchronization) accesses to memory, it has been a natural choice to build syscon on top of regmap. To get familiar with the MFD subsystem, in this chapter, we will begin with an introduction to MFD, where you will learn about its data structures and APIs, and then we will look at device tree binding in order to describe these devices to the kernel. Finally, we will talk about syscon and introduce the simple-mfd driver for a zero conf/hacks subdevice.

This chapter will cover the following topics:

- Introducing the MFD and syscon APIs and data structures

- Device tree binding for MFD devices

- Understanding syscon and simple-mfd

Technical requirements

In order to leverage this chapter, you will need the following:

- C programming skills

- Good knowledge of Linux device driver models

- Linux kernel v4.19.X sources, available at `https://git.kernel.org/pub/scm/linux/kernel/git/stable/linux.git/refs/tags`

Introducing the MFD subsystem and Syscon APIs

Prior to delving into the syscon framework and its APIs, we will cover MFDs. There are peripherals or hardware blocks exposing more than a single functionality by means of subdevices they embed into them and that are handled by separate subsystems in the kernel. That being said, a subdevice is a dedicated entity in a so-called multifunction device, responsible for a specific task, and managed through a reduced set of registers, in the chip's register map. `ADP5520` is a typical example of an MFD device, as it contains a backlight, a keypad, LEDs, and GPIO controllers. Each of these is then considered as a subdevice, and as you can see, each of these falls into a different subsystem. The MFD subsystem, defined in `include/linux/mfd/core.h` and implemented in `drivers/mfd/mfd-core.c`, has been created to deal with these devices, allowing the following features:

- Registering the same device with multiple subsystems

- Multiplexing bus and register access, as there may be some registers shared between subdevices

- Handling IRQs and clocks

Throughout this section, we will study the driver of the `da9055` device from the dialog-semiconductor and located in `drivers/mfd/da9055-core.c` in the kernel source tree. The datasheet for this device can be found at `https://www.dialog-semiconductor.com/sites/default/files/da9055-00-ids3a_20120710.pdf`.

In most cases, MFD device drivers consist of two parts:

- **A core driver** that should be hosted in drivers/mfd, responsible for the main initialization and registering each subdevice as a platform device (along with its platform data) on the system. This driver should provide common services for the subdevice drivers. These services include register access, control, and shared interrupt management. When a platform driver for one of the subsystems is instantiated, the core initializes the chip (which may be specified by the platform data). There can be support for multiple block devices of the same type built into a single kernel image. This is possible thanks to the mechanism of platform data. A platform-specific data abstraction mechanism in the kernel is used to pass configurations to the core, and subsidiary drivers make it possible to support multiple block devices of the same type.

- **The subdevice driver**, which is responsible for handling a specific subdevice registered earlier by the core driver. These drivers are located in their respective subsystem directories. Each peripheral (subsystem device) has a limited view of the device, which is implicitly reduced to the specific set of resources that the peripheral requires in order to function correctly.

> **Important note**
>
> The concept of subdevices in this chapter should not be confused with the concept of the same name in *Chapter 7, Demystifying V4L2 and Video Capture Device Drivers*, which is slightly different, where a subdevice also represents an entity in the video pipeline.

A subdevice is represented in the MFD subsystem by an instance of the struct mfd_cell structure, which you can call a **cell**. A cell is meant to describe a subdevice. The core driver must provide an array of as many cells as there are subdevices in the given peripheral. The MFD subsystem will use the information registered in each structure in the array to create a platform device for each subdevice, along with the platform data associated with each subdevice. In a struct mfd_cell structure, you can specify more advanced things, such as the resources used by the subdevice and suspend-resume operations (to be called from the driver for the subdevice). This structure is presented as follows, with some fields removed for simplicity reasons:

```
/*
 * This struct describes the MFD part ("cell").
 * After registration the copy of this structure will
 * become the platform data of the resulting platform_device
 */
```

```
struct mfd_cell {
    const char *name;
    int id;
    [...]
    int (*suspend)(struct platform_device *dev);
    int (*resume)(struct platform_device *dev);

    /* platform data passed to the sub devices drivers */
    void *platform_data;
    size_t pdata_size;

    /* Device Tree compatible string */
    const char *of_compatible;
    /* Matches ACPI */
    const struct mfd_cell_acpi_match *acpi_match;

    /*
     * These resources can be specified relative to the
     * parent device. For accessing hardware, you should
     * use resources from the platform dev
     */
    int num_resources;
    const struct resource *resources;
    [...]
};
```

> **Important note**
>
> The new platform devices that are created will have the cell structure as their platform data. The real platform data can then be accessed through pdev->mfd_cell->platform_data. A driver can also use mfd_get_cell() in order to retrieve the MFD cell corresponding to a platform device: const struct mfd_cell *cell = mfd_get_cell(pdev);.

The functionality of each member of this structure is self-explanatory. However, the following gives you more details.

The .resources element is an array that represents the resources specific to the subdevice (which is also a platform device), and .num_resources in the number of entries in the array. These are defined as it was done using platform_data, and you probably want to name them for easy retrieval. The following is an example from an MFD driver whose original core source file is drivers/mfd/da9055-core.c:

```
static struct resource da9055_rtc_resource[] = {
    {
        .name = „ALM",
        .start = DA9055_IRQ_ALARM,
        .end = DA9055_IRQ_ALARM,
        .flags = IORESOURCE_IRQ,
    },
    {
        .name = "TICK",
        .start = DA9055_IRQ_TICK,
        .end = DA9055_IRQ_TICK,
        .flags = IORESOURCE_IRQ,
    },
};

static const struct mfd_cell da9055_devs[] = {
    ...
    {
        .of_compatible = "dlg,da9055-rtc",
        .name = "da9055-rtc",
        .resources = da9055_rtc_resource,
        .num_resources = ARRAY_SIZE(da9055_rtc_resource),
    },
    ...
};
```

The following example shows how to retrieve the resource from the subdevice driver, in this case, which is implemented in drivers/rtc/rtc-da9055.c:

```
static int da9055_rtc_probe(struct platform_device *pdev)
{
    [...]
```

```
    alm_irq = platform_get_irq_byname(pdev, "ALM");
    if (alm_irq < 0)
        return alm_irq;


    ret = devm_request_threaded_irq(&pdev->dev, alm_irq, NULL,
                                    da9055_rtc_alm_irq,
                                    IRQF_TRIGGER_HIGH |
                                    IRQF_ONESHOT,
                                    "ALM", rtc);
    if (ret != 0)
        dev_err(rtc->da9055->dev,
                "irq registration failed: %d\n", ret);


    [...]
}
```

Actually, you should use `platform_get_resource()`, `platform_get_resource_byname()`, `platform_get_irq()`, and `platform_get_irq_byname()` to retrieve the resources.

When using `.of_compatible`, the function has to be a child of the MFD (see the *Device tree binding for MFD devices* section). You should statically fill an array of this structure, containing as many entries as there are subdevices on your device:

```
static struct resource da9055_rtc_resource[] = {
    {
        .name = „ALM",
        .start = DA9055_IRQ_ALARM,
        .end = DA9055_IRQ_ALARM,
        .flags = IORESOURCE_IRQ,
    },
    [...]
};


[...]
static const struct mfd_cell da9055_devs[] = {
    {
        .of_compatible = "dlg,da9055-gpio",
```

```
            .name = "da9055-gpio",
        },
        {
            .of_compatible = "dlg,da9055-regulator",
            .name = "da9055-regulator",
            .id = 1,
        },
        [...]
        {
            .of_compatible = "dlg,da9055-rtc",
            .name = "da9055-rtc",
            .resources = da9055_rtc_resource,
            .num_resources = ARRAY_SIZE(da9055_rtc_resource),
        },
        {
            .of_compatible = "dlg,da9055-watchdog",
            .name = "da9055-watchdog",
        },
    };
```

After the array of `struct mfd_cell` is filled, it has to be passed to the `devm_mfd_add_devices()` function, as follows:

```
int devm_mfd_add_devices(
                struct device *dev,
                int id,
                const struct mfd_cell *cells,
                int n_devs,
                struct resource *mem_base,
                int irq_base,
                struct irq_domain *domain)
```

This method's arguments are explained as follows:

- `dev` is the generic struct device structure of the MFD chip. It will be used to set the subdevice's parent.

- `id`: Since subdevices are created as platform devices, they should be given an ID. This field should be set with `PLATFORM_DEVID_AUTO` for automatic ID allocation, in which case `mfd_cell.id` of the corresponding cell is ignored. Otherwise, you should use `PLATFORM_DEVID_NONE`.

- `cells` is a pointer to a list (an array actually) of `struct mfd_cell` structures that describe subdevices.

- `n_dev` is the number of `struct mfd_cell` entries to use in the array to create platform devices. To create as many platform devices as there are cells in the array, you should use the `ARRAY_SIZE()` macro.

- `mem_base`: If not `NULL`, its `.start` field will be used as the base of each resource of type `IORESOURCE_MEM` of each MFD cell in the previously mentioned array. The following is an excerpt of the `mfd_add_device()` function showing this:

```
for (r = 0; r < cell->num_resources; r++) {
    res[r].name = cell->resources[r].name;
    res[r].flags = cell->resources[r].flags;

    /* Find out base to use */
    if ((cell->resources[r].flags & IORESOURCE_MEM) &&
        mem_base) {
        res[r].parent = mem_base;
        res[r].start =
            mem_base->start + cell->resources[r].start;
        res[r].end =
            mem_base->start + cell->resources[r].end;
    } else if (cell->resources[r].flags & IORESOURCE_IRQ)
    {
    [...]
```

- irq_base: This parameter is ignored if the domain is set. Otherwise, it behaves like mem_base but for each resource of type IORESOURCE_IRQ. The following is an excerpt of the mfd_add_device() function showing this:

```
        } else if (cell->resources[r].flags & IORESOURCE_IRQ)
{
        if (domain) {
            /* Unable to create mappings for IRQ ranges. */
            WARN_ON(cell->resources[r].start !=
                                cell->resources[r].end);
            res[r].start = res[r].end =
                irq_create_mapping(
                        domain,cell->resources[r].start);
        } else {
            res[r].start =
                irq_base + cell->resources[r].start;

            res[r].end =
                irq_base + cell->resources[r].end;
        }
    } else {
    [...]
```

- domain: For MFD chips that also play the role of IRQ controller for their subdevices, this parameter will be used as the IRQ domain to create IRQ mappings for these subdevices. It works this way: for each resource r of type IORESOURCE_IRQ in each cell, the MFD core will create a new resource, res, of the same type (actually, an IRQ resource, whose res.start and res.end fields are set with the IRQ mapping in this domain that corresponds to the initial resource's .start field: res[r].start = res[r].end = irq_create_mapping(domain, cell->resources[r].start);). New IRQ resources are then assigned to the platform device of the current cell and correspond to its virqs. Please have a look at the preceding excerpt, in the previous parameter description. Note that this parameter can be NULL.

Let's now see how to put this all together with an excerpt of the da9055 MFD driver:

```
#define DA9055_IRQ_NONKEY_MASK 0x01
#define DA9055_IRQ_ALM_MASK 0x02
#define DA9055_IRQ_TICK_MASK 0x04
```

```
#define DA9055_IRQ_ADC_MASK 0x08
#define DA9055_IRQ_BUCK_ILIM_MASK 0x08


/*
 * PMIC IRQ
 */
#define DA9055_IRQ_ALARM 0x01
#define DA9055_IRQ_TICK 0x02
#define DA9055_IRQ_NONKEY 0x00
#define DA9055_IRQ_REGULATOR 0x0B
#define DA9055_IRQ_HWMON 0x03

struct da9055 {
    struct regmap *regmap;
    struct regmap_irq_chip_data *irq_data;
    struct device *dev;
    struct i2c_client *i2c_client;

    int irq_base;
    int chip_irq;
};
```

In the preceding excerpt, the driver defined some constants, along with a private data structure, whose meaning will be clear as and when you read the code. After, the IRQs are defined for the register map core, as follows:

```
static const struct regmap_irq da9055_irqs[] = {
    [DA9055_IRQ_NONKEY] = {
        .reg_offset = 0,
        .mask = DA9055_IRQ_NONKEY_MASK,
    },
    [DA9055_IRQ_ALARM] = {
        .reg_offset = 0,
        .mask = DA9055_IRQ_ALM_MASK,
    },
    [DA9055_IRQ_TICK] = {
        .reg_offset = 0,
```

```
            .mask = DA9055_IRQ_TICK_MASK,
        },
        [DA9055_IRQ_HWMON] = {
            .reg_offset = 0,
            .mask = DA9055_IRQ_ADC_MASK,
        },
        [DA9055_IRQ_REGULATOR] = {
            .reg_offset = 1,
            .mask = DA9055_IRQ_BUCK_ILIM_MASK,
        },
};

static const struct regmap_irq_chip da9055_regmap_irq_chip = {
    .name = "da9055_irq",
    .status_base = DA9055_REG_EVENT_A,
    .mask_base = DA9055_REG_IRQ_MASK_A,
    .ack_base = DA9055_REG_EVENT_A,
    .num_regs = 3,
    .irqs = da9055_irqs,
    .num_irqs = ARRAY_SIZE(da9055_irqs),
};
```

In the preceding excerpt, da9055_irqs is an array of elements of type regmap_irq, which describes a generic regmap IRQ. It is assigned to da9055_regmap_irq_chip, which is of type regmap_irq_chip and represents the regmap IRQ chip. Both are part of the regmap IRQ data structures set. Finally, the probe method is implemented, as follows:

```
static int da9055_i2c_probe(struct i2c_client *client,
                            const struct i2c_device_id *id)
{
    int ret;
    struct da9055_pdata *pdata = dev_get_platdata(da9055->dev);

    uint8_t clear_events[3] = {0xFF, 0xFF, 0xFF};
    [...]
    ret =
```

```
        devm_regmap_add_irq_chip(
            &client->dev, da9055->regmap,
            da9055->chip_irq, IRQF_TRIGGER_LOW | IRQF_ONESHOT,
            da9055->irq_base, &da9055_regmap_irq_chip,
            &da9055->irq_data);
    if (ret < 0)
            return ret;

    da9055->irq_base = regmap_irq_chip_get_base(
                    da9055->irq_data);

    ret = devm_mfd_add_devices(
                    da9055->dev, -1,
                    da9055_devs, ARRAY_SIZE(da9055_devs),
                    NULL, da9055->irq_base,
                    regmap_irq_get_domain(da9055->irq_data));
    if (ret)
        goto err;
    [...]
}
```

In the preceding probe method, da9055_regmap_irq_chip (defined earlier) is given as a parameter to regmap_add_irq_chip() in order to add a valid regmap IRQ controller to the IRQ core. This function returns 0 on success. Moreover, it also returns a fully configured regmap_irq_chip_data structure through its last parameter, which can be used later as the runtime data structure for the controller. This regmap_irq_chip_data structure will contain the IRQ domain associated with the previously added IRQ controller. This IRQ domain is finally given as a parameter to devm_mfd_add_devices(), along with the array of MFD cells and its size in terms of the number of cells.

> **Important note**
>
> Do note that `devm_mfd_add_devices()` is actually the resource-managed version of `mfd_add_devices()`, which has the following function call sequence:

```
mfd_add_devices()-> mfd_add_device()-> platform_device_alloc()
                                    -> platform_device_add_data()
                                    -> platform_device_add_resources()
                                    -> platform_device_add()
```

There are I2C chips where both the chip itself and internal subdevices have different I2C addresses. Such I2C subdevices can't be probed as I2C clients because the MFD core only instantiates a platform device given an MFD cell. This issue is addressed by the following:

- Creating a dummy I2C client given the subdevice's I2C address and the MFD chip's adapter. This actually corresponds to the adapter (bus) managing the MFD device. This can be achieved using `i2c_new_dummy()`. The returned I2C client should be saved for later use – for example, with `i2c_unregister_device()`, which should be called when the module is being unloaded.

- If a subdevice needs its own regmap, then this regmap has to be built on top of its dummy I2C client.

- Storing either the I2C client only (for later removal) or with the regmap in a private data structure that can be assigned to the underlying platform device.

To summarize the preceding steps, let's walk through the driver of a real MFD device, the max8925 (which is mainly a power management IC, but is also made up of a large group of subdevices). Our code is a summary (dealing with two subdevices only) of the original one, with function names modified for the sake of readability. That being said, the original driver can be found in `drivers/mfd/max8925-i2c.c` in the kernel source tree.

Let's jump to our excerpt, starting with the context data structure definition, as follows:

```
struct priv_chip {
    struct device *dev;
    struct regmap *regmap;

    /* chip client for the parent chip, let's say the PMIC */
    struct i2c_client *client;

    /* chip client for subdevice 1, let's say an rtc */
```

```
    struct i2c_client *subdev1_client;

    /* chip client for subdevice 2 let's say a gpio controller
     */
    struct i2c_client *subdev2_client;

    struct regmap *subdev1_regmap;
    struct regmap *subdev2_regmap;
    unsigned short subdev1_addr; /* subdevice 1 I2C address */
    unsigned short subdev2_addr; /* subdevice 2 I2C address */
};

const struct regmap_config chip_regmap_config = {
    [...]
};
const struct regmap_config subdev_rtc_regmap_config = {
    [...]
};
const struct regmap_config subdev_gpiochip_regmap_config = {
    [...]
};
```

In the preceding excerpt, the driver defines the context data structure, `struct priv_chip`, which contains subdevice regmaps, and then initializes the MFD device regmap configuration as well as the subdevice's own configuration. Then, the `probe` method is defined, as follows:

```
static int my_mfd_probe(struct i2c_client *client,
                        const struct i2c_device_id *id)
{
    struct priv_chip *chip;
    struct regmap *map;

    chip = devm_kzalloc(&client->dev,
                        sizeof(struct priv_chip), GFP_KERNEL);
    map = devm_regmap_init_i2c(client, &chip_regmap_config);
    chip->client = client;
    chip->regmap = map;
```

```
    chip->dev = &client->dev;
    dev_set_drvdata(chip->dev, chip);
    i2c_set_clientdata(chip->client, chip);


    chip->subdev1_addr = client->addr + 1;
    chip->subdev2_addr = client->addr + 2;


    /* subdevice 1, let's say an RTC */
    chip->subdev1_client = i2c_new_dummy(client->adapter,
                                            chip->subdev1_addr);
    chip->subdev1_regmap =
        devm_regmap_init_i2c(chip->subdev1_client,
                            &subdev_rtc_regmap_config);
    i2c_set_clientdata(chip->subdev1_client, chip);


    /* subdevice 2, let's say a gpio controller */
    chip->subdev2_client = i2c_new_dummy(client->adapter,
                                            chip->subdev2_addr);

    chip->subdev2_regmap =
        devm_regmap_init_i2c(chip->subdev2_client,
                            &subdev_gpiochip_regmap_config);
    i2c_set_clientdata(chip->subdev2_client, chip);
    /* mfd_add_devices() is called somewhere */
    [...]
}
```

For the sake of readability, the preceding excerpt omits an error check. Additionally, the following code shows how to remove the dummy I2C clients:

```
static int my_mfd_remove(struct i2c_client *client)
{
    struct priv_chip *chip = i2c_get_clientdata(client);

    mfd_remove_devices(chip->dev);
    i2c_unregister_device(chip->subdev1_client);
    i2c_unregister_device(chip->subdev2_client);
    return 0;
}
```

Finally, the following simplified code shows how the subdevice driver can grab the pointer to either of the regmap data structures set up in the MFD driver:

```
static int subdev_rtc_probe(struct platform_device *pdev)
{
    struct priv_chip *chip = dev_get_drvdata(pdev->dev.parent);
    struct regmap *rtc_regmap = chip->subdev1_regmap;
    int ret;

    [...]

    if (!rtc_regmap) {
        dev_err(&pdev->dev, "no regmap!\n");
        ret = -EINVAL;
        goto out;
    }
    [...]
}
```

Though we have most of the knowledge required to develop MFD device drivers, it is necessary to integrate this with the device tree in order to have a better (that is, not hardcoded) description of our MFD device. This is what we will discuss in the next section.

Device tree binding for MFD devices

Even though we have the necessary tools and inputs to write our own MFD driver, it is important for the underlying MFD device to have its description defined in the device tree, since this lets the MFD core know what our MFD device is made of and how to deal with it. Moreover, the device tree remains the right place to declare devices, whether they are MFD or not. Please keep in mind that its purpose is only to describe devices on the system. As subdevices are children of the MFD device into which they are built (there is a parent-and-child bond of belonging), it is good practice to declare these subdevice nodes beneath their parent node, as in the following example. Moreover, the resources used by the subdevices are sometimes part of the resources of the parent device. So, it enforces the idea of putting the subdevice node beneath the main device node. In each subdevice node, the compatible property should match either both the subdevice's `cell.of_compatible` field and one of the `.compatible` string entries in the subdevice's `platform_driver.of_match_table` array, or both the subdevice's `cell.name` field and the subdevice's `platform_driver.name` field:

> **Important note**
>
> The subdevice's `cell.of_compatible` and `cell.name` fields are those declared in the subdevice's `mfd_cell` structure in the MFD core driver.

```
&i2c3 {
    pinctrl-names = "default";
    pinctrl-0 = <&pinctrl_i2c3>;
    clock-frequency = <400000>;
    status = "okay";

    pmic0: da9062@58 {
        compatible = "dlg,da9062";
        reg = <0x58>;
        pinctrl-names = "default";
        pinctrl-0 = <&pinctrl_pmic>;
        interrupt-parent = <&gpio6>;
        interrupts = <11 IRQ_TYPE_LEVEL_LOW>;
        interrupt-controller;

        regulators {
            DA9062_BUCK1: buck1 {
```

```
                    regulator-name = "BUCK1";
                    regulator-min-microvolt = <300000>;
                    regulator-max-microvolt = <1570000>;
                    regulator-min-microamp = <500000>;
                    regulator-max-microamp = <2000000>;
                    regulator-boot-on;
            };
            DA9062_LDO1: ldo1 {
                    regulator-name = "LDO_1";
                    regulator-min-microvolt = <900000>;
                    regulator-max-microvolt = <3600000>;
                    regulator-boot-on;
            };
        };

        da9062_rtc: rtc {
            compatible = "dlg,da9062-rtc";
        };

        watchdog {
            compatible = "dlg,da9062-watchdog";
        };

        onkey {
            compatible = "dlg,da9062-onkey";
            dlg,disable-key-power;
        };
    };
};
```

In the preceding device tree sample, the parent node (da9062, a **PMIC, Power Management Integrated Circuit**) is declared under its bus node. The regulated output of this PMIC is declared as children of the PMIC node. Here, again, everything is normal. Now, each subdevice is declared as a standalone device node under its parent (da9092, actually) node. Let's focus on the subdevice's compatible properties and use onkey as an example. The MFD cell of this node is declared in the MFD core driver (whose source file is drivers/mfd/da9063-core.c), as follows:

```
static struct resource da9063_onkey_resources[] = {
    {
        .name = "ONKEY",
        .start = DA9063_IRQ_ONKEY,
        .end = DA9063_IRQ_ONKEY,
        .flags = IORESOURCE_IRQ,d
    },
};

static const struct mfd_cell da9062_devs[] = {
    [...]
    {
        .name = "da9062-onkey",
        .num_resources = ARRAY_SIZE(da9062_onkey_resources),
        .resources = da9062_onkey_resources,
        .of_compatible = "dlg,da9062-onkey",
    },
};
```

Now, this onekey platform driver structure is declared (along with its .of_match_table entry) in the driver (whose source file is drivers/input/misc/da9063_onkey.c), as follows:

```
static const struct of_device_id
da9063_compatible_reg_id_table[] = {
    { .compatible = "dlg,da9063-onkey", .data = &da9063_regs },
    { .compatible = "dlg,da9062-onkey", .data = &da9062_regs },
    { },
};
MODULE_DEVICE_TABLE(of, da9063_compatible_reg_id_table);
```

```
[...]

static struct platform_driver da9063_onkey_driver = {
    .probe = da9063_onkey_probe,
    .driver = {
        .name = DA9063_DRVNAME_ONKEY,
        .of_match_table = da9063_compatible_reg_id_table,
    },
};
```

You can see that both `compatible` strings match the node's `compatible` string in the device's node. On the other hand, we can see that the same platform driver may be used for two or more (sub)devices. Using name matching would be confusing, then. That is why you would use a device tree for declaration and a `compatible` string for matching. So far, we have learned how the MFD subsystem deals with the device and vice versa. In the next section, we will extend these concepts to syscon and simple-mfd, two frameworks that help with MFD driver development.

Understanding Syscon and simple-mfd

Syscon stands for **system controller**. SoCs sometimes have a set of MMIO registers dedicated to miscellaneous features that don't relate to a specific IP. Clearly, there can't be a functional driver for this as these registers are neither representative nor cohesive enough to represent a specific type of device. The syscon driver handles this kind of situation. Syscon permits other nodes to access this register space through the regmap mechanism. It is actually just a set of wrapper APIs for regmap. When you request access to syscon, the regmap is created, if it doesn't exist yet.

The header required for using the syscon API is `<linux/mfd/syscon.h>`. As this API is based on regmap, you must also include `<linux/regmap.h>`. The syscon API is implemented in `drivers/mfd/syscon.c` in the kernel source tree. Its main data structure is `struct syscon`, though this structure is not to be used directly:

```
struct syscon {
    struct device_node *np;
    struct regmap *regmap;
    struct list_head list;
};
```

In the preceding structure, np is a pointer to the node acting as syscon. It is also used for syscon lookup by the device node. regmap is the regmap associated with this syscon, and list is used for implementing a kernel linked-lists mechanism, used to link all the syscons in the system together to the system-wide list, syscon_list, defined in drivers/mfd/syscon.c. This linked-list mechanism allows walking through the whole syscon list, either for a match by node or for a match by regmap.

Syscons are declared exclusively from within the device tree, by adding "syscon" to the compatible strings list in the device node that should act as Syscon. During early-boot, each node having syscon in its compatible string list will have its reg memory region IO-mapped and bound to an MMIO regmap, according to a default regmap configuration, syscon_regmap_config, as follows:

```
static const struct regmap_config syscon_regmap_config = {
    .reg_bits = 32,
    .val_bits = 32,
    .reg_stride = 4,
};
```

The syscon that is created is then added to the syscon framework-wide syscon_list, protected by the syscon_list_slock spinlock, as follows:

```
static DEFINE_SPINLOCK(syscon_list_slock);
static LIST_HEAD(syscon_list);
static struct syscon *of_syscon_register(struct device_node
                                         *np)
{
    struct syscon *syscon;
    struct regmap *regmap;
    void __iomem *base;
    [...]

    if (!of_device_is_compatible(np, "syscon"))
        return ERR_PTR(-EINVAL);
    [...]

    spin_lock(&syscon_list_slock);
    list_add_tail(&syscon->list, &syscon_list);
    spin_unlock(&syscon_list_slock);
```

```
    return syscon;
}
```

Syscon binding requires the following mandatory properties:

- `compatible`: This property value should be `"syscon"`.

- `reg`: This is the register region that can be accessed from syscon.

The following are optional properties, used to mangle the default `syscon_regmap_config` regmap config:

- `reg-io-width`: The size (or width, in terms of bytes) of the IO accesses that should be performed on the device

- `hwlocks`: Reference to a phandle of a hardware spinlock provider node

An example is shown in the following, an excerpt from the kernel docs, whose full version is available in `Documentation/devicetree/bindings/mfd/syscon.txt` in the kernel sources:

```
gpr: iomuxc-gpr@20e0000 {
    compatible = "fsl,imx6q-iomuxc-gpr", "syscon";
    reg = <0x020e0000 0x38>;
    hwlocks = <&hwlock1 1>;
};


hwlock1: hwspinlock@40500000 {
    ...
    reg = <0x40500000 0x1000>;
    #hwlock-cells = <1>;
};
```

From within the device tree, you can reference a syscon node in three different ways: either by phandle (specified in the device node of this driver), by its path, or by searching it using a specific compatible value, after which the driver can interrogate the node (or associated OS driver of this regmap) to determine the location of the registers, and finally, access the registers directly. You can use one of the following syscon APIs in order to grab a pointer to the regmap associated with a given syscon node:

```
struct regmap * syscon_node_to_regmap (struct device_node *np);
struct regmap * syscon_regmap_lookup_by_compatible(const char
                                        *s);
```

```
struct regmap * syscon_regmap_lookup_by_pdevname(const char
                                                      *s);
struct regmap * syscon_regmap_lookup_by_phandle(
                        struct device_node *np,
                        const char *property);
```

The preceding APIs have the following descriptions:

- `syscon_regmap_lookup_by_compatible()`: Given one of the compatible strings of the syscon device node, this function returns the associated regmap, or creates one if it does not exist yet, before returning it.

- `syscon_node_to_regmap()`: Given a syscon device node as a parameter, this function returns the associated regmap, or creates one if it does not exist yet, before returning it.

- `syscon_regmap_lookup_by_phandle()`: Given a phandle property holding an identifier of a syscon node, this function returns the regmap corresponding to this syscon node.

Before showing an example of using the preceding APIs, let's introduce the following platform device node, for which we will write the `probe` function. To better understand `syscon_node_to_regmap()`, let's declare this node as a child of the previous `gpr` node:

```
gpr: iomuxc-gpr@20e0000 {
    compatible = "fsl,imx6q-iomuxc-gpr", "syscon";
    reg = <0x020e0000 0x38>;

    my_pdev: my_pdev {
        compatible = "company,regmap-sample";
        regmap-phandle = <&gpr>;
        [...]
    };
};
```

Now that the device tree node is defined, we can focus on the code of the driver, implemented as follows and using the functions enumerated earlier:

```
static struct regmap *by_node_regmap;
static struct regmap *by_compat_regmap;
static struct regmap *by_pdevname_regmap;
```

```
static struct regmap *by_phandle_regmap;

static int my_pdev_regmap_sample(struct platform_device *pdev)
{
    struct device_node *np = pdev->dev.of_node;
    struct device_node *syscon_node;
    [...]
    syscon_node = of_get_parent(np);
    if (!syscon_node)
        return -ENODEV;


    /* If we have a pointer to the syscon device node,
    we use it */
    by_node_regmap = syscon_node_to_regmap(syscon_node);
    of_node_put(syscon_node);
    if (IS_ERR(by_node_regmap)) {
        pr_err("%s: could not find regmap by node\n",
        __func__);
        return PTR_ERR(by_node_regmap);
    }


    /* or we have one of the compatible string of the syscon
    node */
    by_compat_regmap =
        syscon_regmap_lookup_by_compatible("fsl,
        imx6q-iomuxc-gpr");
    if (IS_ERR(by_compat_regmap)) {
        pr_err("%s: could not find regmap by compatible\n",
        __func__);
        return PTR_ERR(by_compat_regmap);
    }


    /* Or a phandle property pointing to the syscon device node
     */
    by_phandle_regmap =
        syscon_regmap_lookup_by_phandle(np, "fsl,tempmon");
    if (IS_ERR(map)) {
```

```
        pr_err("%s: could not find regmap by phandle\n",
            __func__);
        return PTR_ERR(by_phandle_regmap);
    }


    /*
     * It is the extrem and rare case fallback
     * As of Linux kernel v4.18, there is only one driver
     * using this, drivers/tty/serial/clps711x.c
     */
    char pdev_syscon_name[9];
    int index = pdev->id;
    sprintf(syscon_name, "syscon.%i", index + 1);
    by_pdevname_regmap =
        syscon_regmap_lookup_by_pdevname(syscon_name);
    if (IS_ERR(by_pdevname_regmap)) {
        pr_err("%s: could not find regmap by pdevname\n",
            __func__);
        return PTR_ERR(by_pdevname_regmap);
    }


    [...]
    return 0;
}
```

In the preceding example, if we consider that syscon_name contains the platform device name for the gpr device, then the by_node_regmap, by_compat_regmap, by_pdevname_regmap, and by_phandle_regmap variables will all point to the same syscon regmap. However, the purpose here is just to explain the concept. my_pdev could have been the sibling (or whatever relationship) node of gpr. Using it here as its child was done for the sake of understanding the concept and the code and showing that either API has its place, depending on the situation. Now that we are familiar with the syscon framework, let's see how it can be used along with simple-mfd.

Introducing simple-mfd

For MMIO-based MFD devices, there may be no need to configure subdevices prior to adding them to the system. As this configuration is done from within the MFD core driver, the only goal of this MFD core driver would be to populate the system with platform subdevices. As a lot of these MMIO-based MFD devices exist, there would be a lot of redundant code. The simple MFD, which is a simple DT binding, addresses this.

When the `simple-mfd` string is added to the list of compatible strings of a given device node (considered here as the MFD device), it will make the **OF (open firmware)** core spawn child devices (subdevices, actually) for all subnodes of that MFD device, using the `for_each_child_of_node()` iterator. simple-mfd is implemented in `drivers/of/platform.c` as an alias of simple-bus, and its documentation is located in `Documentation/devicetree/bindings/mfd/mfd.txt` in the kernel source tree.

Used in conjunction with syscon to create the regmap, it helps to avoid writing an MFD driver, and the developer can put their effort into writing subdevice drivers. The following is an example:

```
snvs: snvs@20cc000 {
    compatible = "fsl,sec-v4.0-mon", "syscon", "simple-mfd";
    reg = <0x020cc000 0x4000>;

    snvs_rtc: snvs-rtc-lp {
        compatible = "fsl,sec-v4.0-mon-rtc-lp";
        regmap = <&snvs>;
        offset = <0x34>;
        interrupts = <GIC_SPI 19 IRQ_TYPE_LEVEL_HIGH>,
                     <GIC_SPI 20 IRQ_TYPE_LEVEL_HIGH>;
    };

    snvs_poweroff: snvs-poweroff {
        compatible = "syscon-poweroff";
        regmap = <&snvs>;
        offset = <0x38>;
        value = <0x60>;
        mask = <0x60>;
        status = "disabled";
    };
```

```
snvs_pwrkey: snvs-powerkey {
    compatible = "fsl,sec-v4.0-pwrkey";
    regmap = <&snvs>;
    interrupts = <GIC_SPI 4 IRQ_TYPE_LEVEL_HIGH>;
    linux,keycode = <KEY_POWER>;
    wakeup-source;
};
[...]
};
```

In the preceding device tree excerpt, `snvs` is the main device. It is made up of a power control subdevice (represented by a register subregion in the main device register region), an `rtc` subdevice, as well as a power key, and so on. The whole definition can be found in `arch/arm/boot/dts/imx6qdl.dtsi`, which is the SoC vendor `dtsi` for the i.MX6 chip series. The respective drivers can be found in the kernel source by grepping (searching for) the content of their `compatible` properties. To summarize, for each subnode in the `snvs` node, the MFD core will create a corresponding device along with its regmap, which would correspond to their memory region from within the main device's memory region.

This section shows the way to ease into MFD driver development when it comes to MMIO devices. Though SPI/I2C devices do not fall into this category, it covers almost 95% of MMIO-based MFD devices.

Summary

This chapter dealt with MFD devices, along with the syscon and regmap APIs. Here, we discussed how MFD devices work and how deep regmap is embedded into syscon. Having reached the end of this chapter, we can assume that you are able to develop regmap-enabled IRQ controllers, as well as to design and use syscon to share register regions between devices. The next chapter will deal with the common clock framework and how this framework is organized, its implementation, how to use it, and how to add your own clocks.

4
Storming the Common Clock Framework

From the beginning, embedded systems have always needed clock signals in order to orchestrate their inner workings, either for synchronization or for power management (for example, enabling clocks when the device is in active use or adjusting the clock depending on some criteria, such as the system load). Therefore, Linux has always had a clock framework. There has only ever been programming interface declaration support for software management of the system clock tree, and each platform had to implement this API. Different **System on Chips (SoCs)** had their own implementation. This was okay for a while, but people soon found that their hardware implementations were quite similar. The code also became bushy and redundant, which meant it was necessary to use platform-dependent APIs to get/set the clock.

This was rather an uncomfortable situation. Then, the **common clock framework (CCF)** came in, allowing software to manage clocks available on the system in a hardware-independent manner. The CCF is an interface that allows us to control various clock devices (most of time, these are embedded in SoCs) and offers a uniform API that can be used to control them (enabling/disabling, getting/setting the rate, gating/un-gating, and so on). In this chapter, the concept of a clock does not refer to **Real-Time Clocks (RTCs)**, nor timekeeping devices, which are other kinds of devices that have their own subsystems in the kernel.

The main idea behind the CCF is to unify and abstract the similar code that's spread in different SoC clock drivers. This standardized approach introduced the concept of a clock provider and a clock consumer in the following manner:

- Providers are Linux kernel drivers that connect with the framework and provide access to hardware, thereby providing (making these available to consumers) the clock tree (thanks to which one can dump the whole clock tree nowadays) according to the SoC datasheet.

- Consumers are Linux kernel drivers or subsystems that access the framework through a common API.

- That being said, a driver can be both a provider and a consumer (it would then either consume one or more clocks it provides, or one or more clocks provided by others).

In this chapter, we will introduce CCF data structures, and then focus on writing clock provider drivers (regardless of the clock type) before introducing the consumer API. We will do this by covering the following topics:

- CCF data structures and interfaces
- Writing a clock provider device driver
- Clock consumer device drivers and APIs

Technical requirements

The following are the technical requirements for this chapter:

- Advanced computer architecture knowledge and C programming skills
- Linux kernel v4.19.X sources, available at `https://git.kernel.org/pub/scm/linux/kernel/git/stable/linux.git/refs/tags`

CCF data structures and interfaces

In the old kernel days, each platform had to implement a basic API defined in the kernel (to grab/release the clock, set/get the rate, enable/disable the clock, and so on) that could be used by consumer drivers. Since the implementation of these specific APIs was done by each machine's code, this resulted in a similar file in each machine directory, with similar logic to implement the clock provider functions. This had several drawbacks, among which there was a lot of redundant code inside them. Later, the kernel abstracted this common code in the form of a clock provider (`drivers/clk/clk.c`), which became what we now call the CCF core.

Before playing with the CCF, its support needs to be pulled into the kernel by means of the `CONFIG_COMMON_CLK` option. The CCF itself is divided into two halves:

- **The Common Clock Framework core**: This is the core of the framework and is not supposed to be modified when you add a new driver and provide the common definition of `struct clk`, which unifies the framework-level code and the traditional platform-dependent implementation that used to be duplicated across a variety of platforms. This half also allows us to wrap the consumer interface (also called the **clk implementation**) on top of `struct clk_ops`, which must be provided by each clock provider.

- **The hardware-specific half**: This targets the clock device that must be written for each new hardware clock. This requires the driver to provide `struct clk_ops` that corresponds to the callbacks that are used to let us operate on the underlying hardware (these are invoked by the clock's core implementation), as well as the corresponding hardware-specific structures that wrap and abstract the clock hardware.

The two halves are tied together by the `struct clk_hw` structure. This structure helps us with implementing our own hardware clock type. In this chapter, this is referenced as `struct clk_foo`. Since `struct clk_hw` is also pointed to within `struct clk`, it allows for navigation between the two halves.

Now, we can introduce CCF data structures. The CCF is built on top of common heterogeneous data structures (in `include/linux/clk-provider.h`) that help keep this framework as generic as possible. These are as follows:

- `struct clk_hw`: This structure abstracts the hardware clock line and is used in the provider code only. It ties the two halves introduced previously and allows navigation to occur between them. Moreover, this hardware clock's base structure allows platforms to define their own hardware-specific clock structure, along with their own clock operation callbacks, as long as they wrap an instance of the `struct clk_hw` structure.

- `struct clk_ops`: This structure represents the hardware-specific callbacks that can operate on a clock line; that is, the hardware. This is why all of the callbacks in this structure accept a pointer to a `struct clk_hw` as the first parameter, though only a few of these operations are mandatory, depending on the clock type.

- `struct clk_init_data`: This holds `init` data that's common to all clocks that are shared between the clock provider and the common clock framework. The clock provider is responsible for preparing this static data for each clock in the system, and then handing it to the core logic of the clock framework.

- `struct clk`: This structure is the consumer representation of a clock since each consumer API relies on this structure.

- `struct clk_core`: This is the CCF representation of a clock.

> **Important note**
>
> Discerning the difference between `struct clk_hw` and `struct clk` allows us to move closer to a clear split between the consumer and provider clk APIs.

Now that we have enumerated the data structures of this framework, we can go through them and learn how they are implemented and what they are used for.

Understanding struct clk_hw and its dependencies

`struct clk_hw` is the base structure for every clock type in the CCF. It can be seen as a handle for traversing from a `struct clk` to its corresponding hardware-specific structure. The following is the body of `struct clk_hw`:

```
struct clk_hw {
    struct clk_core *core;
    struct clk *clk;
```

```
      const struct clk_init_data *init;
};
```

Let's take a look at the fields in the preceding structure:

- core: This structure is internal to the framework core. It also internally points back to this struct clk_hw instance.

- clk: This is a per-user struct clk instance that can operate with the clk API. It is assigned and maintained by the clock framework and provided to the clock consumer when needed. Whenever the consumer initiates access to the clock device (that is, clk_core) in the CCF through clk_get, it needs to obtain a handle, which is clk.

- init: This is a pointer to struct clk_init_data. In the process of initializing the underlying clock provider driver, the clk_register() interface is called to register the clock hardware. Prior to this, you need to set some initial data, and this initial data is abstracted into a struct clk_init_data data structure. During the initialization process, the data from clk_init_data is used to initialize the clk_core data structure that corresponds to clk_hw. When the initialization is completed, clk_init_data has no meaning.

struct clk_init_data is defined as follows:

```
struct clk_init_data {
    const char *name;
    const struct clk_ops *ops;
    const char * const *parent_names;
    u8 num_parents;
    unsigned long flags;
};
```

It holds initialization data that's common to all clocks and is shared between the clock provider and the common clock framework. Its fields are as follows:

- name, which denotes the name of the clock.

- ops is a set of operation functions related to the clock. This will be described later in the *Providing clock ops* section. Its callbacks are to be provided by the clock provider driver (in order to allow driving hardware clocks), and will be invoked by drivers through the clk_* consumer API.

- parent_names contains the names of all the parent clocks of the clock. This is an array of strings that holds all possible parents.

- `num_parents` is the number of parents. It should correspond to the number of entries in the preceding array.

- `flags` represent the framework-level flags of the clock. We will explain this in detail later in the *Providing clock ops* section, since these flags actually modify some ops.

> **Important note**
>
> `struct clk` and `struct clk_core` are private data structures and are defined in `drivers/clk/clk.c`. The `struct clk_core` structure abstracts a clock device to the CCF layer in such a way that each actual hardware clock device (`struct clk_hw`) corresponds to a `struct clk_core`.

Now that we are done with `struct clk_hw`, which is the centerpiece of the CCF, we can learn how to register a clock provider with the system.

Registering/unregistering the clock provider

The clock provider is responsible for exposing the clocks it provides in the form of a tree, sorting them out, and initializing the interface through the provider or the clock framework's core during system initialization.

In the early kernel days (before the CCF), clock registration was unified by the `clk_register()` interface. Now that we have `clk_hw`-based (provider) APIs, we can get rid of `struct clk`-based APIs while registering clocks. Since it's recommended that clock providers use the new `struct clk_hw`-based API, the appropriate registration interface to consider is `devm_clk_hw_register()`, which is the managed version of `clk_hw_register()`. However, for historical reasons, the old `clk`-based API name is still maintained, and you may find several drivers using it. A resource managed version has even been implemented called `devm_clk_register()`. We're only discussing this old API is to let you understand the existing code, not to help you implement new drivers:

```
struct clk *clk_register(struct device *dev, struct clk_hw *hw)
int clk_hw_register(struct device *dev, struct clk_hw *hw)
```

Based on this `clk_hw_register()` interface, the kernel also provides other more convenient registration interfaces (which will be introduced later), depending on the clock type to be registered. It is responsible for registering the clock to the kernel and returning a `struct clk_hw` pointer representing the clock.

It accepts a pointer to a `struct clk_hw` (since `struct clk_hw` is the provider side representation of a clock) and must contain some of the information of the clock to be registered. This will be populated with further data by the kernel. The implementation logic for this is as follows:

- Assigning the `struct clk_core` space (`clk_hw->core`):

  --Initializing the field's name, `ops`, `hw`, `flags`, `num_parents`, and `parents_names` of `clk` according to the information provided by the `struct clk_hw` pointer.

  --Calling the kernel interface, `__clk_core_init()`, on it to perform subsequent initialization operations, including building the clock tree hierarchy.

- Assigning the `struct clk` space (`clk_hw->clk`) by means of the internal kernel interface, `clk_create_clk()`, and returning this `struct clk` variable.

- Even though `clk_hw_register()` wraps `clk_register()`, you should not use `clk_register()` directly as it returns `struct clk`. This may lead to confusion and breaks the strict separation between the provider and consumer interfaces.

The following is the implementation of `clk_hw_register` in `drivers/clk/clk.c`:

```
int clk_hw_register(struct device *dev, struct clk_hw *hw)
{
    return PTR_ERR_OR_ZERO(clk_register(dev, hw));
}
```

You should check the return value of `clk_hw_register()` prior to executing further steps. Since the CCF framework is responsible for establishing the tree structure of the entire abstract clock tree and maintaining its data, it does this by means of two static linked lists that are defined in `drivers/clk/clk.c`, as follows:

```
static HLIST_HEAD(clk_root_list);
static HLIST_HEAD(clk_orphan_list);
```

Whenever you call `clk_hw_register()` (which internally calls `__clk_core_init()` in order to initialize the clock) on a clock hw, if there is a valid parent for this clock, it will end up in the `children` list of the parent. On other hand, if `num_parent` is 0, it is placed in `clk_root_list`. Otherwise, it will hang inside `clk_orphan_list`, meaning that it has no valid parent. Moreover, every time a new `clk` is clk_init'd, CCF will walk through `clk_orphan_list` (the list of orphan clocks) and re-parent any that are children of the clock currently being initialized. This is how CCF keeps the clock tree consistent with the hardware topology.

On the other hand, `struct clk` is the consumer-side instance of a clock device. Basically, all user access to the clock device creates an access handle of the `struct clk` type. When different users access the same clock device, although the same `struct clk_core` instance is being used under the hood, the handles they access (`struct clk`) are different.

> **Important note**
>
> You should keep in mind that `clk_hw_register` (or its ancestor, `clk_register()`) plays with `struct clk_core` under the hood since this is the CCF representation of a clock.

The CCF manages `clk` entities by means of a globally linked list declared in `drivers/clk/clkdev.c`, along with a mutex to protect its access, as follows:

```
static LIST_HEAD(clocks);
static DEFINE_MUTEX(clocks_mutex);
```

This comes from the era where the device tree was not heavily used. Back then, the clock consumer obtained clk by name (the name of the clk). This was used to identify clocks. Knowing that the purpose of `clk_register()` is just to register to the common clock framework, there was no way for the consumer to know how to locate the clk. So, for the underlying clock provider driver, in addition to calling the `clk_register()` function to register to the common clock framework, `clk_register_clkdev()` also had to be called immediately after `clk_register()` in order to bind the clock with a name (otherwise, the clock consumer wouldn't know how to locate the clock). Therefore, the kernel used `struct clk_lookup`, as its name says, to look up the available clock in case a consumer requested a clock (by name, of course).

This mechanism is still valid and supported in the kernel. However, in order to enforce separation between the provider and consumer code using a hw-based API, `clk_register()` and `clk_register_clkdev()` should be replaced with `clk_hw_register()` and `clk_hw_register_clkdev()` in your code, respectively.

In other words, let's say you have the following code:

```
/* Not be used anymore, introduced here for studying purpose */
int clk_register_clkdev(struct clk *clk,
                    const char *con_id, const char *dev_id)
```

This should be replaced with the following code:

```
/* recommended interface */
int clk_hw_register_clkdev(struct clk_hw *hw,
                        const char *con_id,
                        const char *dev_id)
```

Going back to the `struct clk_lookup` data structure, let's take a look at its definition:

```
struct clk_lookup {
    struct list_head node;
    const char *dev_id;
    const char *con_id;
    struct clk *clk;
    struct clk_hw *clk_hw;
};
```

In the preceding data structure, `dev_id` and `con_id` are used to identify/find the appropriate `clk`. This `clk` is the corresponding underlying clock. `node` is the list entry that will hang inside the global clocks list, as shown in the low-level `__clkdev_add()` function in the following excerpt:

```
static void __clkdev_add(struct clk_lookup *cl)
{
    mutex_lock(&clocks_mutex);
    list_add_tail(&cl->node, &clocks);
    mutex_unlock(&clocks_mutex);
}
```

The preceding `__clkdev_add()` function is indirectly called from within `clk_hw_register_clkdev()`, which actually wraps `clk_register_clkdev()`. Now that we've introduced the device tree, things have changed. Basically, each clock provider became a node in DTS; that is, each `clk` has a device node in the device tree that corresponds to it. In this case, instead of bundling `clk` and a name, it is better to bundle `clk` and your device nodes by means of a new data structure, `struct of_clk_provider`. This specific data structure is as follows:

```
struct of_clk_provider {
    struct list_head link;
    struct device_node *node;
    struct clk *(*get)(struct of_phandle_args *clkspec,
                       void *data);
    struct clk_hw *(*get_hw)(struct of_phandle_args *clkspec,
                             void *data);
    void *data;
};
```

In the preceding structure, the following takes place:

- `link` hangs in the `of_clk_providers` global list.

- `node` represents the DTS node of the clock device.

- `get_hw` is a callback for the decoding clock. For devices (consumers), it is called through `clk_get()` to return the clock associated with the node or NULL.

- `get` is there for the old clk-based APIs for historical and compatibility reasons.

However, nowadays, due to the frequent and common use of the device tree, for the underlying provider driver, the original `clk_hw_register()` + `clk_hw_register_clkdev()` (or its old clk-based implementation, `clk_register()` + `clk_register_clkdev()`) combination becomes a combination of `clk_hw_register` + `of_clk_add_hw_provider` (formerly `clk_register` + `of_clk_add_provider` – this can be found in old and non-`clk_hw`-based drivers). Also, a new globally linked list, `of_clk_providers`, has been introduced in the CCF to help manage the correspondence between all DTS nodes and clocks, along with a mutex to protect this list:

```
static LIST_HEAD(of_clk_providers);
static DEFINE_MUTEX(of_clk_mutex);
```

Although the `clk_hw_register()` and `clk_hw_register_clkdev()` function names are quite similar, the goals of these two functions differ. With the former, the clock provider can register a clock in the common clock framework. On the other hand, `clk_hw_register_clkdev()` registers a `struct clk_lookup` in the common clock framework, as its name suggests. This operation is mainly for finding clk. If you have a device tree-only platform, you no longer need all the calls to `clk_hw_register_clkdev()` (unless you have a strong reason to), so you should rely on one call to `of_clk_add_provider()`.

Important note

Clock providers are recommended to use the new `struct clk_hw`-based API as this allows us to move closer to a clear split of consumer and provider clk APIs.

`clk_hw_*` interfaces are provider interfaces that should be used in clock provider drivers, while `clk_*` is for the consumer side. Whenever you encounter a `clk_*`-based API in provider code, note that this driver should be updated to support the new hw-based interface.

Some drivers still use both functions (`clk_hw_register_clkdev()` and `of_clk_add_hw_provider()`) in order to support both clock lookup methods, such as SoC clock drivers, but you should not use both unless you have a reason to do so.

So far, we have spent time discussing clock registration. However, it might be necessary to unregister a clock, either because the underlying clock hardware goes off the system or because things went wrong during hardware initialization. Clock unregistration APIs are fairly straightforward:

```
void clk_hw_unregister(struct clk_hw *hw)
void clk_unregister(struct clk *clk)
```

The former targets `clk_hw`-based clocks, while the second targets clk-based ones. When it comes to managed variants, unless the Devres core handles unregistration, you should use the following APIs:

```
void devm_clk_unregister(struct device *dev, struct clk *clk)
void devm_clk_hw_unregister(struct device *dev, struct clk_hw
*hw)
```

In both case, `dev` represents the underlying device structure associated with the clock.

With that, we have finished looking at clock registration/unregistration. That being said, one of the main purposes of the driver is to expose device resources to potential consumers, and this applies to clock devices as well. In the next section, we'll learn how to expose clock lines to consumers.

Exposing clocks to others (in detail)

Once the clocks have been registered with CCF, the next step consists of registering this clock provider so that other devices can consume its clock lines. In the old kernel days (when the device tree was not heavily used), you had to expose clocks to the consumer by calling `clk_hw_register_clkdev()` on each clock line, which resulted in registering a lookup structure for the given clock line. Nowadays, the device tree is used for this purpose by calling the `of_clk_add_hw_provider()` interface, as well as a certain number of arguments:

```
int of_clk_add_hw_provider(
    struct device_node *np,
    struct clk_hw *(*get)(struct of_phandle_args *clkspec,
                          void *data),
    void *data)
```

Let's take a look at the arguments in this function:

- np is the device node pointer associated with the clock provider.
- get is a callback for the decoding clock. We will discuss this callback in detail in the next section.
- data is the context pointer for the given get callback. This is usually a pointer to the clock(s) that need to be associated with the device node. This is useful for decoding.

This function returns 0 on a success path. It does the opposite to `of_clk_del_provider()`, which consists of removing the provider from the global list and freeing its space:

```
void of_clk_del_provider(struct device_node *np)
```

Its resource managed version, `devm_of_clk_add_hw_provider()`, can also be used to get rid of the deletion function.

The clock provider device tree node and its associated mechanisms

For a quite some time now, the device tree is the preferred method to describe (declare) devices on a system. The common clock framework does not escape this rule. Here, we will try to figure out how clocks are described from within the device tree and related driver code. To achieve this, we'll need to consider the following device tree excerpt:

```
clocks {
    /* Provider node */
    clk54: clk54 {
        #clock-cells = <0>;
        compatible = 'fixed-clock';
        clock-frequency = <54000000>;
        clock-output-names = 'osc';
    };
};
[...]
i2c0: i2c-master@d090000 {
    [...]
    /* Consumer node */
    cdce706: clock-synth@69 {
        compatible = 'ti,cdce706';
        #clock-cells = <1>;
        reg = <0x69>;
        clocks = <&clk54>;
        clock-names = 'clk_in0';
    };
};
```

Keep in mind that clocks are assigned to consumers through the clocks property, and that a clock provider can be a consumer as well. In the preceding excerpt, clk54 is a fixed clock; we won't go into the details here. cdce706 is a clock provider that also consumes clk54 (given as a phandle in the clocks property).

The most important piece of information that clock provider nodes need to specify is the #clock- cells property, which determines the length of a clock specifier: when it is 0, this means that only the phandle property of this provider needs to be given to the consumer. When it is 1 (or greater), this means that the phandle property has multiple outputs and needs to be provided with additional information, such as an ID indicating what output needs to be used. This ID is directly represented by an immediate value. It is better to define the ID of all clocks in the system in a header file. The device tree can include this header file, such as clocks = <&clock CLK_SPI0>, where CLK_SPI0 is a macro defined in a header file.

Now, let's have a look at clock-output-names. This is an optional but recommended property and should be a list of strings that correspond to the names of the output (that is, provided) clock lines.

Take a look at the following provider node excerpt:

```
osc {
    #clock-cells = <1>;
    clock-output-names = 'ckout1', 'ckout2';
};
```

The preceding node defines a device that's providing two clock output lines named ckout1 and ckout2, respectively. Consumer nodes should never use these names directly to reference these clock lines. Instead, they should use an appropriate clock specifier (referencing clocks by index in respect to #clock-cells of the provider) that allows them to name their input clock line with respect to the device's needs:

```
device {
    clocks = <&osc 0>, <&osc 1>;
    clock-names = 'baud', 'register';
};
```

This device consumes the two clock lines provided by osc and names its input lines according to its needs. We will discuss consumer nodes at the end of this chapter.

When a clock line is assigned to a consumer device and when this consumer's driver calls clk_get() (or similar interfaces that are used to grab a clock), this interface calls of_clk_get_by_name(), which, in turn, calls __of_clk_get(). The function of interest here is __of_clk_get(). It is defined in drivers/clk/clkdev.c as follows:

```
static struct clk * of_clk_get(struct device_node *np,
                               int index,
```

```
                                      const char *dev_id,
                                      const char *con_id)
{
    struct of_phandle_args clkspec;
    struct clk *clk;
    int rc;

    rc = of_parse_phandle_with_args(np, 'clocks',
                                    '#clock-cells',
                                    index, &clkspec);
    if (rc)
        return ERR_PTR(rc);

    clk = of_clk_get_from_provider(&clkspec, dev_id, con_id);
    of_node_put(clkspec.np);
    return clk;
}
```

> **Important note**
>
> It is totally normal for this function to return a pointer to `struct clk` instead of a pointer to `struct clk_hw` as this interface operates from the consumer side.

The magic here comes from `of_parse_phandle_with_args()`, which parses lists of `phandle` and its arguments, and then calls `__of_clk_get_from_provider()`, which we will describe later.

Understanding the of_parse_phandle_with_args() API

The following is the prototype of `of_parse_phandle_with_args`:

```
int of_parse_phandle_with_args(const struct device_node *np,
                               const char *list_name,
                               const char *cells_name,
                               int index,
                               struct of_phandle_args *out_
args)
```

This function returns 0 on success and fills out_args; it returns an appropriate errno value on error. Let's take a look at its arguments:

- np is a pointer to a device tree node containing a list. In our case, it will be the node corresponding to the consumer.

- list_name is the property name that contains a list. In our case, it is clocks.

- cells_name is the property name that specifies the argument count of phandle. In our case, it is #clock-cells. It helps us grab an argument (other cells) after the phandle property in the specifier.

- index is the index of the phandle property and is used to parse out the list.

- out_args is an optional and output parameter that's filled on the success path. This parameter is of the of_phandle_args type and is defined as follows:

```
#define MAX_PHANDLE_ARGS 16
struct of_phandle_args {
    struct device_node *np;
    int args_count;
    uint32_t args[MAX_PHANDLE_ARGS];
};
```

In struct of_phandle_args, the np element is the pointer to the node that corresponds to the phandle property. In the case of the clock specifier, it will be the device tree node of the clock provider. The args_count element corresponds to the number of cells after the phandle in the specifier. It is can be used to walk through args, which is an array containing the arguments in question.

Let's look at an example of using of_parse_phandle_with_args(), given the following DTS excerpt:

```
phandle1: node1 {
    #gpio-cells = <2>;
};
phandle2: node2 {
    #list-cells = <1>;
};
node3 {
    list = <&phandle1 1 2 &phandle2 3>;
};
/* or */
```

```
node3 {
    list = <&phandle1 1 2>, <&phandle2 3>;
}
```

Here, `node3` is a consumer. To get a `device_node` pointer to the `node2` node, you can call `of_parse_phandle_with_args(node3, 'list', '#list-cells', 1, &args);`. Since `&phandle2` is at index 1 (starting from 0) in the list, we specified 1 in the `index` parameter.

In the same way, to get the associated `device_node` of the `node1` node, you can call `of_parse_phandle_with_args(node3, 'list', '#gpio-cells', 0, &args);`. For this second case, if we look at the `args` output parameter, we will see that `args->np` corresponds to `node3`, the value of `args->args_count` is 2 (as this specifier requires 2 parameters), the value of `args->args[0]` is 1, and the value of `args->args[1]` is 2, which would correspond to the 2 argument in the specifier.

> **Important note**
>
> For further reading about the device tree API, take a look at `of_parse_phandle_with_fixed_args()` and the other interfaces provided by the device tree core code in `drivers/of/base.c`.

Understanding the __of_clk_get_from_provider() API

The next function call in `__of_clk_get()` is `__of_clk_get_from_provider()`. The reason why I'm providing its prototype is that you must not use this in your code. However, this function simply walks through the clock providers (in the `of_clk_providers` list) and when the appropriate provider is found, it calls the underlying callback given as the second parameter to `of_clk_add_provider()` to decode the underlying clock. Here, the clock specifier returned by `of_parse_phandle_with_args()` is given as a parameter. As you may recall when you have to expose a clock provider to other devices, we had to use `of_clk_add_hw_provider()`. As a second parameter, this interface accepts a callback used by the CCF to decode the underlying clock whenever the consumer calls `clk_get()`. The structure of this callback is as follows:

```
struct clk_hw *(*get_hw)(struct of_phandle_args *clkspec, void
*data)
```

This callback should return the underlying `clock_hw` according to its parameters. `clkspec` is the clock specifier returned by `of_parse_phandle_with_args()`, while `data` is the context data given as the third parameter to `of_clk_add_hw_provider()`. Remember, `data` is usually a pointer to the clock(s) to be associated with the node. To see how this callback is internally called, we need to have a look at the definition of the `__of_clk_get_from_provider()` interface, which is defined as follows:

```
struct clk * of_clk_get_from_provider(struct
                                      of_phandle_args *clkspec,
                                      const char *dev_id,
                                      const char *con_id)
{
    struct of_clk_provider *provider;
    struct clk *clk = ERR_PTR(-EPROBE_DEFER);
    struct clk_hw *hw;

    if (!clkspec)
        return ERR_PTR(-EINVAL);

    /* Check if we have such a provider in our array */
    mutex_lock(&of_clk_mutex);
    list_for_each_entry(provider, &of_clk_providers, link) {
        if (provider->node == clkspec->np) {
            hw = of_clk_get_hw_from_provider(provider, clkspec);
            clk = clk_create_clk(hw, dev_id, con_id);
        }

        if (!IS_ERR(clk)) {
            if (! clk_get(clk)) {
                clk_free_clk(clk);
                clk = ERR_PTR(-ENOENT);
            }
            break;
        }
    }
    mutex_unlock(&of_clk_mutex);
```

```
    return clk;
}
```

Clock decoding callbacks

If we had to summarize the mechanisms behind getting a clock from the CCF, we would say that, when a consumer calls clk_get(), the CCF internally calls __of_clk_get(). This is given as the first parameter of the device_node property of this consumer so that the CCF can grab the clock specifier and find the device_node property (by means of of_parse_phandle_with_args()) that corresponds to the provider. It then returns this in the form of of_phandle_args. This of_phandle_args corresponds to the clock specifier and is given as a parameter to __of_clk_get_from_provider(), which simply compares the device_node property of the provider in of_phandle_args (that is, of_phandle_args->np) to those that exist in of_clk_providers, which is the list of device tree clock providers. Once a match is found, the corresponding of_clk_provider->get() callback of this provider is called and the underlying clock is returned.

> **Important note**
>
> If __of_clk_get() fails, this means there was no way to find a valid clock for the given device node. This may also mean that the provider did not register its clocks with the device tree interface. Therefore, when of_clk_get() fails, the CCF code calls clk_get_sys(), which is a fall back to using a lookup for a clock based on its name that's not on the device tree anymore. This is the real logic behind clk_get().

This of_clk_provider->get() callback often relies on the context data given as a parameter to of_clk_add_provider() so that the underlying clock is returned. Though it is possible to write your own callback (which should respect the prototype that was already introduced in the previous section), the CCF framework provides two generic decoding callbacks that cover the majority of cases. These are of_clk_src_onecell_get() and of_clk_src_simple_get(), and both have the same prototype:

```
struct clk_hw *of_clk_hw_simple_get(struct
                                    of_phandle_args *clkspec,
                                    void *data);
struct clk_hw *of_clk_hw_onecell_get(struct
                                     of_phandle_args *clkspec,
                                     void *data);
```

`of_clk_hw_simple_get()` is used for simple clock providers, where no special context data structure except for the clock itself is needed, such as the clock-gpio driver (in `drivers/clk/clk-gpio.c`). This callback simply returns the data given as a context data parameter as-is, meaning that this parameter should be a clock. It is defined in `drivers/clk/clk.c` as follows:

```
struct clk_hw *of_clk_hw_simple_get(struct
                                    of_phandle_args *clkspec,
                                    void *data)
{
    return data;
}
EXPORT_SYMBOL_GPL(of_clk_hw_simple_get);
```

On the other hand, `of_clk_hw_onecell_get()` is a bit more complex as it requires a special data structure called `struct clk_hw_onecell_data`. This can be defined as follows:

```
struct clk_hw_onecell_data {
    unsigned int num;
    struct clk_hw *hws[];
};
```

In the preceding structure, hws is an array of pointers to `struct clk_hw`, and num is the number of entries in this array.

Important note

In old clock provider drivers that do not implement clk_hw-based APIs yet, you may see `struct clk_onecell_data`, `of_clk_add_provider()`, `of_clk_src_onecell_get()`, and `of_clk_add_provider()` instead of the data structures and interfaces that have been introduced in this book.

That being said, to keep a hand on the clocks stored in this data structure, it is recommended to wrap them inside your context data structure, as shown in the following example from `drivers/clk/sunxi/clk-sun9i-mmc.c`:

```
struct sun9i_mmc_clk_data {
    spinlock_t    lock;
    void iomem        *membase;
```

```
    struct clk    *clk;
    struct reset_control       *reset;
    struct clk_hw_onecell_data     clk_hw_data;
    struct reset_controller_dev        rcdev;
};
```

You should then dynamically allocate space for these clocks according to the number of clocks that should be stored:

```
int sun9i_a80_mmc_config_clk_probe(struct
                                    platform_device *pdev)
{
    struct device_node *np = pdev->dev.of_node;
    struct sun9i_mmc_clk_data *data;
    struct clk_hw_onecell_data *clk_hw_data;
    const char *clk_name = np->name;
    const char *clk_parent;
    struct resource *r;
    [...]
    data = devm_kzalloc(&pdev->dev, sizeof(*data), GFP_KERNEL);
    if (!data)
        return -ENOMEM;
    clk_hw_data = &data->clk_hw_data;
    clk_hw_data->num = count;
    /* Allocating space for clk_hws, and 'count' is the number
     *of entries
     */
    clk_hw_data->hws =
    devm_kcalloc(&pdev->dev, count, sizeof(struct clk_hw *),
                 GFP_KERNEL);
    if (!clk_hw_data->hws)
        return -ENOMEM;
    /* A clock provider may be a consumer from another
     * provider as well
     */
    data->clk = devm_clk_get(&pdev->dev, NULL);
    clk_parent = __clk_get_name(data->clk);
    for (i = 0; i < count; i++) {
```

```
        of_property_read_string_index(np, 'clock-output-names',
                                i, &clk_name);
        /* storing each clock in its location */
        clk_hw_data->hws[i] =
        clk_hw_register_gate(&pdev->dev, clk_name,
                    clk_parent, 0,
                    data->membase + SUN9I_MMC_WIDTH * i,
                    SUN9I_MMC_GATE_BIT, 0, &data->lock);
        if (IS_ERR(clk_hw_data->hws[i])) {
            ret = PTR_ERR(clk_hw_data->hws[i]);
            goto err_clk_register;
        }
    }

    ret =
        of_clk_add_hw_provider(np, of_clk_hw_onecell_get,
                        clk_hw_data);
    if (ret)
        goto err_clk_provider;
    [...]
    return 0;
}
```

> **Important note**
>
> At the time of writing, the preceding excerpt, which has been taken from the sunxi A80 SoC MMC config clocks/resets driver, still use the clk-based API (along with the `struct clk`, `clk_register_gate()`, and `of_clk_add_src_provider()` interfaces) instead of the `clk_hw` one. Therefore, for learning purposes, I've modified this excerpt so that it uses the recommended `clk_hw` API.

As you can see, the context data that's given during clock registration is `clk_hw_data`, which is of the `clk_hw_onecell_data` type. Moreover, `of_clk_hw_onecell_get` is given as a clock decoder callback function. This helper simply returns the clock at the index that was given as an argument in the clock specifier (which is of the `of_phandle_args` type). Take a look at its definition to get a better understanding:

```
struct clk_hw * of_clk_hw_onecell_get(struct
                                      of_phandle_args *clkspec,
                                      void *data)
{
    struct clk_hw_onecell_data *hw_data = data;
    unsigned int idx = clkspec->args[0];

    if (idx >= hw_data->num) {
        pr_err('%s: invalid index %u\n', func , idx);
        return ERR_PTR(-EINVAL);
    }

    return hw_data->hws[idx];
}
EXPORT_SYMBOL_GPL(of_clk_hw_onecell_get);
```

Of course, depending on your needs, feel free to implement your own decoder callback, similar to the one in the `max9485` audio clock generator, whose driver is `drivers/clk/clk-max9485.c` in the kernel source's tree.

In this section, we have learned about the device tree aspects of clock providers. We have learned how to expose a device's clock source lines, as well as how to assign those clock lines to consumers. Now, the time has come to introduce the driver side, which also consists of writing code for its clock providers.

Writing a clock provider driver

While the purpose of a device tree is to describe the hardware at hand (the clock provider, in this case), it is worth noting that the code used to manage the underlying hardware needs to be written. This section deals with writing code for clock providers so that once their clock lines have been assigned to consumers, they behave the way they were designed to. When writing clock device drivers, it is a good practice to embed the full `struct clk_hw` (not a pointer) into your private and bigger data structure, since it is given as the first parameter to each callback in `clk_ops`. This lets you define a custom `to_<my-data-structure>` helper upon the `container_of` macro, which gives you back a pointer to your private data structure, as follows:

```
/* forward reference */
struct max9485_driver_data;

struct max9485_clk_hw {
    struct clk_hw hw;
    struct clk_init_data init;
    u8 enable_bit;
    struct max9485_driver_data *drvdata;
;
struct max9485_driver_data {
    struct clk *xclk;
    struct i2c_client *client;
    u8 reg_value;
    struct regulator *supply;
    struct gpio_desc *reset_gpio;
    struct max9485_clk_hw hw[MAX9485_NUM_CLKS];
};

static inline struct max9485_clk_hw *to_max9485_clk(struct
                                            clk_hw *hw)
{
    return container_of(hw, struct max9485_clk_hw, hw);
}
```

In the preceding example, max9485_clk_hw abstracts the hw clock (as it contains struct clk_hw). Now, from the driver's point of view, each struct max9485_clk_hw represents a hw clock, allowing us to define another bigger structure that will be used as the driver data this time: the max9485_driver_data struct. You will notice some cross-referencing in the preceding structures, notably in struct max9485_clk_hw, which contains a pointer to struct max9485_driver_data, and struct max9485_driver_data, which contains a max9485_clk_hw array. This allows us to grab the driver data from within any clk_ops callback, as follows:

```
static unsigned long
max9485_clkout_recalc_rate(struct clk_hw *hw,
                                unsigned long parent_rate)
{
    struct max9485_clk_hw *max_clk_hw = to_max9485_clk(hw);
    struct max9485_driver_data *drvdata = max_clk_hw->drvdata;

    [...]
    return 0;
}
```

Moreover, as shown in the following excerpt, it is a good practice to statically declare the clock lines (abstracted by max9485_clk_hw in this case), as well as the associated ops. This is because, unlike private data (which may change from one device to another), this information never changes, regardless of the number of clock chips of the same type that are present on the system:

```
static
const struct max9485_clk max9485_clks[MAX9485_NUM_CLKS] = {
    [MAX9485_MCLKOUT] = {
        .name = 'mclkout',
        .parent_index = -1,
        .enable_bit = MAX9485_MCLK_ENABLE,
        .ops = {
            .prepare        = max9485_clk_prepare,
            .unprepare = max9485_clk_unprepare,
        },
    },
    [MAX9485_CLKOUT] = {
        .name = 'clkout',
```

```
        .parent_index = -1,
        .ops = {
            .set_rate = max9485_clkout_set_rate,
            .round_rate    = max9485_clkout_round_rate,
            .recalc_rate   = max9485_clkout_recalc_rate,
        },
    },
    [MAX9485_CLKOUT1] = {
        .name = 'clkout1',
        .parent_index = MAX9485_CLKOUT,
        .enable_bit = MAX9485_CLKOUT1_ENABLE,
        .ops = {
            .prepare   = max9485_clk_prepare,
            .unprepare = max9485_clk_unprepare,
        },
    },
    [MAX9485_CLKOUT2] = {
        .name = 'clkout2',
        .parent_index = MAX9485_CLKOUT,
        .enable_bit = MAX9485_CLKOUT2_ENABLE,
        .ops = {
            .prepare   = max9485_clk_prepare,
            .unprepare = max9485_clk_unprepare,
        },
    },
};
```

Though ops are embedded in the abstraction data structure, they could have been declared separately, as in the `drivers/clk/clk-axm5516.c` file in the kernel sources. On the other hand, it is better to dynamically allocate the driver data structure as it would be easier for it to be private to the driver, thus allowing private data per declared device, as shown in the following excerpt:

```
static int max9485_i2c_probe(struct i2c_client *client,
                             const struct i2c_device_id *id)
{
    struct max9485_driver_data *drvdata;
```

```
    struct device *dev = &client->dev;
    const char *xclk_name;
    int i, ret;

    drvdata = devm_kzalloc(dev, sizeof(*drvdata), GFP_KERNEL);
    if (!drvdata)
        return -ENOMEM;
    [...]

    for (i = 0; i < MAX9485_NUM_CLKS; i++) {
        int parent_index = max9485_clks[i].parent_index;
        const char *name;
        if (of_property_read_string_index
            (dev->of_node, 'clock-output-names', i, &name) == 0)
{
            drvdata->hw[i].init.name = name;
        } else {
            drvdata->hw[i].init.name = max9485_clks[i].name;
        }

        drvdata->hw[i].init.ops = &max9485_clks[i].ops;
        drvdata->hw[i].init.num_parents = 1;
        drvdata->hw[i].init.flags = 0;
        if (parent_index > 0) {
            drvdata->hw[i].init.parent_names =
                        &drvdata->hw[parent_index].init.name;
            drvdata->hw[i].init.flags |= CLK_SET_RATE_PARENT;
        } else {
            drvdata->hw[i].init.parent_names = &xclk_name;
        }

        drvdata->hw[i].enable_bit = max9485_clks[i].enable_bit;
        drvdata->hw[i].hw.init = &drvdata->hw[i].init;
        drvdata->hw[i].drvdata = drvdata;

        ret = devm_clk_hw_register(dev, &drvdata->hw[i].hw);
```

```
        if (ret < 0)
            return ret;
    }


    return
      devm_of_clk_add_hw_provider(dev, max9485_of_clk_get,
                                    drvdata);
}
```

In the preceding excerpt, the driver calls `clk_hw_register()` (this is actually `devm_clk_hw_register()`, which is the managed version) in order to register each clock with the CCF. Now that we have looked at the basics of a clock provider driver, we will learn how to allow interactions with the clock line thanks to a set of operations that can be exposed in the driver.

Providing clock ops

`struct clk_hw` is the base hardware clock structure on top of which the CCF builds other clock variant structures. As a quick callback, the common clock framework provides the following base clocks:

- **fixed-rate**: This type of clock can't have its rate changed and is always running.

- **gate**: This acts as a gate to a clock source as is its parent. Obviously, it can't have its rate changed as it is just a gate.

- **mux**: This type of clock cannot gate. It has two or more clock inputs: its parents. It allows us to select a parent among those it is connected to. Moreover, it allows us to get the rate from the selected parent.

- **fixed-factor**: This clock type can't gate/ungate but does divide and multiply the parent rate by its constants.

- **divider**: This type of clock cannot gate/ungate. However, it divides the parent clock rate by using a divider that can be selected from among the various arrays that are provided at registration.

- **composite**: This is a combination of three of the base clocks we described earlier: mux, rate, and gate. It allows us to reuse those base clocks to build a single clock interface.

You may be wondering how the kernel (that is, the CCF) knows what the type of a given clock is when giving clk_hw as a parameter to the clk_hw_register() function. Actually, the CCF does not know this, and does not have to know anything. This is the aim of the clk_hw->init.ops field, which is of the struct clk_ops type. According to the callback functions set in this structure, you can guess what type of clock it is facing. The following is a detailed presentation of this set of operation functions for the clock in a struct clk_ops:

```
struct clk_ops {
    int     (*prepare)(struct clk_hw *hw);
    void    (*unprepare)(struct clk_hw *hw);
    int     (*is_prepared)(struct clk_hw *hw);
    void    (*unprepare_unused)(struct clk_hw *hw);
    int     (*enable)(struct clk_hw *hw);
    void    (*disable)(struct clk_hw *hw);
    int     (*is_enabled)(struct clk_hw *hw);
    void    (*disable_unused)(struct clk_hw *hw);
    unsigned long (*recalc_rate)(struct clk_hw *hw,
                            unsigned long parent_rate);
    long    (*round_rate)(struct clk_hw *hw, unsigned long rate,
                        unsigned long *parent_rate);
    int     (*determine_rate)(struct clk_hw *hw,
                            struct clk_rate_request *req);
    int     (*set_parent)(struct clk_hw *hw, u8 index);
    u8      (*get_parent)(struct clk_hw *hw);
    int     (*set_rate)(struct clk_hw *hw, unsigned long rate,
                        unsigned long parent_rate);
[...]
    void    (*init)(struct clk_hw *hw);
};
```

For clarity, some fields have been removed.

Each `prepare*`/`unprepare*`/`is_prepared` callback is allowed to sleep and therefore must not be called from an atomic context, while each `enable*`/`disable*`/`is_enabled` callback may not — and must not – sleep. Let's take a look at this code in more detail:

- `prepare` and `unprepare` are optional callbacks. What has been done in `prepare` should be undone in `unprepare`.

- `is_prepared` is an optional callback that tells is whether the clock is prepared or not by querying the hardware. If omitted, the clock framework core will do the following:

  --Maintain a prepare counter (incremented by one when the `clk_prepare()` consumer API is called, and decremented by one when `clk_unprepare()` is called).

  --Based on this counter, it will determine whether the clock is prepared.

- `unprepare_unused`/`disable_unused`: These callbacks are optional and used in the `clk_disable_unused` interface only. This interface is provided by the clock framework core and called (in `drivers/clk/clk.c: late_initcall_sync(clk_disable_unused)`) in the system-initiated late call in order to unprepare/ungate/close unused clocks. This interface will call the corresponding `.unprepare_unused` and `.disable_unused` functions of each unused clock on the system.

- `enable`/`disable`: Enables/disables the clock atomically. These functions must run atomically and must not sleep. For `enable`, for example, it should return **only** when the underlying clock is generating a valid clock signal that can be used by consumer nodes.

- `is_enabled` has the same logic as `is_prepared`.

- `recalc_rate`: This is an optional callback that queries the hardware to recalculate the rate of the underlying clock, given the parent rate as an input parameter. The initial rate is 0 if this op is omitted.

- `round_rate`: This callback accepts a target rate (in Hz) as input and should return the closest rate actually supported by the underlying clock. The parent rate is an input/output parameter.

- `determine_rate`: This callback is given a targeted clock rate as a parameter and returns the closest one supported by the underlying hardware.

- `set_parent`: This concerns clocks with multiple inputs (multiple possible parents). This callback accepts changing the input source when given the index as a parameter (as a u8) of the parent to be selected. This index should correspond to a parent that's valid in either the `clk_init_data.parent_names` or `clk_init_data.parents` arrays of the clock. This callback should return 0 on a success path or `-EERROR` otherwise.

- `get_parent` is a mandatory callback for clocks with multiple (at least two) inputs (multiple `parents`). It queries the hardware to determine the parent of the clock. The return value is a u8 that corresponds to the parent index. This index should be valid in either the `clk_init_data.parent_names` or `clk_init_data.parents` arrays. In other words, this callback translates the parent value that's read from the hardware into an array index.

- `set_rate`: Changes the rate of the given clock. The requested rate should be the return value of the `.round_rate` call in order to be valid. This callback should return 0 on a success path or `-EERROR` otherwise.

- `init` is a platform-specific clock initialization hook that will be called when the clock is registered to the kernel. For now, no basic clock type implements this callback.

> **Tip**
>
> Since `.enable` and `.disable` must not sleep (they are called with spinlocks held), clock providers in discrete chips that are connected to sleepable buses (such as SPI or I2C) cannot be controlled with spinlocks held and should therefore implement their enable/disable logic in the prepare/unprepare hooks. The general API will directly call the corresponding operation function. This is one of the reasons why, from the consumer side (the clk-based API), a call to `clk_enable` must be preceded by a call to `clk_prepare()` and a call to `clock_disable()` should be followed by `clock_unprepare()`.

Last but not the least, the following difference should be noticed as well:

> **Important note**
>
> SoC-internal clocks can be seen as fast clocks (controlled via simple MMIO register writes), and can therefore implement `.enable` and `.disable`, while SPI/I2C-based clocks can be seen as slow clocks and should implement `.prepare` and `.unprepare`.

These functions are not mandatory for all clocks. Depending on the clock type, some may be mandatory, while others may not be. The following array summarizes which `clk_ops` callbacks are mandatory for which clock type, based on their hardware capabilities:

```
+-----------------+------+-------------+---------------+-------------+------+
|                 | gate | change rate | single parent | multiplexer | root |
+=================+======+=============+===============+=============+======+
|.prepare         |      |             |               |             |      |
+-----------------+------+-------------+---------------+-------------+------+
|.unprepare       |      |             |               |             |      |
+-----------------+------+-------------+---------------+-------------+------+
+-----------------+------+-------------+---------------+-------------+------+
|.enable          | y    |             |               |             |      |
+-----------------+------+-------------+---------------+-------------+------+
|.disable         | y    |             |               |             |      |
+-----------------+------+-------------+---------------+-------------+------+
|.is_enabled      | y    |             |               |             |      |
+-----------------+------+-------------+---------------+-------------+------+
+-----------------+------+-------------+---------------+-------------+------+
|.recalc_rate     |      | y           |               |             |      |
+-----------------+------+-------------+---------------+-------------+------+
|.round_rate      |      | y **        |               |             |      |
+-----------------+------+-------------+---------------+-------------+------+
|.determine_rate  |      | y **        |               |             |      |
+-----------------+------+-------------+---------------+-------------+------+
|.set_rate        |      | y           |               |             |      |
+-----------------+------+-------------+---------------+-------------+------+
+-----------------+------+-------------+---------------+-------------+------+
|.set_parent      |      |             | n             | y           | n    |
+-----------------+------+-------------+---------------+-------------+------+
|.get_parent      |      |             | n             | y           | n    |
+-----------------+------+-------------+---------------+-------------+------+
+-----------------+------+-------------+---------------+-------------+------+
|.recalc_accuracy |      |             |               |             |      |
+-----------------+------+-------------+---------------+-------------+------+
+-----------------+------+-------------+---------------+-------------+------+
|.init            |      |             |               |             |      |
+-----------------+------+-------------+---------------+-------------+------+
```

Figure 4.1 – Mandatory clk_ops callbacks for clock types

In the preceding array, the ** marker means either `round_rate` or `determine_rate` is required.

In the preceding array, **y** means mandatory, while **n** means the concerned callback is either invalid or otherwise unnecessary. Empty cells should be considered as either optional or that they must be evaluated on a case-by-case basis.

Clock flags in clk_hw.init.flags

Since we have already introduced the clock ops structure, we will now introduce the different flags (defined in `include/linux/clk-provider.h`) and see how they affect the behavior of some of the callbacks in this structure:

```
/*must be gated across rate change*/
#define CLK_SET_RATE_GATE  BIT(0)
/*must be gated across re-parent*/
#define CLK_SET_PARENT_GATE    BIT(1)
/*propagate rate change up one level */
#define CLK_SET_RATE_PARENT    BIT(2)
/* do not gate even if unused */
#define CLK_IGNORE_UNUSED    BIT(3)
/*Basic clk, can't do a to_clk_foo()*/
#define CLK_IS_BASIC BIT(5)
/*do not use the cached clk rate*/
#define CLK_GET_RATE_NOCACHE BIT(6)
/* don't re-parent on rate change */
#define CLK_SET_RATE_NO_REPARENT BIT(7)
/* do not use the cached clk accuracy */
#define CLK_GET_ACCURACY_NOCACHE BIT(8)
/* recalc rates after notifications */
#define CLK_RECALC_NEW_RATES BIT(9)
/* clock needs to run to set rate */
#define CLK_SET_RATE_UNGATE BIT(10)
/* do not gate, ever */
#define CLK_IS_CRITICAL    BIT(11)
```

The preceding code shows the different framework-level flags that can be set in the `clk_hw->init.flags` field. You can specify multiple flags by OR'ing them. Let's take a look at them in more detail:

- `CLK_SET_RATE_GATE`: When you change the rate of the clock, it must be gated (disabled). This flag also ensures there's rate change and rate glitch protection; when a clock has the `CLK_SET_RATE_GATE` flag set and it has been prepared, the `clk_set_rate()` request will fail.

- `CLK_SET_PARENT_GATE` : When you change the parent of the clock, it must be gated.

- `CLK_SET_RATE_PARENT`: Once you've changed the rate of the clock, the change must be passed to the upper parent. This flag has two effects:

  --When a clock consumer calls `clk_round_rate()` (which the CCF internally maps to `.round_rate`) to get an approximate rate, if the clock does not provide the `.round_rate` callback, the CCF will immediately return the cached rate of the clock if `CLK_SET_RATE_PARENT` is not set. However, if this flag is set still without `.round_rate` provided, then the request is routed to the clock parent. This means the parent is queried and `clk_round_rate()` is called to get the value that the parent clock can provide that's closest to the targeted rate.

  --This flag also modifies the behavior of the `clk_set_rate()` interface (which the CCF internally maps to `.set_rate`). If set, any rate change request will be forwarded upstream (passed to the parent clock).
  That is to say, if the parent clock can get an approximate rate value, then by changing the parent clock rate, you can get the required rate. This flag is usually set on the clock gate and mux. Use this flag with care.

- `CLK_IGNORE_UNUSED`: Ignore the disable unused call. This is primarily useful when there's a driver that doesn't claim clocks properly, but the bootloader leaves them on. It is the equivalent of the `clk_ignore_unused` kernel boot parameters but for a single clock. It's not expected to be used in normal cases, but for bring up and debug, it's very useful to have the option to not gate (not disable) unclaimed clocks that are still on.

- `CLK_IS_BASIC`: This is no longer used.

- `CLK_GET_RATE_NOCACHE`: There are chips where the clock rate can be changed by internal hardware without the Linux clock framework being aware of that change at all. This flag makes sure the clk rate from the Linux clock tree always matches the hardware settings. In other words, the get/set rate does not come from the cache and is calculated at the time.

> **Important note**
>
> While dealing with the gate clock type, note that a gated clock is a disabled clock, while an ungated clock is an enabled clock. See `https://elixir.bootlin.com/linux/v4.19/source/drivers/clk/clk.c#L931` and `https://elixir.bootlin.com/linux/v4.19/source/drivers/clk/clk.c#L862` for more details.

Now that we are familiar with clock flags, as well as the way those flags may modify the behavior of clock-related callbacks, we can walk through each clock type and learn how to provide their associated ops.

Fixed-rate clock case study and its ops

This is the simplest type of clock. Therefore, we will use this to build some of the strong guidelines we must respect while writing clock drivers. The frequency of this type of clock cannot be adjusted as it is fixed. Moreover, this type of clock cannot be switched, cannot choose its parent, and does not need to provide a `clk_ops` callback function.

The clock framework uses the `struct clk_fixed_rate` structure (described as follows) to abstract this type of clock hardware:

```
Struct clk_fixed_rate {
        struct clk_hw hw;
        unsigned long fixed_rate;
        u8 flags; [...]
};


#define to_clk_fixed_rate(_hw) \
           container_of(_hw, struct clk_fixed_rate, hw)
```

In the preceding structure, hw is the base structure and ensures there's a link between the common and hardware-specific interfaces. Once given to the `to_clk_fixed_rate` macro (which is based on `container_of`), you should get a pointer to `clk_fixed_rate`, which wraps this hw. `fixed_rate` is the constant (fixed) rate of the clock device. `flags` represents framework-specific flags.

Let's have a look at the following excerpt, which simply registers two fake fixed-rate clock lines:

```
#include <linux/clk.h>
#include <linux/clk-provider.h>
```

```c
#include <linux/init.h>
#include <linux/of_address.h>
#include <linux/platform_device.h>
#include <linux/reset-controller.h>

static struct clk_fixed_rate clk_hw_xtal = {
    .fixed_rate = 24000000,
    .hw.init = &(struct clk_init_data){
        .name = 'xtal',
        .num_parents = 0,
        .ops = &clk_fixed_rate_ops,
    },
};

static struct clk_fixed_rate clk_hw_pll = {
    .fixed_rate = 45000000,
    .hw.init = &(struct clk_init_data){
        .name = 'fixed_pll',
        .num_parents = 0,
        .ops = &clk_fixed_rate_ops,
    },
};

static struct clk_hw_onecell_data fake_fixed_hw_onecell_data =
{
    .hws = {
        [CLKID_XTAL]     = &clk_hw_xtal.hw,
        [CLKID_PLL_FIXED]    = &clk_hw_pll.hw,
        [CLK_NR_CLKS]  = NULL,
    },
    .num = CLK_NR_CLKS,
};
```

With that, we have defined our clocks. The following code shows how to register these clocks on the system:

```
static int fake_fixed_clkc_probe(struct platform_device *pdev)
{
    int ret, i;
    struct device *dev = &pdev->dev;
    for (i = CLKID_XTAL; i < CLK_NR_CLKS; i++) {
        ret = devm_clk_hw_register(dev,
                        fake_fixed_hw_onecell_data.hws[i]);
        if (ret)
            return ret;
    }
    return devm_of_clk_add_hw_provider(dev,
                        of_clk_hw_onecell_get,
                        &fake_fixed_hw_onecell_data);
}


static const
struct of_device_id fake_fixed_clkc_match_table[] = {
    { .compatible = 'l.abcsmart,fake-fixed-clkc' },
    { }
};


static struct platform_driver meson8b_driver = {
    .probe = fake_fixed_clkc_probe,
    .driver      = {
        .name    = 'fake-fixed-clkc',
        .of_match_table = fake_fixed_clkc_match_table,
    },
};
```

General simplification considerations

In the previous excerpt, we used `clk_hw_register()` to register the clock. This interface is the base registration interface and can be used to register any type of clock. Its main parameter is a pointer to the `struct clk_hw` structure that's embedded in the underlying clock-type structure.

Clock initialization and registration through a call to `clk_hw_register()` requires populating the `struct clk_init_data` (thus implementing `clk_ops`) object, which gets bundled with `clk_hw`. As an alternative, you can use a hardware-specific (that is, clock-type-dependent) registration function. Here, the kernel is responsible for building the appropriate `init` data from arguments given to the function according to the clock type, before internally calling `clk_hw_register(...)`. With this alternative, the CCF will provide appropriate `clk_ops` according to the clock hardware type.

Generally, the clock provider does not need to use nor allocate the base clock type directly, which in this case is `struct clk_fixed_rate`. This is because the kernel clock framework provides dedicated interfaces for this purpose. In a real-life scenario (where there's a fixed clock), this dedicated interface would be `clk_hw_register_fixed_rate()`:

```
struct clk_hw *
    clk_hw_register_fixed_rate(struct device *dev,
                               const char *name,

                               const char *parent_name,
                               unsigned long flags,

                               unsigned long fixed_rate)
```

The `clk_register_fixed_rate()` interface uses the clock's name, `parent_name`, and `fixed_rate` as parameters to create a clock with a fixed frequency. `flags` represents the framework-specific flags, while `dev` is the device that is registering the clock. The `clk_ops` property of the clock is also provided by the clock framework and does not require the provider to care about it. The kernel clock ops data structure for this kind of clock is `clk_fixed_rate_ops`. It is defined in `drivers/clk/clk-fixed-rate.c` as follows:

```
static unsigned long
    clk_fixed_rate_recalc_rate(struct clk_hw *hw,
                               unsigned long parent_rate)

{

    return to_clk_fixed_rate(hw)->fixed_rate;

}
```

```
static unsigned long
    clk_fixed_rate_recalc_accuracy(struct clk_hw *hw,
                                    unsigned long parent_ accuracy)
{
    return to_clk_fixed_rate(hw)->fixed_accuracy;
}

const struct clk_ops clk_fixed_rate_ops = {
    .recalc_rate = clk_fixed_rate_recalc_rate,
    .recalc_accuracy = clk_fixed_rate_recalc_accuracy,
};
```

`clk_register_fixed_rate()` returns a pointer to the underlying `clk_hw` structure of the fixed-rate clock. The code can then use the `to_clk_fixed_rate` macro the grab a pointer to the original clock-type structure.

However, you can still use the low-level `clk_hw_register()` registration interface and reuse some of the CCF provided ops callbacks. The fact that the CCF provides an appropriate ops structure for your clock does not mean you should use it as-is. You may not wish to use the clock-type-dependent registration interface (using `clock_hw_register()` instead) and instead use one or more of the individual ops provided by the CCF. This does not just apply to adjustable clocks, as per the following example, but to all other clock types that we will discuss in this book.

Let's have a look at an example from `drivers/clk/clk-stm32f4.c` for a clock divider driver:

```
static unsigned long stm32f4_pll_div_recalc_rate(
                                    struct clk_hw *hw,
                                    unsigned long parent_rate)
{
    return clk_divider_ops.recalc_rate(hw, parent_rate);
}

static long stm32f4_pll_div_round_rate(struct clk_hw *hw,
                                    unsigned long rate,
                                    unsigned long *prate)
{
    return clk_divider_ops.round_rate(hw, rate, prate);
```

```
}

static int stm32f4_pll_div_set_rate(struct clk_hw *hw,
                                    unsigned long rate,
                                    unsigned long parent_rate)
{
    int pll_state, ret;
    struct clk_divider *div = to_clk_divider(hw);
    struct stm32f4_pll_div *pll_div = to_pll_div_clk(div);
    pll_state = stm32f4_pll_is_enabled(pll_div->hw_pll);

    if (pll_state)
        stm32f4_pll_disable(pll_div->hw_pll);
    ret = clk_divider_ops.set_rate(hw, rate, parent_rate);
    if (pll_state)
        stm32f4_pll_enable(pll_div->hw_pll);
    return ret;
}

static const struct clk_ops stm32f4_pll_div_ops = {
    .recalc_rate = stm32f4_pll_div_recalc_rate,
    .round_rate = stm32f4_pll_div_round_rate,
    .set_rate = stm32f4_pll_div_set_rate,
};
```

In the preceding excerpt, the driver only implements the .set_rate ops and reuses the .recalc_rate and .round_rate properties of the CCF-provided clock divider ops known as clk_divider_ops.

Fixed clock device binding

This type of clock can also be natively and directly supported by DTS configuration without the need to write any code. This device tree-based interface is generally used to provide dummy clocks. There are cases where some devices in the device tree may require clock nodes to describe their own clock inputs. For example, the *mcp2515* SPI to CAN converter needs to be provided with a clock to let it know the frequency of the quartz it is connected to. For such a dummy clock node, the compatible property should be `fixed-clock`. An example of this is as follows:

```
/* fixed crystal dedicated to mpc251x */
clocks {
    /* fixed crystal dedicated to mpc251x */
    clk8m: clk@1 {
        compatible = 'fixed-clock';
        reg=<0>;
        #clock-cells = <0>;
        clock-frequency = <8000000>;
        clock-output-names = 'clk8m';
    };
};


/* consumer */
can1: can@1 {
    compatible = 'microchip,mcp2515';
    reg = <0>;
    spi-max-frequency = <10000000>;
    clocks = <&clk8m>;
};
```

The clock framework's core will directly extract the clock information provided by DTS and will automatically register it to the kernel without any driver support. `#clock-cells` is 0 here because only one fixed rate line is provided, and in this case, the specifier only needs to be the `phandle` of the provider.

PWM clock alternative

Because of the lack of output clock sources (clock pads), some board designers (rightly or wrongly) use PWM output pads as the clock source for external components. This kind of clock is only instantiated from the device tree. Moreover, since PWM binding requires specifying the period of the PWM signal, pwm-clock falls into the fixed-rate clock category. An example of such an instantiation can be seen in the following code, which is an excerpt from imx6qdl-sabrelite.dtsi:

```
mipi_xclk: mipi_xclk {
    compatible = 'pwm-clock';
    #clock-cells = <0>;
    clock-frequency = <22000000>;
    clock-output-names = 'mipi_pwm3';
    pwms = <&pwm3 0 45>; /* 1 / 45 ns = 22 MHz */
    status = 'okay';
};


ov5640: camera@40 {
    compatible = 'ovti,ov5640';
    pinctrl-names = 'default';
    pinctrl-0 = <&pinctrl_ov5640>;
    reg = <0x40>;
    clocks = <&mipi_xclk>;
    clock-names = 'xclk';
    DOVDD-supply = <&reg_1p8v>;
    AVDD-supply = <&reg_2p8v>;
    DVDD-supply = <&reg_1p5v>;
    reset-gpios = <&gpio2 5 GPIO_ACTIVE_LOW>;
    powerdown-gpios = <&gpio6 9 GPIO_ACTIVE_HIGH>;
    [...]
};
```

As you can see, the compatible property should be pwm-clock, while #clock-cells should be <0>. This clock-type driver is located at drivers/clk/clk-pwm.c, and further reading about this can be found at Documentation/devicetree/bindings/clock/pwm-clock.txt.

Fixed-factor clock driver and its ops

This type of clock divides and multiplies the parent rate by constants (hence it being a fixed-factor clock driver). This clock cannot gate:

```
struct clk_fixed_factor {
    struct clk_hw      hw;
    unsigned int       mult;
    unsigned int       div;
};


#define to_clk_fixed_factor(_hw) \
            container_of(_hw, struct clk_fixed_factor, hw)
```

The frequency of the clock is determined by the frequency of the parent clock, multiplied by `mult`, and then divided by `div`. It is actually a **fixed multiplier and divider** clock. The only way for a fixed-factor clock to have its rate changed would be to change its parent rate. In this case, you need to set the `CLK_SET_RATE_PARENT` flag. Since the frequency of the parent clock can be changed, the fixed-factor clock can also have its frequency changed, so callbacks such as `.recalc_rate`/`.set_rate`/`.round_rate` are also provided. That being said, since the set rate request will be propagated upstream if the `CLK_SET_RATE_PARENT` flag is set, the `.set_rate` callback of such a clock needs to return 0 to ensure its call is a valid **nop (no-operation)**:

```
static int clk_factor_set_rate(struct clk_hw *hw,
                               unsigned long rate,
                               unsigned long parent_rate)
{
    return 0;
}
```

For such clocks, you're better off using the clock framework provider helper ops known as `clk_fixed_factor_ops`, which is defined and implemented in `drivers/clk/clk-fixed-factor.c` as follows:

```
const struct clk_ops clk_fixed_factor_ops = {
    .round_rate = clk_factor_round_rate,
    .set_rate = clk_factor_set_rate,
    .recalc_rate = clk_factor_recalc_rate,
};
EXPORT_SYMBOL_GPL(clk_fixed_factor_ops);
```

The advantage of using this is that you don't need to care about ops anymore since the kernel has already set everything up for you. Its `round_rate` and `recalc_rate` callbacks even take care of the `CLK_SET_RATE_PARENT` flag, which means we can adhere to our simplification path. Moreover, you're better off using the clock framework helper interface to register such a clock; that is, `clk_hw_register_fixed_factor()`:

```
struct clk_hw *
    clk_hw_register_fixed_factor(struct device *dev,
                                 const char *name,

                                 const char *parent_name,
                                 unsigned long flags,

                                 unsigned int mult,
                                 unsigned int div)
```

This interface internally sets up a `struct clk_fixed_factor` that it allocates dynamically, and then returns a pointer to the underlying `struct clk_hw`. You can use this with the `to_clk_fixed_factor` macro to grab a pointer to the original fixed-factor clock structure. The ops that's assigned to the clock is `clk_fixed_factor_ops`, as discussed previously. In addition, this type of interface is similar to the fixed-rate clock. You do not need to provide a driver. You only need to configure the device tree.

Device tree binding for fixed-factor clocks

You can find binding documentation for such simple fixed factor rate clocks at `Documentation/devicetree/bindings/clock/fixed-factor-clock.txt`, in the kernel sources. The required properties are as follows:

- `#clock-cells`: This will be set to 0 according to the common clock binding.

- `compatible`: This will be `'fixed-factor-clock'`.

- `clock-div`: Fixed divider.

- `clock-mult`: Fixed multiplier.

- `clocks`: The `phandle` of the parent clock.

Here's an example:

```
clock {
    compatible = 'fixed-factor-clock';
    clocks = <&parentclk>;
    #clock-cells = <0>;
    clock-div = <2>;
    clock-mult = <1>;
};
```

Now that the fixed-factor clock has been addressed, the next logical step would be to look at the gateable clock, another simple clock type.

Gateable clock and its ops

This type of clock can only be switched, so only providing .enable/.disable callbacks makes sense here:

```
struct clk_gate {
    struct clk_hw hw;
    void   iomem *reg;
    u8     bit_idx;
    u8     flags;
    spinlock_t   *lock;
};


#define to_clk_gate(_hw) container_of(_hw, struct clk_gate, hw)
```

Let's take a look at the preceding structure in more detail:

- reg: This represents the register address (virtual address; that is, MMIO) for controlling the clock switch.

- bit_idx: This is the control bit of the clock switch (this can be 1 or 0 and sets the state of the gate).

- `clk_gate_flags`: This represents the gate-specific flags of the gate clock. These are as follows:

  `--CLK_GATE_SET_TO_DISABLE`: This is the clock switch's control mode. If set, writing 1 turns off the clock, and writing 0 turns on the clock.

  `--CLK_GATE_HIWORD_MASK`: Some registers use the concept of `reading-modifying-writing` to operate at the bit level, while other registers only support **hiword mask**. **Hiword mask** is a concept in which (in a 32-bit register) changing a bit at a given index (between 0 and 15) consists of changing the corresponding bit in the 16 lower bits (0 to 15) and masking the same bit index in the 16 higher bits (16 to 31, hence hiword or High Word) in order to indicate/validate the change.

  For example, if bit b1 needs to be set as a gate, it also needs to indicate the change by setting the hiword mask (b1 << 16). This means that the gate settings are truly in the lower 16 bits of the register, while the mask of gate bits is in the higher 16 bits of this same register. When setting this flag, `bit_idx` should be no higher than 15.

- `lock`: This is the spinlock that should be used if the clock switch requires mutual exclusion.

As you have probably guessed, this structure assumes that the clock gate register is mmio. As for the previous clock type, it is better to use the provided kernel interface to deal with such a clock; that is, `clk_hw_register_gate()`:

```
struct clk_hw *
    clk_hw_register_gate(struct device *dev, const char *name,
                         const char *parent_name,
                         unsigned long flags,
                         void iomem *reg, u8 bit_idx,
                         u8 clk_gate_flags, spinlock_t *lock);
```

Some of the parameters of this interface are the same ones we described regarding the clock-type structure. Moreover, the following are extra arguments that need to be described:

- `dev` is the device that is registering the clock.

- `name` is the name of the clock.

- `parent_name` is the name of the parent clock, which should be NULL if it has no parent.

- `flags` represents the framework-specific flags for this clock. It is common to set the `CLK_SET_RATE_PARENT` flag for gate clocks that have a parent so that rate change requests are propagated up one level.

- `clk_gate_flags` corresponds to the `.flags` in the clock-type structure.

This interface returns a pointer to the underlying `struct clh_hw` of the clock gate structure. Here, you can use the `to_clk_gate` helper macro to grab the original clock gate structure.

While setting up this clock and prior to its registration, the clock framework assigns the `clk_gate_ops` ops to it. This is actually the default ops for the gate clock. It relies on the fact that the clock is controlled through mmio registers:

```
const struct clk_ops clk_gate_ops = {
    .enable = clk_gate_enable,
    .disable = clk_gate_disable,
    .is_enabled = clk_gate_is_enabled,
};
EXPORT_SYMBOL_GPL(clk_gate_ops);
```

The entire gate clock API is defined in `drivers/clk/clk-gate.c`. Such a clock driver can be found in `drivers/clk/clk-asm9260.c`, while its device tree binding can be found in `Documentation/devicetree/bindings/clock/alphascale,acc.txt`, in the kernel source tree.

I2C/SPI-based gate clock

Not just mmio peripherals can provide gate clocks. There are also discrete chips behind I2C/SPI buses that can provide such clocks. Obviously, you cannot rely on the structure (`struct clk_gate`) or the interface helper (`clk_hw_register_gate()`) that we introduced earlier to develop drivers for such chips. The main reasons for this are as follows:

- The aforementioned interface and data structure assume that the clock gate register control is mmio, which is definitely not the case here.

- The standard gate clock ops are `.enable` and `.disable`. However, these callbacks don't need to sleep as they are called with spinlocks held, but we all know that I2C/SPI register accesses may sleep.

Both of these restrictions have workarounds:

- Instead of using the gate-specific clock framework helper, you can want to use the low-level `clk_hw_register()` interface to control the parameters of the clock, from its flags to its ops.

- You can implement the `.enable`/`.disable` logic in the `.prepare`/`.unprepare` callbacks. Remember, `.prepare`/`.unprepare` ops may sleep. This is guaranteed to work as it is a requirement for the consumer side to call `clk_prepare()` prior to calling `clk_enable()`, and then to follow a call to `clk_disable()` by a call to `clk_unprepare()`. By doing so, any consumer call to `clk_enable()` (mapped to the provider's `.enable` callback) will immediately return. However, since it is always preceded by a consumer call to `clk_prepare()` (mapped to the `.prepare` callback), we can be sure that our clock will be ungated. The same goes for `clk_disable` (mapped to the `.disable` callback), which guarantees that our clock will be gated.

This clock driver implementation can be found in `drivers/clk/clk-max9485.c`, while its device tree binding can found in `Documentation/devicetree/bindings/clock/maxim,max9485.txt`, in the kernel source tree.

GPIO gate clock alternative

This is a basic clock that can be enabled and disabled through a gpio output. `gpio-gate-clock` instances can only be instantiated from the device tree. For this, the `compatible` property should be `gpio-gate-clock` and `#clock-cells` should be `<0>` as shown in the following excerpt from `imx6qdl-sr-som-ti.dtsi`:

```
clk_ti_wifi: ti-wifi-clock {
    compatible = 'gpio-gate-clock';
    #clock-cells = <0>;
    clock-frequency = <32768>;
    pinctrl-names = 'default';
    pinctrl-0 = <&pinctrl_microsom_ti_clk>;
    enable-gpios = <&gpio5 5 GPIO_ACTIVE_HIGH>;
};

pwrseq_ti_wifi: ti-wifi-pwrseq {
    compatible = 'mmc-pwrseq-simple';
    pinctrl-names = 'default';
    pinctrl-0 = <&pinctrl_microsom_ti_wifi_en>;
```

```
        reset-gpios = <&gpio5 26 GPIO_ACTIVE_LOW>;
        post-power-on-delay-ms = <200>;
        clocks = <&clk_ti_wifi>;
        clock-names = 'ext_clock';
};
```

This clock-type driver is located in `drivers/clk/clk-gpio.c`, and further reading can be found in `Documentation/devicetree/bindings/clock/gpio-gate-clock.txt`.

Clock multiplexer and its ops

A clock multiplexer has multiple input clock signals or parents, among which only one can be selected as output. Since this type of clock can choose from among multiple parents, the `.get_parent/.set_parent/.recalc_rate` callbacks should be implemented. A mux clock is represented in the CCF by an instance of `struct clk_mux`, which looks as follows:

```
struct clk_mux {
    struct clk_hw hw;
    void __iomem *reg;
    u32    *table;
    u32    mask;
    u8     shift;
    u8     flags;
    spinlock_t   *lock;
};

#define to_clk_mux(_hw) container_of(_hw, struct clk_mux, hw)
```

Let's take a look at the elements shown in the preceding structure:

- `table` is an array of register values corresponding to the parent index.

- `mask` and `shift` are used to modify the `reg` bit field prior to getting the appropriate value.

- reg is the mmio register used for parent selection. By default, when the register's value is 0, it corresponds to the first parent, and so on. If there are exceptions, various flags can be used, as well as another interface.

- flags represents the unique flags of the mux clock, which are as follows:

  --CLK_MUX_INDEX_BIT: The register value is a power of 2. We will look at how this works shortly.

  --CLK_MUX_HIWORD_MASK: This uses the concept of the hiword mask, which we explained earlier.

  --CLK_MUX_INDEX_ONE: The register value does not start from 0, instead starting at 1. This means that the final value should be incremented by one.

  --CLK_MUX_READ_ONLY: Some platforms have read-only clock muxes that are preconfigured at reset and cannot be changed at runtime.

  --CLK_MUX_ROUND_CLOSEST : This flag uses the parent rate that is closest to the desired frequency.

- lock, if provided, is used to protect access to the register.

The CCF helper that's used to register such a clock is clk_hw_register_mux(). This looks as follows:

```
struct clk_hw *
    clk_hw_register_mux(struct device *dev, const char *name,
                        const char * const *parent_names,
                        u8 num_parents, unsigned long flags,
                        void iomem *reg, u8 shift, u8 width,
                        u8 clk_mux_flags, spinlock_t *lock)
```

Some of the parameters in the preceding registration interface were introduced when we described the mux clock structure. The remaining parameters are as follows:

- parent_names: This is an array of strings that describes all possible parent clocks.

- num_parents: This specifies the number of parent clocks.

While registering such a clock, depending on the CLK_MUX_READ_ONLY flag being set or not, the CCF assigns different clock ops. If set, clk_mux_ro_ops is used. This clock ops only implements the .get_parent ops as there would be no way to change the parent. If this is not set, clk_mux_ops is used. This ops implements .get_parent, .set_parent, and .determine_rate, as follows:

```
if (clk_mux_flags & CLK_MUX_READ_ONLY)
    init.ops = &clk_mux_ro_ops;
else
    init.ops = &clk_mux_ops;
```

These clock ops are defined as follows:

```
const struct clk_ops clk_mux_ops = {
    .get_parent = clk_mux_get_parent,
    .set_parent = clk_mux_set_parent,
    .determine_rate = clk_mux_determine_rate,
};
EXPORT_SYMBOL_GPL(clk_mux_ops);

const struct clk_ops clk_mux_ro_ops = {
    .get_parent = clk_mux_get_parent,
};
EXPORT_SYMBOL_GPL(clk_mux_ro_ops);
```

In the preceding code, there is a .table field. This is used to provide a set of values according to the parent index. However, the preceding registration interface, clk_hw_register_mux(), does not provide us with any way to feed this table.

Due to this, there is another variant available in the CCF that allows us to pass the table:

```
struct clk *
    clk_register_mux_table(struct device *dev,
                           const char *name,
                           const char **parent_names,
                           u8 num_parents,
                           unsigned long flags,
                           void iomem *reg, u8 shift, u32 mask,
    u8 clk_mux_flags, u32 *table, spinlock_t *lock);
```

The interface registers a mux to control an irregular clock through a table. Whatever the registration interface is, the same internal ops are used. Now, let's pay special attention to the most important ones; that is, .set_parent and .get_parent:

- clk_mux_set_parent: When this is called, if table is not NULL, it gets a register value from the index in table. If table is NULL and the CLK_MUX_INDEX_BIT flag is set, this means the register value is a power of 2 according to index. This value is then obtained with val = 1 << index; if CLK_MUX_INDEX_ONE is set, this value is incremented by one. If table is NULL and CLK_MUX_INDEX_BIT is not set, index is used as the default value. In either case, the final value is left-shifted at shift time and OR'ed with a mask prior to us obtaining the real value. This should be written into reg for parent selection:

```
unsigned int
    clk_mux_index_to_val(u32 *table, unsigned int flags,
                         u8 index)
{
    unsigned int val = index;
    if (table) {
        val = table[index];
    } else {
        if (flags & CLK_MUX_INDEX_BIT)
            val = 1 << index;
        if (flags & CLK_MUX_INDEX_ONE) val++;
    }
    return val;
}


static int clk_mux_set_parent(struct clk_hw *hw,
                              u8 index)
{
    struct clk_mux *mux = to_clk_mux(hw);
    u32 val =
        clk_mux_index_to_val(mux->table, mux->flags,
                             index);
    unsigned long flags = 0; u32 reg;

    if (mux->lock)
        spin_lock_irqsave(mux->lock, flags);
```

```
        else
            __acquire(mux->lock);

        if (mux->flags & CLK_MUX_HIWORD_MASK) {
            reg = mux->mask << (mux->shift + 16);
        } else {
            reg = clk_readl(mux->reg);
            reg &= ~(mux->mask << mux->shift);
        }
        val = val << mux->shift; reg |= val;
        clk_writel(reg, mux->reg);

        if (mux->lock)
            spin_unlock_irqrestore(mux->lock, flags);
        else
            __release(mux->lock);

        return 0;
}
```

- `clk_mux_get_parent`: This reads the value in `reg`, shifts it `shift` time to the right and applies (the AND operation) `mask` to it prior to getting the real value. This value is then given to the `clk_mux_val_to_index()` helper, which will return the right index according to the `reg` value. `clk_mux_val_to_index()` first gets the number of parents for the given clock. If `table` is not NULL, this number is used as the upper limit in a loop to walk through `table`. Each iteration will check whether the `table` value at the current position matches `val`. If it does, the current position in the iteration is returned. If no match is found, an error is returned. `ffs()` returns the position of the first (least significant) bit set in the word:

```
int clk_mux_val_to_index(struct clk_hw *hw, u32 *table,
                         unsigned int flags,
                         unsigned int val)
{
    int num_parents = clk_hw_get_num_parents(hw);
    if (table) {
        int i;
```

```
            for (i = 0; i < num_parents; i++)
                if (table[i] == val)
                    return i;
            return -EINVAL;
        }


        if (val && (flags & CLK_MUX_INDEX_BIT))
            val = ffs(val) - 1;
        if (val && (flags & CLK_MUX_INDEX_ONE))
            val--;
        if (val >= num_parents)
            return -EINVAL;
        return val;
    }
EXPORT_SYMBOL_GPL(clk_mux_val_to_index);


static u8 clk_mux_get_parent(struct clk_hw *hw)
{
    struct clk_mux *mux = to_clk_mux(hw);
    u32 val;
    val = clk_readl(mux->reg) >> mux->shift;
    val &= mux->mask;


    return clk_mux_val_to_index(hw, mux->table,
                                mux->flags, val);
}
```

An example of such a driver can be found in `drivers/clk/microchip/clk-pic32mzda.c`.

I2C/SPI-based clock mux

The aforementioned CCF interfaces that are used to handle clock muxes assume that control is provided via mmio registers. However, there are some I2C/SPI-based clock mux chips where you have to rely on the low-level `clk_hw` (using a `clk_hw_register()` registration-based interface) interface and register each clock according to its properties before providing the appropriate ops.

Each mux input clock should be a parent of the mux output, which must have at least
`.set_parent` and `.get_parent` ops. Other ops are also allowed but not mandatory.
A concrete example is the Linux driver for the `Si5351a/b/c` programmable I2C
clock generator from Silicon Labs, available in `drivers/clk/clk-si5351.c` in the
kernel sources. Its device tree binding is available in `Documentation/devicetree/`
`bindings/clock/silabs,si5351.txt`.

> **Important note**
>
> To write such clock drivers, you must learn how `clk_hw_register_`
> `mux` is implemented and base your registration function on it, without the
> mmio/spinlock part, and then provide your own ops according to the clock's
> properties.

GPIO mux clock alternative

The GPIO mux clock can be represented as follows:

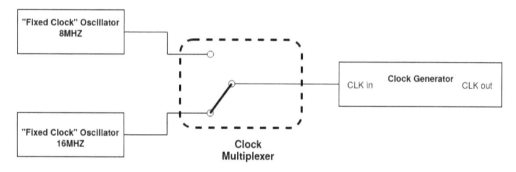

Figure 4.2 – GPIO mux clock

This is a limited alternative to clock multiplexing that only accepts two parents, as
stated in the following excerpt from its drivers, which are available in `drivers/clk/`
`clk-gpio.c`. In this case, the parent selection depends on the value of the gpio being
used:

```
struct clk_hw *clk_hw_register_gpio_mux(struct device *dev,
                            const char *name,
                            const char *
                            const *parent_names,
                            u8 num_parents,
                            struct gpio_desc *gpiod,
                            unsigned long flags)
{
```

```
    if (num_parents != 2) {
        pr_err('mux-clock %s must have 2 parents\n', name);
        return ERR_PTR(-EINVAL);
    }
    return clk_register_gpio(dev, name, parent_names,
                             num_parents,
                             gpiod, flags, &clk_gpio_mux_ops);
}
EXPORT_SYMBOL_GPL(clk_hw_register_gpio_mux);
```

According to its binding, it is only instantiable in the device tree. This binding can be found in Documentation/devicetree/bindings/clock/gpio-mux-clock. txt, in the kernel sources. The following example show how to use it:

```
clocks {
    /* fixed clock oscillators */
    parent1: oscillator22 {
        compatible = 'fixed-clock';
        #clock-cells = <0>;
        clock-frequency = <22579200>;
    };


    parent2: oscillator24 {
        compatible = 'fixed-clock';
        #clock-cells = <0>;
        clock-frequency = <24576000>;
    };


    /* gpio-controlled clock multiplexer */
    mux: multiplexer {
        compatible = 'gpio-mux-clock';
        clocks = <&parent1>, <&parent2>;
        /* parent clocks */
        #clock-cells = <0>;
        select-gpios = <&gpio 42 GPIO_ACTIVE_HIGH>;
    };
};
```

Here, we have looked at the clock multiplexer, which allows us to select a clock source from its APIs and device tree binding. Moreover, we introduced the GPIO-based clock multiplexer alternative, which does not require that we write any code. The next clock type in this series is the divider clock, which, as its name suggests, divides the parent rate by a given ratio.

(Adjustable) divider clock and its ops

This type of clock divides the parent rate and cannot gate. Since you can set the divider ratio, providing a .recalc_rate/.set_rate/.round_rate callback is a must. A clock divider is represented in the kernel as an instance of a struct clk_divider. This can be defined as follows:

```
struct clk_divider {
    struct clk_hw   hw;
    void iomem    *reg;
    u8      shift;
    u8      width;
    u8      flags;
    const struct clk_div_table      *table;
    spinlock_t    *lock;
};


#define to_clk_divider(_hw) container_of(_hw,
                                        struct clk_divider,
                                        hw)
```

Let's take a look at element in this structure:

- hw: The underlying clock_hw structure that defined the provider side.

- reg: This is the register that controls the clock division ratio. By default, the actual divider value is the register value plus one. If there are other exceptions, you can refer to the flags field descriptions for adapting.

- shift: This controls the offset of the bit of the division ratio in the register.

- width: This is the width of the divider bit field. It controls the bit number of the division ratio. For example, if width is 4, this means the division ratio is coded on 4 bits.

- `flags`: This is the divider-clock-specific flag of the clock. Various flags can be used here, some of which are as follows:

  --`CLK_DIVIDER_ONE_BASED`: When set, this means that the divider is the raw value that's read from the register since the default divisor is the value that's read from the register plus one. This also implies 0 is invalid, unless the `CLK_DIVIDER_ALLOW_ZERO` flag is set.

  --`CLK_DIVIDER_ROUND_CLOSEST`: This should be used when we want to be able to round the divider to the closest and best calculated one instead of just rounding up, which is the default behavior.

  --`CLK_DIVIDER_POWER_OF_TWO`: The actual divider value is the register value raised to a power of 2.

  --`CLK_DIVIDER_ALLOW_ZERO`: The divider value can be 0 (no change, depending on hardware support).

  --`CLK_DIVIDER_HIWORD_MASK`: See the *Gateable clock and its ops* section for more details on this flag.

  --`CLK_DIVIDER_READ_ONLY`: This flag shows that the clock has preconfigured settings and instructs the framework not to change anything. This flag also affects the ops that have been assigned to the clock.

  `CLK_DIVIDER_MAX_AT_ZERO`: This allows a clock divider to have a max divisor when it's set to zero. So, if the field value is zero, the divisor value should be 2 bits in width. For example, let's consider a divisor clock with a 2-bit field:

Value	divisor
0	4
1	1
2	2
3	3

- `table`: This is an array of value/divider pairs whose last entry should have `div = 0`. This will be described shortly.

- `lock`: Like in other clock data structures, if provided, it is used to protect access to the register.

- `clk_hw_register_divider()`: This is the most commonly used registration interface for such clocks. It is defined as follows:

```
struct clk_hw *
    clk_hw_register_divider(struct device *dev,
                            const char *name,
                            const char *parent_name,

                            unsigned long flags,
                            void iomem *reg,

                            u8 shift, u8 width,
                            u8 clk_divider_flags,

                            spinlock_t *lock)
```

This function registers a divider clock with the system and returns a pointer to the underlying `clk_hw` field. Here, you can can use the `to_clk_divider` macro to grab a pointer to the wrapper's `clk_divider` structure. Except for `name` and `parent_name`, which represent the name of the clock and the name of its parent, respectively, the other arguments in this function match the fields described in the `struct clk_divider` structure.

You may have noticed that the `.table` field is not being used here. This field is kind of special as it is used for clock dividers whose division ratios are uncommon. Actually, there are clock dividers where each individual clock line has a number of division ratios that are not related to each other's clock lines. Sometimes, there is not even any linearity between each ratio and the register value. For such cases, the best solution is to feed each clock line a table, where each ratio corresponds to its register value. This requires us to introduce a new registration interface that accepts such a table; that is, `clk_hw_register_divider_table`. This can be defined as follows:

```
struct clk_hw *
    clk_hw_register_divider_table(
                            struct device *dev,
                            const char *name,
                            const char *parent_name,

                            unsigned long flags,
                            void iomem *reg,

                            u8 shift, u8 width,
                            u8 clk_divider_flags,

                            const struct clk_div_table *table,

                            spinlock_t *lock)
```

This interface is used to register the clock with an irregular frequency division ratio, compared to the preceding interface. The difference is that the relationship between the value of the divider and the value of the register is determined by a table of the `struct clk_div_table` type. This table structure can be defined as follows:

```
struct clk_div_table {
    unsigned int val;
    unsigned int div;
};
```

In the preceding code, `val` represents the register value, while `div` represents the division ratio. Their relationship can also be changed byusing `clk_divider_flags`. Regardless of what registration interface is used, the `CLK_DIVIDER_READ_ONLY` flag determines the ops to be assigned to the clock, as follows:

```
if (clk_divider_flags & CLK_DIVIDER_READ_ONLY)
    init.ops = &clk_divider_ro_ops;
else
    init.ops = &clk_divider_ops;
```

Both these clock ops are defined in `drivers/clk/clk-divider.c`, as follows:

```
const struct clk_ops clk_divider_ops = {
    .recalc_rate = clk_divider_recalc_rate,
    .round_rate = clk_divider_round_rate,
    .set_rate = clk_divider_set_rate,
};
EXPORT_SYMBOL_GPL(clk_divider_ops);

const struct clk_ops clk_divider_ro_ops = {
    .recalc_rate = clk_divider_recalc_rate,
    .round_rate = clk_divider_round_rate,
};
EXPORT_SYMBOL_GPL(clk_divider_ro_ops);
```

While the former can set the clock rate, the last one cannot.

> **Important note**
>
> Once again, so far, using the clock-type-dependent registration interface provided by the kernel requires your clock to be mmio. Implementing such a clock driver for a non-mmio-based (SPI or I2C-based) clock would require using the low-level hw_clk registration interface and implementing the appropriate ops. An example of such a driver for an I2C-based clock, along with the appropriate ops implemented, can be found in drivers/clk/clk-max9485.c. Its binding can be found in Documentation/devicetree/bindings/clock/maxim,max9485.txt. This is a much more adjustable clock driver than the divider one.

The adjustable clock has no secrets for us anymore. Its APIs and ops have been described, as well as how it deals with irregular ratios. Next, we'll look at our final clock type, which is a mix of all the clock types we have seen so far: the composite clock.

Composite clock and its ops

This clock is used for clock branches that use a combination of mux, divider, and gate components. This is the case on most Rockchip SoCs. The clock framework abstracts such clocks by means of struct clk_composite, which looks as follows:

```
struct clk_composite {
    struct clk_hw        hw;
    struct clk_ops       ops;
    struct clk_hw        *mux_hw;
    struct clk_hw        *rate_hw;
    struct clk_hw        *gate_hw;
    const struct clk_ops     *mux_ops;
    const struct clk_ops     *rate_ops;
    const struct clk_ops     *gate_ops;
};
#define to_clk_composite(_hw) container_of(_hw,
                                    struct clk_composite,
                                    hw)
```

The fields in this data structure are quite self-explanatory, as follows:

- hw, as in other clock structures, is the handle between common and hardware-specific interfaces.
- mux_hw represents the mux clock.
- rate_hw represents the divider clock.
- gate_hw represents the gate clock.
- mux_ops, rate_ops, and gate_ops are the clock ops for mux, rate, and gate, respectively.

Such a clock can be registered through the following interface:

```
struct clk_hw *clk_hw_register_composite(
          struct device *dev, const char *name,
          const char * const *parent_names, int num_parents,
          struct clk_hw *mux_hw,
          const struct clk_ops *mux_ops,
          struct clk_hw *rate_hw,
          const struct clk_ops *rate_ops,
          struct clk_hw *gate_hw,
          const struct clk_ops *gate_ops,
          unsigned long flags)
```

This may look a bit complicated, but if you went through the previous clock, this one will be more or less obvious to you. Take a look at `drivers/clk/sunxi/clk-a10-hosc.c` in the kernel source for an example of a composite clock driver.

Putting it all together – global overview

If you are still confused, then take a look at the following diagram:

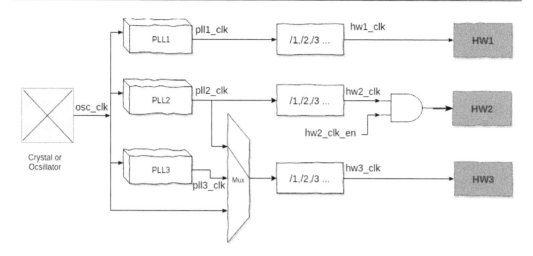

Figure 4.3 – Clock tree example

The preceding clock tree shows an oscillator clock feeding three PLLs – that is, pll1, pll2, and pll3 – as well as a multiplexer. According to the multiplexer (mux), hw3_clk can be derived from either the pll2, pll3, or osc clock.

The following device tree excerpt can be used to model the preceding clock tree:

```
osc: oscillator {
    #clock-cells = <0>;
    compatible = 'fixed-clock';
    clock-frequency = <20000000>;
    clock-output-names = 'osc20M';
};

pll2: pll2 {
    #clock-cells = <0>;
    compatible = 'abc123,pll2-clock';
    clock-frequency = <23000000>; clocks = <&osc>;
    [...]
};

pll3: pll3 {
    #clock-cells = <0>;
    compatible = 'abc123,pll3-clock';
```

```
        clock-frequency = <23000000>; clocks = <&osc>;
        [...]
};


hw3_clk: hw3_clk {
    #clock-cells = <0>;
    compatible = 'abc123,hw3-clk';
    clocks = <&pll2>, <&pll3>, <&osc>;
    clock-output-names = 'hw3_clk';
};
```

When it comes to the source code, the following excerpt shows how to register hw_clk3 as a mux (a clock multiplexer) and points out the parent relationship of pll2, pll3, and osc:

```
of_property_read_string(node, 'clock-output-names', &clk_name);
parent_names[0] = of_clk_get_parent_name(node, 0);
parent_names[1] = of_clk_get_parent_name(node, 1);
parent_names[2] = of_clk_get_parent_name(node, 2); /* osc */


clk = clk_register_mux(NULL, clk_name, parent_names,
                       ARRAY_SIZE(parent_names), 0, regs_base,
                       offset_bit, one_bit, 0, NULL);
```

A downstream clock provider should use of_clk_get_parent_name() to obtain its parent clock name. For a block with multiple outputs, of_clk_get_parent_name() can return a valid clock name, but only when the clock-output-names property is present.

Now, we can look at the clock tree summary via the CCF *sysfs* interface, /sys/kernel/debug/clk/clk_summary. This can be seen in the following excerpt:

```
$ mount -t debugfs none /sys/kernel/debug
# cat /sys/kernel/debug/clk/clk_summary
[...]
```

With that, we are done with the clock producer side. We have learned about its APIs and discussed its declaration in the device tree. Moreover, we have learned how to dump their topology from *sysfs*. Now, let's look at clock consumer APIs.

Introducing clock consumer APIs

Clock producer device drivers are useless without consumers at the other end to leverage the clock lines that have been exposed. The main purpose of such drivers is to assign their clock source lines to consumers. These clock lines are then used for several purposes, and the Linux kernel provides consequent APIs and helpers to achieve the required goal. Consumer drivers need to include `<linux/clk.h>` in their code for its APIs to be used. Moreover, nowadays, the clock consumer interface entirely relies on the device tree, meaning that consumers should be assigned clocks they need from the device tree. The consumer binding should follow the provider's since the consumer specifier is determined by the provider's `#clock-cells` property. Take a look at the following UART node description, which requires two clock lines:

```
uart1: serial@02020000 {
    compatible = 'fsl,imx6sx-uart', 'fsl,imx21-uart';
    reg = <0x02020000 0x4000>;
    interrupts = <GIC_SPI 26 IRQ_TYPE_LEVEL_HIGH>;
    clocks = <&clks IMX6SX_CLK_UART_IPG>,
             <&clks IMX6SX_CLK_UART_SERIAL>;
    clock-names = 'ipg', 'per';
    dmas = <&sdma 25 4 0>, <&sdma 26 4 0>;
    dma-names = 'rx', 'tx';
    status = 'disabled';
};
```

This represents a device with two clock inputs. The preceding node excerpt allows us to introduce the device tree binding for the clock consumer, which should have, at the very least, the following properties:

- The `clocks` property is where you should specify the source clock lines for a device with respect to the `#clock-cells` property of the provider.

- `clock-names` is the property used to name clocks in the same way they are listed in `clocks`. In other words, this property should be used to list the input name(s) for the clock(s) with respect to the consuming node. This name(s) should reflect the consumer input signal name(s) and can/must be used in the code (see `[devm_]` `clk_get()`) so that it matches the corresponding clock.

> **Important note**
> Clock consumer nodes must never directly reference the provider's `clock-output-names` property.

The consumer has a reduced and portable API based on whatever the underlying hardware clock is. Next, we'll take a look at the common operations that are performed by consumer drivers, along with their associated APIs.

Grabbing and releasing clocks

The following functions allow us to grab and release a clock, given its `id`:

```
struct clk *clk_get(struct device *dev, const char *id);
void clk_put(struct clk *clk);
struct clk *C(struct device *dev, const char *id)
```

`dev` is the device using this clock, while `id` is the name given to the clock in the device tree. On success, `clk_get` returns a pointer to a `struct clk`. This can be given to any other `clk-consumer` API. `clk_put` actually releases the clock line. The first two APIs in the preceding code are defined in `drivers/clk/clkdev.c`. However, other clock consumer APIs are defined in `drivers/clk/clk.c`. `devm_clk_get` is simply the managed version of `clk_get`.

Preparing/unpreparing clocks

To prepare a clock for use, you can use `clk_prepare()`, as follows:

```
void clk_prepare(struct clk *clk);
void clk_unprepare(struct clk *clk);
```

These functions may sleep, which means they cannot be called from within an atomic context. It is worth always calling `clk_prepare()` before `clock_enable()`. This may be useful if the underlying clock is behind a slow bus (SPI/I2C) since such clock drivers must implement their enable/disable (which must not sleep) code from within the prepare/unprepare ops (which are allowed to sleep).

Enabling/disabling

When it comes to gating/ungating the clock, you can use the following API:

```
int clk_enable(struct clk *clk);
void clk_disable(struct clk *clk);
```

`clk_enable` must not sleep and actually ungates the clock. It returns 0 on success or an error otherwise. `clk_disable` does the reverse. To enforce the fact of calling prepare prior to calling enable, the clock framework provide the `clk_prepare_enable` API, which internally calls both. The opposite can be done with `clk_disable_unprepare`:

```
int clk_prepare_enable(struct clk *clk)
void clk_disable_unprepare(struct clk *clk)
```

Rate functions

For clocks whose rates can be changed, we can use the following function to get/set the rate of the clock:

```
unsigned long clk_get_rate(struct clk *clk);
int clk_set_rate(struct clk *clk, unsigned long rate);
long clk_round_rate(struct clk *clk, unsigned long rate);
```

`clk_get_rate()` returns 0 if `clk` is NULL; otherwise, it will return the rate of the clock; that is, the cached rate. However, if the CLK_GET_RATE_NOCACHE flag is set, a new calculation will be done (by means of `recalc_rate()`) to return the real clock rate. On the other hand, `clk_set_rate()` will set the rate of the clock. However, its rate parameter can't take any value. To see if the rate you are targeting is supported or allowed by the clock, you should use `clk_round_rate()`, along with the clock pointer and the target rate in Hz, as follows.

```
rounded_rate = clk_round_rate(clkp, target_rate);
```

This is the return value of `clk_round_rate()` that must be given to `clk_set_rate()`, as follows:

```
ret = clk_set_rate(clkp, rounded_rate);
```

Changing the clock rate may fail in the following cases:

- The clock is drawing its source from a fixed-rate clock source (for example, `OSC0`, `OSC1`, `XREF`, and so on).
- The clock is in use by multiple modules/children, which would mean that `usecount` is greater than 1.
- The clock source is in use by more than one child.

Note that parent rates are returned if `.round_rate()` is not implemented.

Parent functions

There are clocks that are the children of other clocks, thus creating a parent/child relationship. To either get/set the parent of a given clock, you can use the following functions:

```
int clk_set_parent(struct clk *clk, struct clk *parent);
struct clk *clk_get_parent(struct clk *clk);
```

`clk_set_parent()` actually sets the parent of the given clock, while `clk_get_parent()` returns the current parent.

Putting it all together

To summarize this, take a look at the following excerpt of the i.MX serial driver (`drivers/tty/serial/imx.c`), which deals with the preceding device node:

```
sport->clk_per = devm_clk_get(&pdev->dev, 'per');
if (IS_ERR(sport->clk_per)) {
    ret = PTR_ERR(sport->clk_per);
    dev_err(&pdev->dev, 'failed to get per clk: %d\n', ret);
    return ret;
}
sport->port.uartclk = clk_get_rate(sport->clk_per);
/*
 * For register access, we only need to enable the ipg clock.
```

```
    */
    ret = clk_prepare_enable(sport->clk_ipg);
    if (ret)
        return ret;
```

In the preceding code excerpt, we see can how the driver grabs the clock and its current rate, and then enables it.

Summary

In this chapter, we walked through the Linux Common Clock Framework. We introduced both the provider and consumer sides, as well as the user space interface. We then discussed the different clock types and learned how to write the appropriate Linux drivers for each.

The next chapter deals with ALSA SoC, the Linux kernel framework for audio. This framework heavily relies on the clock framework to, for example, sample audio.

Section 2: Multimedia and Power Saving in Embedded Linux Systems

This section walks you through the most widely used Linux kernel multimedia subsystems, V4L2 and ALSA SoC, in a simple but energy-efficient way, helped by the Linux kernel power management subsystem.

This section contains the following chapters:

5
ALSA SoC Framework – Leveraging Codec and Platform Class Drivers

Audio is an analog phenomenon that can be produced in all sorts of ways. Voice and audio have been communication media since the beginning of humanity. Almost every kernel provides audio support to userspace applications as an interaction mechanism between computers and humans. To achieve this, the Linux kernel provides a set of APIs known as **ALSA**, which stands for **Advanced Linux Sound Architecture**.

ALSA was designed for desktop computers, not taking into account embedded world constraints. This added a lot of drawbacks when it came to dealing with embedded devices, such as the following:

- Strong coupling between codec and CPU code, leading to difficulties in porting and code duplication.

- No standard way to handle notifications about users' audio-related behavior. In mobile scenarios, users' audio-related behaviors are frequent, so a special mechanism is needed.

- In the original ALSA architecture, power efficiency was not considered. But for embedded devices (most of the time, battery-backed), this is a key point, so there needs to be a mechanism.

This is where ASoC comes into the picture. The purpose of the **ALSA System on Chip** (**ASoC**) layer is to provide better ALSA support for embedded processors and various codecs.

ASoC is a new architecture designed to solve the aforementioned problems and comes with the following advantages:

- An independent codec driver to reduce coupling with the CPU

- More convenient configuration of the audio data interface between the CPU and codec **Dynamic Audio Power Management** (**DAPM**), dynamically controlling power consumption (more information can be found here: `https://www.kernel.org/doc/html/latest/sound/soc/dapm.html`)

- Reduced pop and click and increased platform-related controls

To achieve the aforementioned features, ASoC divides the embedded audio system into three reusable component drivers, namely the **machine class**, **platform class**, and **codec class**. Among them, the platform and codec classes are *cross-platform*, and the machine class is *board*-specific. In this chapter and the next chapter, we will walk through these component drivers, dealing with their respective data structures and how they are implemented.

Here, we will present the Linux ASoC driver architecture and the implementation of its different parts, looking specifically at the following:

- Introduction to ASoC

- Writing codec class drivers

- Writing platform class drivers

Technical requirements

- Strong knowledge of device tree concepts

- Familiarity with the **Common Clock Framework** (**CCF**) (discussed in *Chapter 4, Storming the Common Clock Framework*)

- Familiarity with the regmap API

- Strong knowledge of the Linux kernel DMA framework

- Linux kernel v4.19.X sources, available at `https://git.kernel.org/pub/scm/linux/kernel/git/stable/linux.git/refs/tags`

Introduction to ASoC

From an architectural point of view, the ASoC subsystem elements and their relationship can be represented as follows:

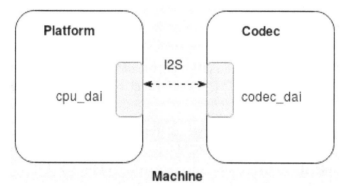

Figure 5.1 – ASoC architecture

The preceding diagram summarizes the new ASoC architecture, in which the machine entity wraps both platform and codec entities.

In the ASoC implementation prior to kernel v4.18, there was strict separation between SoC audio codec devices (represented by `struct snd_soc_codec` these days) and SoC platform interfaces (represented by `struct snd_soc_platform`) and their respective digital audio interfaces. However, there was an increasing amount of similar code between codecs, platforms, and other components. This led to a new and generic approach, the concept of the **component**. All drivers have been moved over to this new generic component, platform code has been removed, and everything has been refactored so that nowadays, we only talk about `struct snd_soc_component` (which may refer to either a codec or a platform) and `struct snd_soc_component_driver` (which refers to their respective audio interface drivers).

Now that we have introduced the ASoC concept, we can get deeper into the details, discussing digital audio interfaces first.

ASoC Digital Audio Interface

The **Digital Audio Interface** (**DAI**) is the bus controller that actually carries audio data from one end (the SoC, for example) to the other end (the codec). ASoC currently supports most of the DAIs found on SoC controllers and portable audio codecs today, such as AC97, I2S, PCM, S/PDIF, and TDM.

> **Important note**
>
> An I2S module supports six different modes, the most useful of which are I2S and TDM.

ASoC sub-elements

As we have seen earlier, an ASoC system is divided into three elements, each having a dedicated driver, described as follows:

- **Platform**: This refers to the audio DMA engine of the SoC (AKA the platform), such as i.MX, Rockchip, and STM32. Platform class drivers can be subdivided into two parts:

  --**CPU DAI driver**: In embedded systems, it usually refers to the audio bus controller (I2S, S/PDIF, AC97, and PCM bus controllers, sometimes integrated into a bigger module, the **Serial Audio Interface** (**SAI**)) of the CPU. It is responsible for transporting audio data from the bus Tx FIFO to the codec in the case of playback (the recording is in opposite direction, from the codec to the bus Rx FIFO) . The platform driver defines DAIs and registers them with the ASoC core.

  --**PCM DMA driver**: The PCM driver helps perform DMA operations by overriding the function pointers exposed by the `struct snd_soc_component_driver` (see the `struct snd_pcm_ops` element) structure. The PCM driver is platform-agnostic and interacts only with the SOC DMA engine upstream APIs. The DMA engine then interacts with the platform-specific DMA driver to get the correct DMA settings.

It is responsible for carrying the audio data in the **DMA buffer** to the bus (or port) Tx FIFO. The logic of this part is more complicated. The next sections will elaborate on it.

- **Codec**: Codec literally means codec, but there are many features in the chip. Common ones are AIF, DAC, ADC, Mixer, PGA, Line-in, and Line-out. Some high-end codec chips also have an echo canceller, noise suppression, and other components. The codec is responsible for the conversion of analog signals from sound sources into digital signals that the processor can operate (for capture operations) or the conversion of digital signals from sound sources (the CPU) to analog signals that humans can recognize in the case of playback. If necessary, it makes the corresponding adjustment to the audio signal and controls the path between the audio signals since there may be a different flow path for each audio signal in the chip.

- **Machine**: This is the system-level representation (the board actually), linking both audio interfaces (`cpu_dai` and `codec_dai`). This link is abstracted in the kernel by instances of `struct snd_soc_dai_link`. After configuring the link, the machine driver registers (by means of `devm_snd_soc_register_card()`) a `struct snd_soc_card` object, which is the Linux kernel abstraction of a sound card. Whereas platform and codec drivers are generally reusable, the machine has its specific hardware features that are almost non-reusable. The so-called hardware characteristics refer to the link between DAIs; an open amplifier through a GPIO; detecting the plugin through a GPIO; using a clock such as MCLK/eternal OSC as the reference clock source of the I2S CODEC module, and so on.

From the preceding description, we can produce the following ASoC scheme and its relationships:

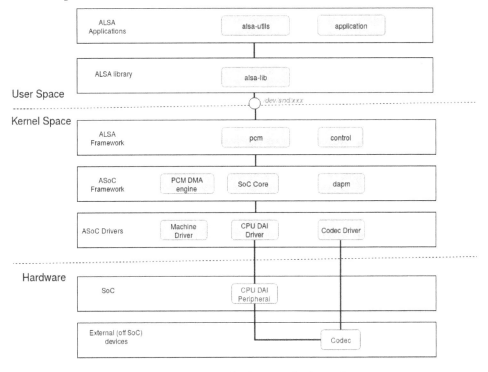

Figure 5.2 – Linux audio layers and relationships

The preceding diagram is a snapshot of the interaction between Linux kernel audio components. Now that we are familiar with ASoC concepts, we can move on to its first device driver class, which deals with codec devices.

Writing codec class drivers

In order to be coupled together, machine, platform, and codec entities need dedicated drivers. The codec class driver is the most basic. It implements code that should leverage the codec device and expose its hardware properties so that user space tools such as amixer can play with it. The codec class driver is and should be platform-independent. The same codec driver can be used whatever the platform. Since it targets a specific codec, it should contain audio controls, audio interface capabilities, a codec DAPM definition, and I/O functions. Each codec driver must fulfill the following specifications:

- Provide an interface to other modules by defining DAI and PCM configurations.

- Provide codec control IO hooks (using I2C or SPI or both APIs).

- Expose additional **kcontrols (kernel controls)** as needed for userspace utilities to dynamically control module behavior.

- Optionally, define DAPM widgets and establish DAPM routes for dynamic power switching and also provide DAC digital mute control.

The codec driver includes the codec device (component, actually) itself and DAIs components, which are used during the binding with the platform. It is platform-independent. By means of devm_snd_soc_register_component(), the codec driver registers a struct snd_soc_component_driver object (which is actually the instance of the codec driver that contains pointers to the codec's routes, widgets, controls, and a set of codec-related function callbacks) along with one or more struct snd_soc_dai_driver, which is an instance of the codec DAI driver that may contain an audio stream, for example:

```
struct snd_soc_component_driver {
    const char *name;
    /* Default control and setup, added after probe() is run */
    const struct snd_kcontrol_new *controls;
    unsigned int num_controls;
    const struct snd_soc_dapm_widget *dapm_widgets;
    unsigned int num_dapm_widgets;
    const struct snd_soc_dapm_route *dapm_routes;
    unsigned int num_dapm_routes;

    int (*probe)(struct snd_soc_component *);
    void (*remove)(struct snd_soc_component *);
    int (*suspend)(struct snd_soc_component *);
    int (*resume)(struct snd_soc_component *);
    unsigned int (*read)(struct snd_soc_component *,
                    unsigned int);
    int (*write)(struct snd_soc_component *, unsigned int,
                    unsigned int);

    /* pcm creation and destruction */
    int (*pcm_new)(struct snd_soc_pcm_runtime *);
    void (*pcm_free)(struct snd_pcm *);
    /* component wide operations */
    int (*set_sysclk)(struct snd_soc_component *component,
```

```
                    int clk_id,
                    int source, unsigned int freq, int dir);
    int (*set_pll)(struct snd_soc_component *component,
                    int pll_id, int source,
                    unsigned int freq_in,
                    unsigned int freq_out);
    int (*set_jack)(struct snd_soc_component *component,
                    struct snd_soc_jack *jack, void *data);
    [...]
    const struct snd_pcm_ops *ops;
    [...]
    unsigned int non_legacy_dai_naming:1;
};
```

This structure must also be provided by the platform driver. However, in the ASoC core, the only element in this structure that is mandatory is name, since it is used for matching the component. The following are the meanings of the elements in the structure:

- name: The name of this component is mandatory for both codec and platform. Other elements in the structure may not be needed on the platform side.

- probe: Component driver probe function, executed (in order to complete the component initialization if necessary) when this component driver is probed by the machine driver (actually, when the machine driver registers a card made of this component with the ASoC core: see snd_soc_instantiate_card()).

- remove: When the component driver is unregistered (which occurs when the sound card to which this component driver is bound is unregistered).

- suspend and resume: Power management callbacks, invoked during system suspend or resume stages.

- controls: Controls interface pointers, such as controlling volume adjustment, channel selection, and so on, mostly for codecs.

- set_pll: Sets function pointers for phase-locked loops.

- read: The function to read the codec register.

- write: The function to write into codec registers.

- num_controls: The number of controls in controls, that is, the number of snd_kcontrol_new objects.

- dapm_widgets: The dapm widget pointer.

- `num_dapm_widgets` : The number of dapm part pointers.

- `dapm_routes`: dapm route pointers.

- `num_dapm_routes` : The number of dapm route pointers.

- `set_sysclk`: Sets clock function pointers.

- `ops`: Platform DMA-related callbacks, only necessary when providing this structure from within the platform driver (ALSA only); however, with ASoC, this field is set up by the ASoC core for you by means of a dedicated ASoC DMA-related API when using the generic PCM DMA engine framework.

So far, we have introduced the `struct snd_soc_component_driver` data structure in the context of the codec class driver. Remember, this structure abstracts both codec and platform devices and will be discussed in the platform driver context as well. Still, in the context of the codec class driver, we will need to discuss the `struct snd_soc_dai_driver` data structure, which, along with `struct snd_soc_component_driver`, abstracts a codec or a platform device, along with its DAI driver.

Codec DAI and PCM (AKA DSP) configurations

This section is rather generic and should probably be named **DAI and PCM** (AKA **DSP**) **configurations**, without the word *Codec*. Each codec (or should I say "component") driver must expose the DAIs of the codec (component) as well as their capabilities and operations. This is done by filling and registering as many instances of `struct snd_soc_dai_driver` as there are DAIs on the codec, and that must be exported using the `devm_snd_soc_register_component()` API. This function also takes a pointer to a `struct snd_soc_component_driver`, which is the component driver to which the provided DAI drivers will be bound and will be exported (actually, inserted into the ASoC global list of components, `component_list`, defined in `sound/soc/soc-core.c`) so that it can be registered with the core by the machine driver prior to registering the sound card. This structure covers the clocking, formatting, and ALSA operations for each interface and is defined in `include/sound/soc-dai.h` as follows:

```
struct snd_soc_dai_driver {
    /* DAI description */
    const char *name;

    /* DAI driver callbacks */
    int (*probe)(struct snd_soc_dai *dai);
    int (*remove)(struct snd_soc_dai *dai);
    int (*suspend)(struct snd_soc_dai *dai);
```

```
    int (*resume)(struct snd_soc_dai *dai);
[...]
    /* ops */
    const struct snd_soc_dai_ops *ops;

    /* DAI capabilities */
    struct snd_soc_pcm_stream capture;
    struct snd_soc_pcm_stream playback;
    unsigned int symmetric_rates:1;
    unsigned int symmetric_channels:1;
    unsigned int symmetric_samplebits:1;
[...]
};
```

In the preceding block, only the main elements of the structure have been enumerated for the sake of readability. The following are their meanings:

- name: This is the name of the DAI interface.

- probe: The DAI driver probe function, executed when the component driver to which this DAI driver belongs is probed by the machine driver (actually, when the machine driver registers a card with the ASoC core).

- remove: Invoked when the component driver to which this DAI driver belongs is unregistered.

- suspend and resume: Power management callbacks.

- ops: Points to the struct snd_soc_dai_ops structure, which provides callbacks for configuring and controlling the DAI.

- capture: Points to a struct snd_soc_pcm_stream structure, which represents the hardware parameters for audio capture. This member describes the number of channels, bit rate, data format, and so on supported during audio capture. It does not need to be initialized if the capture feature is not needed.

- playback: The hardware parameter for audio playback. This member describes the number of channels, bit rate, data format, and so on supported during playback. It does not need to be initialized if the audio playback feature is not needed.

Actually, codec and platform drivers must register this structure for every DAI they have. This is what makes this section generic. It is used later by the machine driver to build the link between the codec and the SoC. However, there are other data structures that need to be granted some study time in order to understand how the whole configuration is done: these are `struct snd_soc_pcm_stream` and `struct snd_soc_dai_ops`, described in the next sections.

DAI operations

The operations are abstracted by instances of the `struct snd_soc_dai_ops` structure. This structure contains a set of callbacks that relate to different events regarding the PCM interface (that is, it's most probable that you'll want to prepare the device in some manner before audio transfer starts, so you would put the code to do this into your `prepare` callback) or callbacks that relate the DAI clock and format configurations. This structure is defined as follows:

```
struct snd_soc_dai_ops {
    int (*set_sysclk)(struct snd_soc_dai *dai, int clk_id,
                        unsigned int freq, int dir);
    int (*set_pll)(struct snd_soc_dai *dai, int pll_id,
                    int source,
                        unsigned int freq_in,
                        unsigned int freq_out);
    int (*set_clkdiv)(struct snd_soc_dai *dai, int div_id,
                        int div);
    int (*set_bclk_ratio)(struct snd_soc_dai *dai,
                        unsigned int ratio);


    int (*set_fmt)(struct snd_soc_dai *dai, unsigned int fmt);
    int (*xlate_tdm_slot_mask)(unsigned int slots,
                                unsigned int *tx_mask,
                                unsigned int *rx_mask);
    int (*set_tdm_slot)(struct snd_soc_dai *dai,
                        unsigned int tx_mask,
                        unsigned int rx_mask,
                        int slots, int slot_width);
    int (*set_channel_map)(struct snd_soc_dai *dai,
                        unsigned int tx_num,
                        unsigned int *tx_slot,
                        unsigned int rx_num,
```

```
                              unsigned int *rx_slot);
    int (*get_channel_map)(struct snd_soc_dai *dai,
                              unsigned int *tx_num,
                              unsigned int *tx_slot,

                              unsigned int *rx_num,
                              unsigned int *rx_slot);
    int (*set_tristate)(struct snd_soc_dai *dai, int tristate);
    int (*set_sdw_stream)(struct snd_soc_dai *dai,
                          void *stream,
                          int direction);


    int (*digital_mute)(struct snd_soc_dai *dai, int mute);
    int (*mute_stream)(struct snd_soc_dai *dai, int mute,
                       int stream);


    int (*startup)(struct snd_pcm_substream *,
                   struct snd_soc_dai *);
    void (*shutdown)(struct snd_pcm_substream *,
                     struct snd_soc_dai *);
    int (*hw_params)(struct snd_pcm_substream *,
                     struct snd_pcm_hw_params *,
                     struct snd_soc_dai *);
    int (*hw_free)(struct snd_pcm_substream *,
                   struct snd_soc_dai *);
    int (*prepare)(struct snd_pcm_substream *,
                   struct snd_soc_dai *);


    int (*trigger)(struct snd_pcm_substream *, int,
                   struct snd_soc_dai *);
};
```

Callback functions in this structure can be basically divided into three classes, and the driver can implement some of them according to the actual situation.

The first class gathers **clock configuration callbacks**, usually called by the machine driver. These callbacks are as follows:

- `set_sysclk` sets the main clock of the DAI. If implemented, this callback should derive the best DAI bit and frame clocks from the system or master clock. The machine driver can use the `snd_soc_dai_set_sysclk()` API on `cpu_dai` and/or `codec_dai` in order to invoke this callback.

- `set_pll` sets the PLL parameters. If implemented, this callback should configure and enable PLL to generate the output clock based on the input clock. The machine driver can use the `snd_soc_dai_set_pll()` API on `cpu_dai` and/or `codec_dai` in order to invoke this callback.

- `set_clkdiv` sets the clock division factor. The API from the machine driver to invoke this callback is `snd_soc_dai_set_clkdiv()`.

The second callback class is the **format configuration callbacks** of the DAI, usually called by the machine driver. These callbacks are as follows:

- `set_fmt` sets the format of the DAI. The machine driver can use the `snd_soc_dai_set_fmt()` API to invoke this callback (on either CPU or codec DAIs, or both) in order to configure the DAI hardware audio format.

- `set_tdm_slot`: If the DAI supports **time-division multiplexing** (TDM), it is used to set the TDM slot. The machine driver API to invoke this callback is `snd_soc_dai_set_tdm_slot()`, in order to configure the specified DAI for TDM operations.

- `set_channel_map`: Channel TDM mapping settings. The machine driver invokes this callback for the specified DAI using the `snd_soc_dai_set_channel_map()` API.

- `set_tristate`: Sets the state of the DAI pin, which is needed when using the same pin in parallel with other DAIs. It is invoked from the machine driver by using the `snd_soc_dai_set_tristate()` API.

The last callback class is the normal standard frontend witch gathers PCM correction operations usually invoked by the ASoC core. The concerned callbacks are as follows:

- `startup`: Invoked by ALSA when a PCM sub-stream is opened (when someone has opened the capture/playback device (At device file open for example).

- `shutdown`: This callback should implement code that will undo what has been done during startup.

- hw_params: This is called when setting up the audio stream. The `struct snd_pcm_hw_params` contains the audio characteristics.

- hw_free: Should undo what has been done in `hw_params`.

- prepare: This is called when PCM is *ready*. Please see the following PCM common state change flow in order to understand when this callback is called. DMA transfer parameters are set according to channels, `buffer_bytes`, and so on, which are related to the specific hardware platform.

- trigger: This is called when PCM starts, stops, and pauses. The `int` argument in this callback is a command that may be one of SNDRV_PCM_TRIGGER_START, SNDRV_PCM_TRIGGER_RESUME, or SNDRV_PCM_TRIGGER_PAUSE_RELEASE according to the event. Drivers can use `switch...case` in order to iterate over events.

- (Optional) `digital_mute`: An anti-pop sound called by the ASoC core. It may be invoked by the core when the system is being suspended, for example.

In order to figure out how the preceding callbacks could be invoked by the core, let's have a look at the PCM common state changes flow:

1. **First started**: *off --> standby --> prepare --> on*

2. **Stop**: *on --> prepare --> standby*

3. **Resume**: *standby --> prepare --> on*

Each state in the preceding flow will invoke a callback. All this being said, we can delve into hardware configuration data structures, either for capture or playback operations.

Capture and playback hardware configuration

During capture or playback operations, the DAI setting (such as the channel number) and capabilities should be set in order to allow the underlying PCM stream to be configured. You achieve this by filling an instance of `struct snd_soc_pcm_stream` defined as follows for each operation and for each DAI, in both codec and platform drivers:

```
struct snd_soc_pcm_stream {
    const char *stream_name;
    u64 formats;
    unsigned int rates;
    unsigned int rate_min;
    unsigned int rate_max;
    unsigned int channels_min;
```

```
    unsigned int channels_max;
    unsigned int sig_bits;
};
```

The main members of this structure can be described as follows:

- `stream_name`: The name of the stream, either `"Playback"` or `"Capture"`.

- `formats`: A collection of supported data formats (valid values are defined in `include/sound/pcm.h` prefixed with `SNDRV_PCM_FMTBIT_`), such as `SNDRV_PCM_FMTBIT_S16_LE` or `SNDRV_PCM_FMTBIT_S24_LE`. If multiple formats are supported, each format can be combined, such as `SNDRV_PCM_FMTBIT_S16_LE | SNDRV_PCM_FMTBIT_S20_3LE`.

- `rates`: A set of supported sampling rates (prefixed with `SNDRV_PCM_RATE_` and the whole valid values are defined in `include/sound/pcm.h`), such as `SNDRV_PCM_RATE_44100` or `SNDRV_PCM_RATE_48000`. If multiple sampling rates are supported, each sampling rate can be increased, such as `SNDRV_PCM_RATE_48000 | SNDRV_PCM_RATE_88200`.

- `rate_min`: The minimum supported sample rate.

- `rate_max`: The maximum supported sample rate.

- `channels_min`: The minimum supported number of channels.

The concept of controls

It is common for codec drivers to expose some codec properties that can be altered from the userspace. These are codec controls. When the codec is initialized, all the defined audio controls are registered to the ALSA core. The structure of an audio control is `struct snd_kcontrol_new` defined as `include/sound/control.h`.

In addition to the DAI bus, codec devices are equipped with a control bus, an I2C or SPI bus most of time. In order to not bother about each codec driver to implement its control access routines, the codec control I/O has been standardized. This was where the regmap API originated. You can use regmap to abstract the control interface so that the codec driver does not have to worry about what the current control method is. The audio codec frontend is implemented in `sound/soc/soc-io.c`. This relies on the regmap API, which has already been discussed, in *Chapter 2, Leveraging the Regmap API and Simplifying the Code.*

The codec driver then needs to provide read and write interfaces in order to access the underlying codec registers. These callbacks need to be set in the .read and .write fields of the codec component driver, the struct snd_soc_component_driver. The following are the high-level APIs that can be used to access the component registers:

```
int snd_soc_component_write(struct
                            snd_soc_component *component,
                            unsigned int reg, unsigned int val)
int snd_soc_component_read(struct snd_soc_component *component,
                            unsigned int reg,
                            unsigned int *val)
int snd_soc_component_update_bits(struct
                            snd_soc_component *component,
                            unsigned int reg,
                            unsigned int mask,
                            unsigned int val)
int snd_soc_component_test_bits(struct
                            snd_soc_component *component,
                            unsigned int reg,
                            unsigned int mask,
                            unsigned int value)
```

Each helper in the preceding is self-descriptive. Before we delve into control implementation, notice that the control framework is made of several types:

- A simple switch control, which is a single logical value in a register
- A stereo control, which is the stereo version of the previous simple switch control, controlling two logical values at the same time in the register
- A mixer control, which is a combination of multiple simple controls and whose output is the mix of its inputs
- MUX controls – the same as the aforementioned mixer control, but selecting one among many

From within ALSA, a control is abstracted by means of the struct snd_kcontrol_new structure, defined as follows:

```
struct snd_kcontrol_new {
    snd_ctl_elem_iface_t iface;
    unsigned int device;
    unsigned int subdevice;
```

```
    const unsigned char *name;
    unsigned int index;
    unsigned int access;
    unsigned int count;
    snd_kcontrol_info_t *info;
    snd_kcontrol_get_t *get;
    snd_kcontrol_put_t *put;
    union {
        snd_kcontrol_tlv_rw_t *c;
        const unsigned int *p;
    } tlv;
    [...]
};
```

The following are the descriptions of the fields in the aforementioned data structure:

- The `iface` field specifies the control type. It is of type `snd_ctl_elem_iface_t`, which is an enum of SNDRV_CTL_ELEM_IFACE_XXX, where XXX can be MIXER, PCM, and so on. The list of possible values can be found here: https://elixir.bootlin.com/linux/v4.19/source/include/uapi/sound/asound.h#L848. If the control is closely associated with a specific device on the sound card, you can use HWDEP, PCM, RAWMIDI, TIMER, or SEQUENCER, and specify the device number with the device and subdevice (which is the substream in the device) fields.

- name is the name of the control. This field has an important role that allows controls to be categorized by name. ALSA has somehow standardized some control names, which we discuss in detail in the *Control naming convention* section.

- The `index` field is used to save the number of controls on the card. If there is more than one codec on the sound card, and each codec has a control with the same name, then we can distinguish these controls by `index`. When `index` is 0, this differentiation strategy can be ignored.

- access contains the access right of the control in the form SNDRV_CTL_ELEM_ACCESS_XXX. Each bit represents an access type that can be combined with multiple OR operations. XXX can be either READ, WRITE, or VOLATILE, and so on. Possible bitmasks can be found here: https://elixir.bootlin.com/linux/v4.19/source/include/uapi/sound/asound.h#L858.

- `get` is the callback function used to read the current value of the control and return it to the application in the userspace.

- `put` is the callback function used to set the application's control value to the control.

- The `info` callback function is used to get the details about the control.

- The `tlv` field provides metadata for the control.

Control naming convention

ALSA expects controls to be named in a certain way. In order to achieve this, ALSA has predefined some commonly used sources (such as Master, PCM, CD, Line, and so on), directions (representing the data flow of the control, such as Playback, Capture, Bypass, Bypass Capture, and so on), and functions (according to the function of the control, such as Switch, Volume, Route, and so on). Do note that no definition of the direction means that the control is two-way (playback and capture).

You can refer to the following link for more details on ALSA control naming: `https://www.kernel.org/doc/html/v4.19/sound/designs/control-names.html`.

Control metadata

There are mixer controls that need to provide information in **decibels (dB)**. We can use the `DECLARE_TLV_xxx` macro to define some variables containing this information, then point the control `tlv.p` field to these variables, and finally add the `SNDRV_CTL_ELEM_ACCESS_TLV_READ` flag to the access field.

TLV literally means **Type-Length-Value** (or **Tag-Length-Value**) and is an encoding scheme. This has been adopted by ALSA in order to define dB range/scale containers. For example, `DECLARE_TLV_DB_SCALE` will define information about a mixer control where each step in the control's value changes the dB value by a constant dB amount. Let's take the following example:

```
static DECLARE_TLV_DB_SCALE(db_scale_my_control, -4050, 150,
                            0);
```

According to the definition of this macro in `include/sound/tlv.h`, the preceding example could be expanded into the following:

```
static struct snd_kcontrol_new my_control devinitdata = {
    [...]
    .access =
    SNDRV_CTL_ELEM_ACCESS_READWRITE |
    SNDRV_CTL_ELEM_ACCESS_TLV_READ,
```

```
      [...]
      .tlv.p = db_scale_my_control,
};
```

The first parameter of the macro represents the name of the variable to be defined; the second one represents the minimum value this control can accept, in units of 0.01 dB. The third parameter is the step size of the change, also in steps of 0.01 dB. If a mute operation is performed when the control is at the minimum value, the fourth parameter needs to be set to 1. Please have a look at include/sound/tlv.h to see the available macros.

Upon sound card registration, the snd_ctl_dev_register() function is called in order to save relevant information about the control device and make it available to users.

Defining kcontrols

kcontrols are used by the ASoC core to export audio controls (such as switch, volume, *MUX...) to the userspace. This means, for example, when a userspace application like PulseAudio switches off headphones or switches on speakers when no headphones are plugged in, the action is handled in the kernel by kcontrols. Normal kcontrols are not involved in power management (DAPM). They are specially meant to control non-power-management-based elements such as volume level, gain level, and so on. Once controls have been set up using the appropriate macros, they must be registered with the system control list using the snd_soc_add_component_controls() method, whose prototype is as follows:

```
int snd_soc_add_component_controls(
                    struct snd_soc_component *component,
                    const struct snd_kcontrol_new *controls,
                    unsigned int num_controls);
```

In the preceding prototype, component is the component you add the controls for, controls is the array of controls to add, and num_controls is the number of entries in the array that need to be added.

In order to see how simple this API is, let's consider the following sample, which defines some controls:

```
static const DECLARE_TLV_DB_SCALE(dac_tlv, -12750, 50, 1);
static const DECLARE_TLV_DB_SCALE(out_tlv, -12100, 100, 1);
static const DECLARE_TLV_DB_SCALE(bypass_tlv, -2100, 300, 0);
```

```
static const struct snd_kcontrol_new wm8960_snd_controls[] = {
    [...]
    SOC_DOUBLE_R_TLV("Playback Volume", WM8960_LDAC,
                     WM8960_RDAC, 0,
                     255, 0, dac_tlv),
    SOC_DOUBLE_R_TLV("Headphone Playback Volume", WM8960_LOUT1,
                     WM8960_ROUT1, 0, 127, 0, out_tlv),
    SOC_DOUBLE_R("Headphone Playback ZC Switch", WM8960_LOUT1,
                 WM8960_ROUT1, 7, 1, 0),
    SOC_DOUBLE_R_TLV("Speaker Playback Volume", WM8960_LOUT2,
                     WM8960_ROUT2, 0, 127, 0, out_tlv),
    SOC_DOUBLE_R("Speaker Playback ZC Switch", WM8960_LOUT2,
                 WM8960_ROUT2, 7, 1, 0),
    SOC_SINGLE("Speaker DC Volume", WM8960_CLASSD3, 3, 5, 0),
    SOC_SINGLE("Speaker AC Volume", WM8960_CLASSD3, 0, 5, 0),
    SOC_ENUM("DAC Polarity", wm8960_enum[1]),
    SOC_SINGLE_BOOL_EXT("DAC Deemphasis Switch", 0,
                        wm8960_get_deemph,
                        wm8960_put_deemph),
    [...]
    SOC_SINGLE("Noise Gate Threshold", WM8960_NOISEG, 3, 31,
0),
    SOC_SINGLE("Noise Gate Switch", WM8960_NOISEG, 0, 1, 0),

    SOC_DOUBLE_R_TLV("ADC PCM Capture Volume", WM8960_LADC,
                     WM8960_RADC, 0, 255, 0, adc_tlv),

    SOC_SINGLE_TLV("Left Output Mixer Boost Bypass Volume",
                   WM8960_BYPASS1, 4, 7, 1, bypass_tlv),
};
```

The corresponding code that would register the preceding controls is as follows:

```
snd_soc_add_component_controls(component, wm8960_snd_controls,
                               ARRAY_SIZE(wm8960_snd_controls));
```

The following are ways to define commonly used controls with these preset macro definitions.

SOC_SINGLE(xname, reg, shift, max, invert)

To set up a simple switch, we can use `SOC_SINGLE`. This is the simplest control:

```
#define SOC_SINGLE(xname, reg, shift, max, invert) \
{   .iface = SNDRV_CTL_ELEM_IFACE_MIXER, .name = xname, \
    .info = snd_soc_info_volsw, .get = snd_soc_get_volsw,\
    .put = snd_soc_put_volsw, \
    .private_value = SOC_SINGLE_VALUE(reg, shift, max, invert) }
```

This type of control has only one setting and is generally used for component switches. Descriptions of the parameters defined by the macro are as follows:

- xname: The name of the control.
- reg: The register address corresponding to the control.
- Shift: The offset control bit (from where to apply the change) for this control in the register reg.
- max: The range of values set by the control. Generally speaking, if the control bit has only 1 bit, then max=1, because the possible values are only 0 and 1.
- invert: Whether the set value is inverted.

Let's study the following example:

```
SOC_SINGLE("PCM Playback -6dB Switch", WM8960_DACCTL1, 7, 1,
            0),
```

In the previous example, PCM Playback -6dB Switch is the name of the control. WM8960_DACCTL1 (defined in wm8960.h) is the address of the register in the codec (the WM8960 chip), which allows you to control this switch:

- 7 means the 7th bit in the DACCTL1 register is used to enable/disable the DAC 6dB attenuation.
- 1 means there is only one enable or disable option.
- 0 means the value you set is not inverted.

SOC_SINGLE_TLV(xname, reg, shift, max, invert, tlv_array)

This macro sets up a switch with levels. It is an extension of SOC_SINGLE that is used to define controls that have gain control, such as volume controls, EQ equalizers, and so on. In this example, the left input volume control is from 000000 (-17.25 dB) to 111111(+30 dB). Each step is 0.75 dB, meaning a total of 63 steps:

```
SOC_SINGLE_TLV("Input Volume of LINPUT1",
             WM8960_LINVOL, 0, 63, 0, in_tlv),
```

The scale of in_tlv (which represents the control metadata) is declared like this:

```
static const DECLARE_TLV_DB_SCALE(in_tlv, -1725, 75, 0);
```

In the preceding, -1725 means the control scale starts from -17.25dB. 75 means each step is 0.75dB, and 0 means the step starts from 0. For some volume control cases, the first step is "mute" and the step starts from 1. Thus the 0 in the preceding code should be replaced by 1.

SOC_DOUBLE_R(xname, reg_left, reg_right, xshift, xmax, xinvert)

SOC_DOUBLE_R is a stereo version of SOC_SINGLE. The difference is that SOC_SINGLE only controls one variable, while SOC_DOUBLE can control two similar variables in one register at the same time. We may use this to control the left and right channels at the same time.

Because there is one more channel, the parameter has a shift value corresponding to it. The following is an example:

```
SOC_DOUBLE_R("Headphone ZC Switch", WM8960_LOUT1,
             WM8960_ROUT1, 7, 1, 0),
```

SOC_DOUBLE_R_TLV(xname, reg_left, reg_right, xshift, xmax, xinvert, tlv_array)

SOC_DOUBLE_R_TLV is the stereo version of SOC_SINGLE_TLV. The following is an example of its usage:

```
SOC_DOUBLE_R_TLV("PCM DAC Playback Volume", WM8960_LDAC,
             WM8960_RDAC, 0, 255, 0, dac_tlv),
```

The mixer control

The mixer control is used for routing the control of audio channels. It consists of multiple inputs and one output. Multiple inputs can be freely mixed together to form a mixed output:

```
static const struct snd_kcontrol_new left_speaker_mixer[] = {
    SOC_SINGLE("Input Switch", WM8993_SPEAKER_MIXER, 7, 1, 0),
    SOC_SINGLE("IN1LP Switch", WM8993_SPEAKER_MIXER, 5, 1, 0),
    SOC_SINGLE("Output Switch", WM8993_SPEAKER_MIXER, 3, 1, 0),
    SOC_SINGLE("DAC Switch", WM8993_SPEAKER_MIXER, 6, 1, 0),
};
```

The preceding mixer uses the third, fifth, sixth, and seventh bits of the WM8993_SPEAKER_MIXER register to control the opening and closing of the four inputs.

SOC_ENUM_SINGLE(xreg, xshift, xmax, xtexts)

This macro defines a single enumerated control, where xreg is the register to modify to apply settings, xshift is the control bit(s) offset in the register, xmask is the control bit(s) size, and xtexts is a pointer to the array of strings that describe each setting. This is used when the control options are some texts.

As an example, we can set up the array for the texts as follows:

```
static const char *aif_text[] = {
    "Left" , "Right"
};
```

And then define the enum as follows:

```
static const struct  soc_enum aifin1_enum =
    SOC_ENUM_SINGLE(WM8993_AUDIO_INTERFACE_2, 15, 2, aif_text);
```

Now we are done with the concept of controls, which are used to change the properties of an audio device, we will learn how to leverage it and play with the power properties of an audio device.

The concept of DAPM

Modern sound cards consist of many independent discrete components. Each component has functional units that can be powered independently. The thing is, embedded systems are, most of the time, battery-powered and require the lowest power mode. Managing power domain dependencies by hand could be tedious and error-prone. **Dynamic Audio Power Management** (**DAPM**) targets the lowest use of power at all times in the audio subsystem. DAPM is to be used for things for which there is power control and can be skipped if power management is not necessary. Things only go into DAPM if they have some relevance to power – that is, if they're a thing for which there is power control or if they control the routing of audio through the chip (and therefore let the core decide which parts of the chip need to be powered on).

DAPM lies in the ASoC Core (this means power switching is done from within the kernel) and becomes active as and when audio streams/paths/settings change, making it completely transparent for all userspace applications.

In the previous sections, we introduced the concept of controls and how to deal with them. However, kcontrols on their own are not involved in audio power management. A normal kcontrol has the following characteristics:

- Self-descriptive and cannot describe the connection relationship between each kcontrol.

- Lacks a power management mechanism.

- Lacks a time processing mechanism to respond to audio events such as playing, stopping, powering on, and powering off.

- Lacks a pop-pop sound prevention mechanism, so that it is up to the user program to pay attention to the power-on and power-off sequence for each kcontrol.

- Manual, because all the control involved in an audio path can't be automatically closed. When an audio path is no longer valid, it requires userspace intervention.

DAPM introduced the concept of widgets in order to solve the aforementioned problems. A widget is the basic DAPM unit. Thus, the so-called widget can be understood as a further upgrade and encapsulation of kcontrols.

A widget is a combination of kcontrols and dynamic power management, and also has the link function of the audio path. It can have a dynamic connection relationship with its neighbor widget.

The DAPM framework abstracts a widget by means of the `struct snd_soc_dapm_widget` structure, defined in `include/sound/soc-dapm.h` as follows:

```
struct snd_soc_dapm_widget {
    enum snd_soc_dapm_type id;
    const char *name;
    const char *sname;
[...]
    /* dapm control */
    int reg;       /* negative reg = no direct dapm */
    unsigned char shift;
    unsigned int mask;
    unsigned int on_val;
    unsigned int off_val;
[...]
    int (*power_check)(struct snd_soc_dapm_widget *w);

    /* external events */
    unsigned short event_flags;
    int (*event)(struct snd_soc_dapm_widget*,
                 struct snd_kcontrol *, int);

    /* kcontrols that relate to this widget */
    int num_kcontrols;
    const struct snd_kcontrol_new *kcontrol_news;
    struct snd_kcontrol **kcontrols;
    struct snd_soc_dobj dobj;

    /* widget input and output edges */
    struct list_head edges[2];

    /* used during DAPM updates */
    struct list_head dirty;
[...]
}
```

For the sake of readability, only the relevant fields are listed in the preceding snippet, and the following are their descriptions:

- `id` is of type `enum snd_soc_dapm_type` and represents the type of the widget, such as `snd_soc_dapm_output`, `snd_soc_dapm_mixer`, and so on. The full list is defined in `include/sound/soc-dapm.h`.

- `name` is the name of the widget.

- `shift` and `mask` are used to control the power state of the widget, corresponding to the register address `reg`.

- The `on_val` and `off_val` values represent the values that are to be used to change the current power state of the widget. They respectively correspond to when it is turned on and when it is turned off.

- `event` represents the DAPM event handling callback function pointer. Each widget is associated with a kcontrol object, pointed to by `**kcontrols`.

- `*kcontrol_news` is the array of controls this kcontrol is made of, and `num_kcontrols` is the number of entries in it. These three fields are used to describe the kcontrol control contained in the widget, such as a mixer control or a MUX control.

- `dirty` is used to insert this widget into a dirty list when the state of the widget is changed. This dirty list is then scanned in order to perform the update of the entire path.

Defining widgets

Like the normal kcontrol, the DAPM framework provides us with a large number of auxiliary macros to define a variety of widget controls. These macro definitions can be spread into several fields according to the type of widgets and to the domain in which they are powered. They are as follows:

- **The codec domain**: Such as `VREF` and `VMID`; they provide reference voltage widgets. These widgets are usually controlled in the codec probe/remove callback.

- **The platform/machine domain**: These widgets are usually input/output interfaces for the platform or board (machine actually) that need to be physically connected, such as headphones, speakers, and microphones. That being said, because these interfaces may differ on each board, they are usually configured by the machine driver and respond to asynchronous events, for example, when headphones are inserted. They can also be controlled by userspace applications to turn them on and off in some way.

- **The audio path domain**: Generally, refers to the MUX, mixer, and other widgets that control the audio path inside the codec. These widgets can automatically set their power state according to the connection relationship of the userspace, for example, `alsamixer` and `amixer`.

- **The audio stream domain**: These need to process audio data streams, such as ADCs, DACs, and so on. Enabled and disabled when stream playback/capture is started and stopped respectively, for example, `aplay` and `arecord`.

All DAPM power switching decisions are made automatically according to a machine-specific audio routing map, which consists of the interconnections between every audio component (including internal codec components).

Codec domain definition

There is only one macro provided by the DAPM framework for this domain:

```
/* codec domain */
#define SND_SOC_DAPM_VMID(wname) \
    .id = snd_soc_dapm_vmid, .name = wname,
    .kcontrol_news = NULL, \
    .num_kcontrols = 0}
```

Defining platform domain widgets

The widgets of the platform domain correspond to the signal generator, input pin, output pin, microphone, earphone, speaker, and line input interface respectively. The DAPM framework provides us with a number of auxiliary definition macros for the platform domain widgets. These are defined as the following:

```
#define SND_SOC_DAPM_SIGGEN(wname) \
{   .id = snd_soc_dapm_siggen, .name = wname,
    .kcontrol_news = NULL, \
    .num_kcontrols = 0, .reg = SND_SOC_NOPM }


#define SND_SOC_DAPM_SINK(wname) \
{   .id = snd_soc_dapm_sink, .name = wname,
    .kcontrol_news = NULL, \
    .num_kcontrols = 0, .reg = SND_SOC_NOPM }


#define SND_SOC_DAPM_INPUT(wname) \
{   .id = snd_soc_dapm_input, .name = wname,
    .kcontrol_news = NULL, \
```

```
        .num_kcontrols = 0, .reg = SND_SOC_NOPM }


#define SND_SOC_DAPM_OUTPUT(wname) \
{   .id = snd_soc_dapm_output, .name = wname,
    .kcontrol_news = NULL, \
    .num_kcontrols = 0, .reg = SND_SOC_NOPM }


#define SND_SOC_DAPM_MIC(wname, wevent) \
{   .id = snd_soc_dapm_mic, .name = wname,
    .kcontrol_news = NULL, \
    .num_kcontrols = 0, .reg = SND_SOC_NOPM, .event = wevent, \
    .event_flags = SND_SOC_DAPM_PRE_PMU |
    SND_SOC_DAPM_POST_PMD}


#define SND_SOC_DAPM_HP(wname, wevent) \
{   .id = snd_soc_dapm_hp, .name = wname,
    .kcontrol_news = NULL, \
    .num_kcontrols = 0, .reg = SND_SOC_NOPM, .event = wevent, \
    .event_flags = SND_SOC_DAPM_POST_PMU |
    SND_SOC_DAPM_PRE_PMD}


#define SND_SOC_DAPM_SPK(wname, wevent) \
{   .id = snd_soc_dapm_spk, .name = wname,
    .kcontrol_news = NULL, \
    .num_kcontrols = 0, .reg = SND_SOC_NOPM, .event = wevent, \
    .event_flags = SND_SOC_DAPM_POST_PMU |
    SND_SOC_DAPM_PRE_PMD}


#define SND_SOC_DAPM_LINE(wname, wevent) \
{   .id = snd_soc_dapm_line, .name = wname,
    .kcontrol_news = NULL, \
    .num_kcontrols = 0, .reg = SND_SOC_NOPM, .event = wevent, \
    .event_flags = SND_SOC_DAPM_POST_PMU |
    SND_SOC_DAPM_PRE_PMD}


#define SND_SOC_DAPM_INIT_REG_VAL(wreg, wshift, winvert) \
    .reg = wreg, .mask = 1, .shift = wshift, \
    .on_val = winvert ? 0 : 1, .off_val = winvert ? 1 : 0
```

In the preceding code, most of the fields in these macros are common. The fact that the reg field is set to SND_SOC_NOPM (defined to -1) means that these widgets have no register control bits to control the power state of the widgets. SND_SOC_DAPM_INPUT and SND_SOC_DAPM_OUTPUT are used to define the output and input pins of the codec chip from within the codec driver. From what we can see, the MIC, HP, SPK, and LINE widgets respond to SND_SOC_DAPM_POST_PMU (after widget power-up) and SND_SOC_DAPM_PMD (before widget power-down) events, and these widgets are usually defined in the machine driver.

Defining an audio path domain widget

This kind of widget usually repackages the ordinary kcontrols and extends them with audio path and power management functions. This extension somehow makes this kind of widget DAPM-aware. Widgets in this domain will contain one or more kcontrols that are not the ordinary kcontrols. There are DAPM-enabled kcontrols. These cannot be defined using the standard method, that is, SOC_*-based macro controls. They need to be defined using the definition macros provided by the DAPM framework. We will discuss them in detail later, in the *Defining DAPM kcontrols* section. However, here are the definition macros for these widgets:

```
#define SND_SOC_DAPM_PGA(wname, wreg, wshift, winvert,\
                         wcontrols, wncontrols) \
{    .id = snd_soc_dapm_pga, .name = wname, \
    SND_SOC_DAPM_INIT_REG_VAL(wreg, wshift, winvert), \
    .kcontrol_news = wcontrols, .num_kcontrols = wncontrols}


#define SND_SOC_DAPM_OUT_DRV(wname, wreg, wshift, winvert,\
                            wcontrols, wncontrols) \
{    .id = snd_soc_dapm_out_drv, .name = wname, \
    SND_SOC_DAPM_INIT_REG_VAL(wreg, wshift, winvert), \
    .kcontrol_news = wcontrols, .num_kcontrols = wncontrols}


#define SND_SOC_DAPM_MIXER(wname, wreg, wshift, winvert, \
                          wcontrols, wncontrols)\
{    .id = snd_soc_dapm_mixer, .name = wname, \
    SND_SOC_DAPM_INIT_REG_VAL(wreg, wshift, winvert), \
    .kcontrol_news = wcontrols, .num_kcontrols = wncontrols}


#define SND_SOC_DAPM_MIXER_NAMED_CTL(wname, wreg,
```

```
                                        wshift, winvert, \
                                        wcontrols, wncontrols)\
{    .id = snd_soc_dapm_mixer_named_ctl, .name = wname, \
     SND_SOC_DAPM_INIT_REG_VAL(wreg, wshift, winvert), \
     .kcontrol_news = wcontrols, .num_kcontrols = wncontrols}


#define SND_SOC_DAPM_SWITCH(wname, wreg, wshift, winvert, \
wcontrols) \
{    .id = snd_soc_dapm_switch, .name = wname, \
     SND_SOC_DAPM_INIT_REG_VAL(wreg, wshift, winvert), \
     .kcontrol_news = wcontrols, .num_kcontrols = 1}


#define SND_SOC_DAPM_MUX(wname, wreg, wshift, \
                         winvert, wcontrols) \
{    .id = snd_soc_dapm_mux, .name = wname, \
     SND_SOC_DAPM_INIT_REG_VAL(wreg, wshift, winvert), \
     .kcontrol_news = wcontrols, .num_kcontrols = 1}


#define SND_SOC_DAPM_DEMUX(wname, wreg, wshift, \
                          winvert, wcontrols) \
{    .id = snd_soc_dapm_demux, .name = wname, \
     SND_SOC_DAPM_INIT_REG_VAL(wreg, wshift, winvert), \
     .kcontrol_news = wcontrols, .num_kcontrols = 1}
```

Unlike platform and codec domain widgets, the reg and shift fields need to be assigned, indicating that these widgets have corresponding power control registers. The DAPM framework uses these registers to control the power state of the widgets when scanning and updating the audio path. Their power states are dynamically allocated, powered up when needed (on a valid audio path), and powered down when not needed (on an inactive audio path). These widgets need to perform the same functions as the mixer, MUX, and so on introduced earlier. In fact, this is done by the kcontrol controls they contain. The driver code must define kcontrols before defining the widget, and then pass the wcontrols and num_kcontrols parameters to these auxiliary definition macros.

There is another variant of those macros that exists and that has a pointer to an event handler. Such macros have the _E suffix. These are `SND_SOC_DAPM_PGA_E`, `SND_SOC_DAPM_OUT_DRV_E`, `SND_SOC_DAPM_MIXER_E`, `SND_SOC_DAPM_MIXER_NAMED_CTL_E`, `SND_SOC_DAPM_SWITCH_E`, `SND_SOC_DAPM_MUX_E`, and `SND_SOC_DAPM_VIRT_MUX_E`. You are encouraged to have a look at the kernel source code to see their definitions at `https://elixir.bootlin.com/linux/v4.19/source/include/sound/soc-dapm.h#L136`.

Defining the audio stream domain

These widgets mainly include audio input/output interfaces, ADC/DAC, and clock lines. Starting with the audio interface widgets, these are the following:

```
#define SND_SOC_DAPM_AIF_IN(wname, stname, wslot, wreg, wshift,
winvert) \
{   .id = snd_soc_dapm_aif_in, .name = wname, .sname = stname, \
    SND_SOC_DAPM_INIT_REG_VAL(wreg, wshift, winvert), }


#define SND_SOC_DAPM_AIF_IN_E(wname, stname, wslot, wreg, \
                              wshift, winvert, wevent, wflags) \
{   .id = snd_soc_dapm_aif_in, .name = wname, .sname = stname, \
    SND_SOC_DAPM_INIT_REG_VAL(wreg, wshift, winvert), \
    .event = wevent, .event_flags = wflags }


#define SND_SOC_DAPM_AIF_OUT(wname, stname, wslot, wreg,
wshift, winvert) \
{ .id = snd_soc_dapm_aif_out, .name = wname, .sname = stname, \
      SND_SOC_DAPM_INIT_REG_VAL(wreg, wshift, winvert), }


#define SND_SOC_DAPM_AIF_OUT_E(wname, stname, wslot, wreg, \
                               wshift, winvert, wevent, wflags) \
{ .id = snd_soc_dapm_aif_out, .name = wname, .sname = stname, \
      SND_SOC_DAPM_INIT_REG_VAL(wreg, wshift, winvert), \
      .event = wevent, .event_flags = wflags }
```

In the preceding macro definition list, `SND_SOC_DAPM_AIF_IN` and `SND_SOC_DAPM_AIF_OUT` are respectively the audio interface input and output. The former defines the connection to the host that receives the audio to be passed into the DAC(s) and the latter defines the connection to the host that transmits the audio received from the ADC(s). `SND_SOC_DAPM_AIF_IN_E` and `SND_SOC_DAPM_AIF_OUT_E` are their respective event variants, allowing `wevent` to be called when one of the events enabled in `wflags` occurs.

Now come the ADC/DAC-related widgets, as well as the clock-related one, defined as the following:

```
#define SND_SOC_DAPM_DAC(wname, stname, wreg,
                         wshift, winvert) \
{      .id = snd_soc_dapm_dac, .name = wname, .sname = stname, \
       SND_SOC_DAPM_INIT_REG_VAL(wreg, wshift, winvert) }


#define SND_SOC_DAPM_DAC_E(wname, stname, wreg, wshift, \
                           winvert, wevent, wflags) \
{      .id = snd_soc_dapm_dac, .name = wname, .sname = stname, \
       SND_SOC_DAPM_INIT_REG_VAL(wreg, wshift, winvert), \
       .event = wevent, .event_flags = wflags}


#define SND_SOC_DAPM_ADC(wname, stname, wreg,
                         wshift, winvert) \
{      .id = snd_soc_dapm_adc, .name = wname, .sname = stname, \
       SND_SOC_DAPM_INIT_REG_VAL(wreg, wshift, winvert), }


#define SND_SOC_DAPM_ADC_E(wname, stname, wreg, wshift,\
                           winvert, wevent, wflags) \
{      .id = snd_soc_dapm_adc, .name = wname, .sname = stname, \
       SND_SOC_DAPM_INIT_REG_VAL(wreg, wshift, winvert), \
       .event = wevent, .event_flags = wflags}


#define SND_SOC_DAPM_CLOCK_SUPPLY(wname) \
{      .id = snd_soc_dapm_clock_supply, .name = wname, \
       .reg = SND_SOC_NOPM, .event = dapm_clock_event, \
       .event_flags = SND_SOC_DAPM_PRE_PMU | SND_SOC_DAPM_POST_PMD
}
```

In the preceding list of macros, SND_SOC_DAPM_ADC and SND_SOC_DAPM_DAC are ADC and DAC widgets respectively. The former is used to control the powering up and shutting down of the ADC on an as-needed basis, while the latter targets DAC(s). The former is typically associated with a capture stream on the device, for example, "Left Capture" or "Right Capture," and the latter is typically associated with a playback stream, for example, "Left Playback" or "Right Playback." The register settings define a single register and bit position that, when flipped, will turn the ADC/DAC on or off. You should also notice their event variants, SND_SOC_DAPM_ADC_E and SND_SOC_DAPM_DAC_E respectively. SND_SOC_DAPM_CLOCK_SUPPLY is a supply-widget variant for connection to the clock framework.

There are other widget types for which no definition macro is provided, and that do not end in any of the domains we have introduced so far. These are snd_soc_dapm_dai_in, snd_soc_dapm_dai_out, and snd_soc_dapm_dai_link.

Such widgets are implicitly created upon DAI registration, either from the CPU or the codec driver. In other words, whenever a DAI is registered, the DAPM core will create a widget either of type snd_soc_dapm_dai_in or of type snd_soc_dapm_dai_out according to the streams of the DAI being registered. Usually, both widgets will be connected to widgets with the same stream name in the codec. Additionally, when the machine driver decides to bind codec and CPU DAIs together, this will result in the DAPM framework creating a widget of type snd_soc_dapm_dai_link to describe the power state of the connection.

The concept of a path – a connector between widgets

Widgets are meant to be linked one to the other in order to build a functional audio stream path. That being said, the connection between two widgets needs to be tracked in order to maintain the audio state. To describe the patch between two widgets, the DAPM core uses the struct snd_soc_dapm_path data structure, defined as follows:

```
/* dapm audio path between two widgets */
struct snd_soc_dapm_path {
    const char *name;
    /*
     * source (input) and sink (output) widgets
     * The union is for convenience,
     * since it is a lot nicer to type
     * p->source, rather than p->node[SND_SOC_DAPM_DIR_IN]
     */
    union {
```

```
    struct {
        struct snd_soc_dapm_widget *source;
        struct snd_soc_dapm_widget *sink;
    };
    struct snd_soc_dapm_widget *node[2];
};


/* status */
u32 connect:1; /* source and sink widgets are connected */
u32 walking:1; /* path is in the process of being walked */
u32 weak:1; /* path ignored for power management */
u32 is_supply:1;  /* At least one of the connected widgets
                     is a supply */


int (*connected)(struct snd_soc_dapm_widget *source, struct
                 snd_soc_dapm_widget *sink);


struct list_head list_node[2];
struct list_head list_kcontrol;
struct list_head list;
};
```

This structure abstracts the link between two widgets. Its source field points to the start widget of the connection, whereas its sink field points to the arrival widget of the connection. The input and output (that is, an endpoint) of the widget may be connected to multiple paths. The snd_soc_dapm_path structure of all inputs is hung in the sources list of the widget through the list_node[SND_SOC_DAPM_DIR_IN] field while the snd_soc_dapm_path structure of all outputs is stored in the sinks list of the widget, which is list_node[SND_SOC_DAPM_DIR_OUT]. The connection goes from the source to the sink and the principle is quite simple. Just remember the connection path is this: *the output of the start widget --> the input of the path data structure* and *the output of the path data structure --> Arrival side widget input.*

The list field will end up in the sound card's path list header field upon card registration. This list allows the sound card to track all the available paths it can use. Finally, the connected field is there to let you implement your own custom method to check the current connection state of the path.

> **Important note**
>
> SND_SOC_DAPM_DIR_IN and SND_SOC_DAPM_DIR_OUT are
> enumerators that are 0 and 1 respectively.

You'll probably never want to deal with a path directly. However, this concept has been introduced here for the sake of pedagogy, as it will help us understand the next section.

The concept of a route – widget inter-connections

The concept of a path, introduced earlier in this chapter, was an introduction to this one. From the preceding discussion, we can introduce the concept of a route. A route connection is made of at least the starter widget, the jumper path, the sink widget, and in the DAPM the struct snd_soc_dapm_route structure is used to describe such a connection:

```
struct snd_soc_dapm_route {
    const char *sink;
    const char *control;
    const char *source;

    /* Note: currently only supported for links where source is
     a supply */
    int (*connected)(struct snd_soc_dapm_widget *source,
                     struct snd_soc_dapm_widget *sink);
};
```

In the preceding data structure, sink points to the name string of the arriving widget, source points to the name string of the starting widget, control points to the kcontrol name string responsible for controlling the connection, and connected defines the custom connection check callback. The meaning of this structure is obvious: source is connected to sink via a kcontrol and a connected callback function can be called to check the connection state.

Routes are to be defined using the following scheme:

```
{Destination Widget, Switch, Source Widget},
```

This means Source Widget is connected to Destination Widget via Swtich. This way, the DAPM core will take care of closing the switch whenever the connection needs to be activated, and both source and destination widget will be powered on as well. Sometimes, the connection may be direct. In this case, Switch should be NULL. You'll then have something like the following:

```
{end point, NULL, starting point},
```

You should directly use the name string to describe the connection relationship, all defined routes, and finally, you have to register to the DAPM core. DAPM core will find the corresponding widget according to these names, and dynamically generate the required `snd_soc_dapm_path` to describe the connection between the two widgets. In the next sections, we'll see how to create routes.

Defining DAPM kcontrols

As mentioned in the previous sections, mixers or MUX-type widgets in the audio path domain are made of several kcontrols, which must be defined using DAPM-based macros. DAPM uses these kcontrols to complete the audio path. However, for widgets, this task is more than that. DAPM also dynamically manages the connection relationships of these audio paths so that the power state of these widgets can be controlled according to these connection relationships. If these kcontrols are defined in the usual way, this is not possible, so DAPM provides us with another set of definition macros that define the kcontrols that are included in the widget:

```
#define SOC_DAPM_SINGLE(xname, reg, shift, max, invert) \
{   .iface = SNDRV_CTL_ELEM_IFACE_MIXER, .name = xname, \
    .info = snd_soc_info_volsw, \
    .get = snd_soc_dapm_get_volsw,
    .put = snd_soc_dapm_put_volsw, \
    .private_value = SOC_SINGLE_VALUE(reg, shift, max, invert) }


#define SOC_DAPM_SINGLE_TLV(xname, reg, shift, max, invert,
                              tlv_array) \
{   .iface = SNDRV_CTL_ELEM_IFACE_MIXER, .name = xname, \
    .info = snd_soc_info_volsw, \
    .access = SNDRV_CTL_ELEM_ACCESS_TLV_READ | \
              SNDRV_CTL_ELEM_ACCESS_READWRITE, \
    .tlv.p = (tlv_array), \
    .get = snd_soc_dapm_get_volsw,
    .put = snd_soc_dapm_put_volsw, \
    .private_value = SOC_SINGLE_VALUE(reg, shift, max, invert) }


#define SOC_DAPM_ENUM(xname, xenum) \
{   .iface = SNDRV_CTL_ELEM_IFACE_MIXER, .name = xname, \
    .info = snd_soc_info_enum_double, \
    .get = snd_soc_dapm_get_enum_double, \
```

```
    .put = snd_soc_dapm_put_enum_double, \
    .private_value = (unsigned long)&xenum}

#define SOC_DAPM_ENUM_VIRT(xname, xenum) \
{   .iface = SNDRV_CTL_ELEM_IFACE_MIXER, .name = xname, \
    .info = snd_soc_info_enum_double, \
    .get = snd_soc_dapm_get_enum_virt, \
    .put = snd_soc_dapm_put_enum_virt, \
    .private_value = (unsigned long)&xenum}

#define SOC_DAPM_ENUM_EXT(xname, xenum, xget, xput) \
{   .iface = SNDRV_CTL_ELEM_IFACE_MIXER, .name = xname, \
    .info = snd_soc_info_enum_double, \
    .get = xget, \
    .put = xput, \
    .private_value = (unsigned long)&xenum }

#define SOC_DAPM_VALUE_ENUM(xname, xenum) \
{   .iface = SNDRV_CTL_ELEM_IFACE_MIXER, .name = xname, \
    .info = snd_soc_info_enum_double, \
    .get = snd_soc_dapm_get_value_enum_double, \
    .put = snd_soc_dapm_put_value_enum_double, \
    .private_value = (unsigned long)&xenum }

#define SOC_DAPM_PIN_SWITCH(xname) \
{   .iface = SNDRV_CTL_ELEM_IFACE_MIXER,
    .name = xname " Switch" , \
    .info = snd_soc_dapm_info_pin_switch, \
    .get = snd_soc_dapm_get_pin_switch, \
    .put = snd_soc_dapm_put_pin_switch, \
    .private_value = (unsigned long)xname }
```

It can be seen that `SOC_DAPM_SINGLE` is the DAPM equivalent to `SOC_SINGLE` of the standard control, `SOC_DAPM_SINGLE_TLV` corresponds to `SOC_SINGLE_TLV`, and so on. Compared to the ordinary kcontrols, DAPM's kcontrols just replace the `info`, `get`, and `put` callback functions. The `put` callback function provided by DAPM kcontrols not only updates the state of the control itself but also passes this change to the adjacent DAPM kcontrol. The adjacent DAPM kcontrol will pass this change to its own neighbor DAPM kcontrol, knowing at the end of the audio path, by changing the connection state of one of the widgets, all widgets associated with it are scanned and tested to see if they are still in the active audio path, thus dynamically changing their power state. This is the essence of DAPM.

Creating widgets and routes

The previous section introduced a lot of auxiliary macros. However, it was theoretical and did not explain how to define the widgets we need for a real system, nor how to define the connection relationship of widgets. Here, we take Wolfson's codec chip **WM8960** as an example to understand this process:

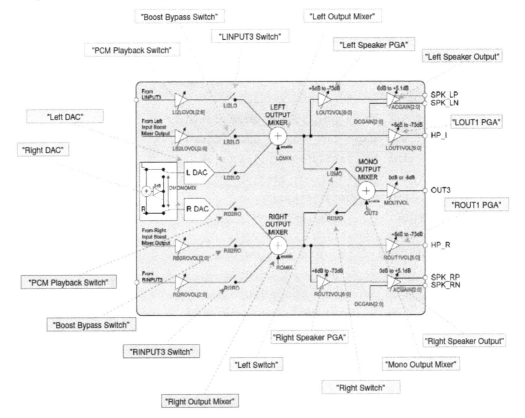

Figure 5.3 – WM8960 internal audio paths and controls

Given the preceding diagram as an example, from the Wolfson WM8960 codec chip, the first step is to use the helper macro to define the DAPM kcontrol required by the widgets:

```
static const struct snd_kcontrol_new wm8960_loutput_mixer[] = {
    SOC_DAPM_SINGLE("PCM Playback Switch", WM8960_LOUTMIX, 8,
                    1, 0),
    SOC_DAPM_SINGLE("LINPUT3 Switch", WM8960_LOUTMIX, 7, 1, 0),
    SOC_DAPM_SINGLE("Boost Bypass Switch", WM8960_BYPASS1, 7,
                    1, 0),
};
static const struct snd_kcontrol_new wm8960_routput_mixer[] = {
    SOC_DAPM_SINGLE("PCM Playback Switch", WM8960_ROUTMIX, 8,
                    1, 0),
    SOC_DAPM_SINGLE("RINPUT3 Switch", WM8960_ROUTMIX, 7, 1, 0),
    SOC_DAPM_SINGLE("Boost Bypass Switch", WM8960_BYPASS2, 7,
                    1, 0),
};
static const struct snd_kcontrol_new wm8960_mono_out[] = {
    SOC_DAPM_SINGLE("Left Switch", WM8960_MONOMIX1, 7, 1, 0),
    SOC_DAPM_SINGLE("Right Switch", WM8960_MONOMIX2, 7, 1, 0),
};
```

In the preceding, we defined the mixer controls for the left and right output channels in wm8960, as well as the mono output mixer: wm8960_loutput_mixer, wm8960_routput_mixer, and wm8960_mono_out.

The second step consists of defining the real widget, including the DAPM control defined in the first step:

```
static const struct snd_soc_dapm_widget wm8960_dapm_widgets[] =
{
    [...]
    SND_SOC_DAPM_INPUT("LINPUT3"),
    SND_SOC_DAPM_INPUT("RINPUT3"),

    SND_SOC_DAPM_SUPPLY("MICB", WM8960_POWER1, 1, 0, NULL, 0),
    [...]
    SND_SOC_DAPM_DAC("Left DAC", "Playback", WM8960_POWER2, 8,
                    0),
    SND_SOC_DAPM_DAC("Right DAC", "Playback", WM8960_POWER2, 7,
```

```
                                0),

    SND_SOC_DAPM_MIXER("Left Output Mixer", WM8960_POWER3, 3,
                        0,
                        &wm8960_loutput_mixer[0],
                        ARRAY_SIZE(wm8960_loutput_mixer)),
    SND_SOC_DAPM_MIXER("Right Output Mixer", WM8960_POWER3, 2,
                        0,
                        &wm8960_routput_mixer[0],
                        ARRAY_SIZE(wm8960_routput_mixer)),


    SND_SOC_DAPM_PGA("LOUT1 PGA", WM8960_POWER2, 6, 0, NULL,
                        0),
    SND_SOC_DAPM_PGA("ROUT1 PGA", WM8960_POWER2, 5, 0, NULL,
                        0),


    SND_SOC_DAPM_PGA("Left Speaker PGA", WM8960_POWER2,
                        4, 0, NULL, 0),
    SND_SOC_DAPM_PGA("Right Speaker PGA", WM8960_POWER2,
                        3, 0, NULL, 0),
    SND_SOC_DAPM_PGA("Right Speaker Output", WM8960_CLASSD1,
                        7, 0, NULL, 0);
    SND_SOC_DAPM_PGA("Left Speaker Output", WM8960_CLASSD1,
                        6, 0, NULL, 0),
    SND_SOC_DAPM_OUTPUT("SPK_LP"),
    SND_SOC_DAPM_OUTPUT("SPK_LN"),
    SND_SOC_DAPM_OUTPUT("HP_L"),
    SND_SOC_DAPM_OUTPUT("HP_R"),
    SND_SOC_DAPM_OUTPUT("SPK_RP"),
    SND_SOC_DAPM_OUTPUT("SPK_RN"),
    SND_SOC_DAPM_OUTPUT("OUT3"),
};


static const struct snd_soc_dapm_widget
wm8960_dapm_widgets_out3[] = {
    SND_SOC_DAPM_MIXER("Mono Output Mixer", WM8960_POWER2, 1,
                        0,
                        &wm8960_mono_out[0],
```

```
                    ARRAY_SIZE(wm8960_mono_out)),
};
```

In this step, a MUX widget is defined for each of the left and right channels as well as the channel selector: these are Left Output Mixer, Right Output Mixer, and Mono Output Mixer. We also define a mixer widget for each of the left and right speakers: SPK_LP, SPK_LN, HP_L, HP_R, SPK_RP, OUT3, and SPK_RN. The specific mixer control is done by wm8960_loutput_mixer, wm8960_routput_mixer, and wm8960_mono_out defined in the previous step. The three widgets have power properties, so when one (or more) of these widgets are in one valid audio path, the DAPM framework can control its power state via bits 7 and/or 8 of their respective registers.

The third step is to define the connection path of these widgets:

```
static const struct snd_soc_dapm_route audio_paths[] = {
    [...]
    {"Left Output Mixer", "LINPUT3 Switch", "LINPUT3"},
    {"Left Output Mixer", "Boost Bypass Switch",
     "Left Boost Mixer"},
    {"Left Output Mixer", "PCM Playback Switch", "Left DAC"},

    {"Right Output Mixer", "RINPUT3 Switch", "RINPUT3"},
    {"Right Output Mixer", "Boost Bypass Switch",
     "Right Boost Mixer"},
    {"Right Output Mixer", "PCM Playback Switch", "Right DAC"},

    {"LOUT1 PGA", NULL, "Left Output Mixer"},
    {"ROUT1 PGA", NULL, "Right Output Mixer"},

    {"HP_L", NULL, "LOUT1 PGA"},
    {"HP_R", NULL, "ROUT1 PGA"},

    {"Left Speaker PGA", NULL, "Left Output Mixer"},
    {"Right Speaker PGA", NULL, "Right Output Mixer"},

    {"Left Speaker Output", NULL, "Left Speaker PGA"},
    {"Right Speaker Output", NULL, "Right Speaker PGA"},

    {"SPK_LN", NULL, "Left Speaker Output"},
```

```
    {"SPK_LP", NULL, "Left Speaker Output"},
    {"SPK_RN", NULL, "Right Speaker Output"},
    {"SPK_RP", NULL, "Right Speaker Output"},
};
```

```
static const struct snd_soc_dapm_route audio_paths_out3[] = {
    {"Mono Output Mixer", "Left Switch", "Left Output Mixer"},
    {"Mono Output Mixer", "Right Switch", "Right Output Mixer"},
    {"OUT3", NULL, "Mono Output Mixer"}
};
```

Through the definition of the first step, we know that `"Left output Mux"` and `"right output Mux"` have three input pins, respectively, `"Boost Bypass Switch"`, `"LINPUT3 Switch"` (or `"RINPUT3 Switch"`), and `"PCM Playback Switch"`. `"Mono Output Mixer"` has only two input select pins, which are `"Left Switch"` and `"Right Switch"`. So, obviously, the meaning of the preceding path definition is as follows:

- `"Left Boost Mixer"` is connected to `"Left Output Mixer"` via `"Boost Bypass Switch"`.

- `"Left DAC"` is connected to `"Left Output Mixer"` via `"PCM Playback Switch"`.

- `"RINPUT3"` is connected to `"Right Output Mixer"` via `"RINPUT3 Switch"`.

- `"Right Boost Mixer"` is connected to `"Right Output Mixer"` via `"Boost Bypass Switch"`.

- `"Right DAC"` is connected to `"Right Output Mixer"` via `"PCM Playback Switch"`.

- `"Left Output Mixer"` is connected to `"LOUT1 PGA"`. However, there is no switch control for this link.

- `"Right Output Mixer"` is connected to `"ROUT1 PGA"`, with no switch controlling this connection.

Not all of the connections have been described, but the idea is there. The fourth step is to register these widgets and paths in the codec-driven probe callback:

```
static int wm8960_add_widgets(struct
                              snd_soc_component *component)
```

```
{
    [...]
    struct snd_soc_dapm_context *dapm =
                        snd_soc_component_get_dapm(component);
    struct snd_soc_dapm_widget *w;

    snd_soc_dapm_new_controls(dapm, wm8960_dapm_widgets,
                        ARRAY_SIZE(wm8960_dapm_widgets));
    snd_soc_dapm_add_routes(dapm, audio_paths,
                        ARRAY_SIZE(audio_paths));
    [...]
    return 0;
}

static int wm8960_probe(struct snd_soc_component *component)
{
    [...]
    snd_soc_add_component_controls(component,
                            wm8960_snd_controls,
                        ARRAY_SIZE(wm8960_snd_controls));
    wm8960_add_widgets(component);
    return 0;
}

static const struct snd_soc_component_driver
    soc_component_dev_wm8960 = {
    .probe = wm8960_probe,
    .set_bias_level = wm8960_set_bias_level,
    .suspend_bias_off = 1,
    .idle_bias_on = 1,
    .use_pmdown_time = 1,
    .endianness = 1,
    .non_legacy_dai_naming = 1,
};

static int wm8960_i2c_probe(struct i2c_client *i2c,
                        const struct i2c_device_id *id)
```

```
{
    [...]
    ret = devm_snd_soc_register_component(&i2c->dev,
                                    &soc_component_dev_wm8960,
                                    &wm8960_dai, 1);


    return ret;
}
```

In the preceding example, controls, widgets, and route registration are deferred into the component driver's probe callback. This helps make sure that these elements are created only when the component is probed by the machine driver. In the machine driver, we can define and register the board-specific widgets and path information in the same way.

Codec component registration

After the codec component has been set up, it has to be registered with the system so that it can be used for what it is designed for. For this purpose, you should use `devm_snd_soc_register_component()`. This function will take care of unregistration/cleaning automatically when needed. The following is its prototype:

```
int devm_snd_soc_register_component(struct device *dev,
                    const struct
                    snd_soc_component_driver *cmpnt_drv,
                    struct snd_soc_dai_driver *dai_drv,
                    int num_dai)
```

The following is an example of codec registration, which is an excerpt from the `wm8960` codec driver. The component driver is first defined as the following:

```
static const struct snd_soc_component_driver
    soc_component_dev_wm8900 = {
    .probe = wm8900_probe,
    .suspend = wm8900_suspend,
    .resume = wm8900_resume,
    [...]
    /* control, widget and route setup */
    .controls    = wm8900_snd_controls,
    .num_controls    = ARRAY_SIZE(wm8900_snd_controls),
    .dapm_widgets    = wm8900_dapm_widgets,
```

```
        .num_dapm_widgets  = ARRAY_SIZE(wm8900_dapm_widgets),
        .dapm_routes       = wm8900_dapm_routes,
        .num_dapm_routes   = ARRAY_SIZE(wm8900_dapm_routes),
};
```

That component driver contains dapm routes and widgets, as well as a set of controls. Then, the codec dai callbacks are provided through the struct snd_soc_dai_ops, as follows:

```
static const struct snd_soc_dai_ops wm8900_dai_ops = {
    .hw_params   = wm8900_hw_params,
    .set_clkdiv  = wm8900_set_dai_clkdiv,
    .set_pll     = wm8900_set_dai_pll,
    .set_fmt     = wm8900_set_dai_fmt,
    .digital_mute   = wm8900_digital_mute,
};
```

Those codec dai callbacks the assigned to the codec dai driver (via the ops field) for registration with the ASoC core as follows:

```
#define WM8900_RATES  (SNDRV_PCM_RATE_8000   |\
                       SNDRV_PCM_RATE_11025  |\

                       SNDRV_PCM_RATE_16000  |\
                       SNDRV_PCM_RATE_22050  |\

                       SNDRV_PCM_RATE_44100  |\
                       SNDRV_PCM_RATE_48000)


#define WM8900_PCM_FORMATS \
    (SNDRV_PCM_FMTBIT_S16_LE | SNDRV_PCM_FMTBIT_S20_3LE | \
    SNDRV_PCM_FMTBIT_S24_LE)


static struct snd_soc_dai_driver wm8900_dai = {
    .name = "wm8900-hifi",
    .playback = {
        .stream_name = "HiFi Playback",
        .channels_min = 1,
        .channels_max = 2,
        .rates = WM8900_RATES,
        .formats = WM8900_PCM_FORMATS,
```

```
    },
    .capture = {
        .stream_name = "HiFi Capture",
        .channels_min = 1,
        .channels_max = 2,
        .rates = WM8900_RATES,
        .formats = WM8900_PCM_FORMATS,
    },
    .ops = &wm8900_dai_ops,
};
static int wm8900_spi_probe(struct spi_device *spi)
{
    [...]
    ret = devm_snd_soc_register_component(&spi->dev,
                                    &soc_component_dev_wm8900,
                                    &wm8900_dai, 1);

    return ret;
}
```

When the machine driver probes this codec, then the probe callback of the codec component driver (wm8900_probe) will be invoked, and they will complete the codec driver initialization. The full version of this codec device driver is sound/soc/codecs/wm8900.c in the Linux kernel source.

Now we are familiar with the codec class driver and its architecture. We have seen how to export codec properties as well, how to build audio routes, and how to implement DAPM features. On its own, the codec driver is quite useless, though it manages the codec device. It needs to be tied to the platform driver, which is the next driver class we are going to learn about.

Writing the platform class driver

The platform driver registers the PCM driver, CPU DAI driver, and their operation functions, pre-allocates buffers for PCM components, and sets playback and capture operations as applicable. In other words, the platform driver contains the audio DMA engine and audio interface drivers (for example, I2S, AC97, and PCM) for that platform.

The platform driver targets the SoC the platform is made of. It concerns the platform's DMA, which is how audio data transits between each block in the SoC, and CPU DAI, which is the path the CPU uses to send/carry audio data to/from the codec. Such a driver has two important data structures: `struct snd_soc_component_driver` and `struct snd_soc_dai_driver`. The former is responsible for DMA data management, and the latter is responsible for the parameter configuration of the DAI. However, both of these data structures have already been described while dealing with codec class drivers. Thus, this part will just deal with additional concepts, related to the platform code.

The CPU DAI driver

Since the platform code has been refactored too, as with the codec driver, the CPU DAI drivers must export an instance of the component driver as well as an instance of the DAI driver, respectively `struct snd_soc_component_driver` and `struct snd_soc_dai_driver`.

On the platform side, most of the work can be done by the core, especially for DMA-related stuff. Thus, it is common for CPU DAI drivers to provide only the name of the interface within the component driver structure and to let the core do the rest. The following is an example from the Rockchip SPDIF driver, implemented in `sound/soc/rockchip/rockchip_spdif.c`:

```
static const struct snd_soc_dai_ops rk_spdif_dai_ops = {
    [...]
};


/* SPDIF has no capture channel */
static struct snd_soc_dai_driver rk_spdif_dai = {
    .probe = rk_spdif_dai_probe,
    .playback = {
        .stream_name = "Playback",
[...]
    },
    .ops = &rk_spdif_dai_ops,
};


/* fill in the name only */
static const struct snd_soc_component_driver rk_spdif_component
= {
    .name = "rockchip-spdif",
```

```
};

static int rk_spdif_probe(struct platform_device *pdev)
{
    struct device_node *np = pdev->dev.of_node;
    struct rk_spdif_dev *spdif;
    int ret;
[...]
    spdif->playback_dma_data.addr = res->start + SPDIF_SMPDR;
    spdif->playback_dma_data.addr_width =
    DMA_SLAVE_BUSWIDTH_4_BYTES;
    spdif->playback_dma_data.maxburst = 4;

    ret = devm_snd_soc_register_component(&pdev->dev,
                                          &rk_spdif_component,
                                          &rk_spdif_dai, 1);
    if (ret) {
        dev_err(&pdev->dev, "Could not register DAI\n");
        goto err_pm_runtime;
    }

    ret = devm_snd_dmaengine_pcm_register(&pdev->dev, NULL, 0);
    if (ret) {
        dev_err(&pdev->dev, "Could not register PCM\n");
        goto err_pm_runtime;
    }

    return 0;
}
```

In the preceding excerpt, `spdif` is the driver state data structure. We can see that only the name is filled in in the component driver, and both the component and DAI drivers are registered as usual by means of `devm_snd_soc_register_component()`. The `struct snd_soc_dai_driver` must be set up according to the actual DAI properties, and the `dai_ops` should be set if necessary. However, a big part of the setup is done by `devm_snd_dmaengine_pcm_register()`, which will set the component driver's PCM ops according to the `dma_data` provided. This is explained in detail in the next section.

The platform DMA driver AKA PCM DMA driver

In a sound ecosystem, we have several types of devices: PCM, MIDI, mixer, sequencer, timer, and so on. Here, PCM does refer to Pulse Code Modulation, but it is a reference to a device that processes sample-based digital audio, that is, not midi and so on. The PCM layer (part of the ALSA core) is responsible for doing all the digital audio work, such as preparing the card for capture or playback, initiating the transfer to and from the device, and so on. In short, if you want to play back or capture sound, you're going to need a PCM.

The PCM driver helps perform DMA operations by overriding the function pointers exposed by the `struct snd_pcm_ops` structure. It is platform agnostic and interacts only with the SOC DMA engine upstream APIs. The DMA engine then interacts with the platform-specific DMA driver to get the correct DMA settings. The `struct snd_pcm_ops` is a structure that contains a set of callbacks that relate to different events regarding the PCM interface.

While dealing with ASoC (not purely ALSA), you'll never need to instantiate this structure as is, as long as you use the generic PCM DMA engine framework. The ASoC core does this for you. Have a look at the following call stack: *snd_soc_register_card -> snd_soc_instantiate_card -> soc_probe_link_dais -> soc_new_pcm*.

The audio DMA interface

Each audio bus driver of the SoC is responsible for providing a DMA interface by means of this API. This is the case, for example, for the audio buses on i.MX-based SoCs, such as ESAI, SAI, SPDIF and SSI whose drivers are located in `sound/soc/fsl/`, respectively `sound/soc/fsl/fsl_esai.c`, `sound/soc/fsl/fsl_sai.c`, `sound/soc/fsl/fsl_spdif.c`, and `sound/soc/fsl/fsl_ssi.c`.

The audio DMA driver is registered via `devm_snd_dmaengine_pcm_register()`. This function registers a `struct snd_dmaengine_pcm_config` for the device. The following is its prototype:

```
int devm_snd_dmaengine_pcm_register(
                        struct device *dev,
                        const struct
                        snd_dmaengine_pcm_config *config,
                        unsigned int flags);
```

In the preceding prototype, `dev` is the parent device for the PCM device, usually `&pdev->dev`. `config` is the platform-specific PCM configuration, which is of type `struct snd_dmaengine_pcm_config`. This structure needs to be described in detail. `flags` represents additional flags describing how to deal with DMA channels. Most of the time, it is `0`. However, possible values are defined in `include/sound/dmaengine_pcm.h` and are all prefixed with `SND_DMAENGINE_PCM_FLAG_`. Frequently used ones are `SND_DMAENGINE_PCM_FLAG_HALF_DUPLEX`, `SND_DMAENGINE_PCM_FLAG_NO_DT`, and `SND_DMAENGINE_PCM_FLAG_COMPAT`. The former indicates that the PCM is half-duplex and the DMA channel is shared between capture and playback. The second one asks the core not to try to request the DMA channels through the device tree. The last one means a custom callback will be used to request the DMA channel. Upon registration, the generic PCM DMA engine framework will build a suitable `snd_pcm_ops` and set the component driver's `.ops` field with it.

The classic DMA operation flow in Linux is as follows:

1. `dma_request_channel`: For allocating the slave channel.

2. `dmaengine_slave_config`: To set slave- and controller-specific parameters.

3. `dma_prep_xxxx`: To get a descriptor for the transaction.

4. `dma_cookie = dmaengine_submit(tx)`: Submit the transaction and grab the DMA cookie.

5. `dma_async_issue_pending(chan)`: To start transmission and wait for a callback notification.

In ASoC, the device tree is used to map the DMA channels to the PCM device. `devm_snd_dmaengine_pcm_register()` requests the DMA channel through `dmaengine_pcm_request_chan_of()`, which is a device-tree-based interface. In order to perform *steps 1* to *3*, the PCM DMA engine core needs to be provided additional information. This can be done either by populating a `struct snd_dmaengine_pcm_config`, which will be given to the registration function or by letting the PCM DMA engine framework retrieve information from the system's DMA engine core. *Steps 4* and *5* are transparently handled by the PCM DMA engine core.

The following is what the `struct snd_dma_engine_pcm_config` looks like:

```
struct snd_dmaengine_pcm_config {
    int (*prepare_slave_config)(
                        struct snd_pcm_substream *substream,
                        struct snd_pcm_hw_params *params,
                        struct dma_slave_config *slave_config);
```

```
    struct dma_chan *(*compat_request_channel)(
                        struct snd_soc_pcm_runtime *rtd,
                        struct snd_pcm_substream *substream);
    [...]

    dma_filter_fn compat_filter_fn;
    struct device *dma_dev;
    const char *chan_names[SNDRV_PCM_STREAM_LAST + 1];
    const struct snd_pcm_hardware *pcm_hardware;
    unsigned int prealloc_buffer_size;
};
```

The preceding data structure mainly deals with DMA channel management, buffer management, and channel configuration:

- `prepare_slave_config`: This callback is used to fill in the DMA `slave_config` (of type `struct dma_slave_config`, which is the DMA slave channel runtime config) for a PCM sub-stream. It will be called from the PCM driver's hwparams callback. Here, you can use `snd_dmaengine_pcm_prepare_slave_config`, which is a generic `prepare_slave_config` callback for platforms that make use of the `snd_dmaengine_dai_dma_data` struct for their DAI DMA data. This generic callback will internally call `snd_hwparams_to_dma_slave_config` to fill in the slave config based on `hw_params`, followed by `snd_dmaengine_set_config_from_dai_data` to fill in the remaining fields based on the DAI DMA data.

  When using the generic callback approach, you should call `snd_soc_dai_init_dma_data()` (given the DAI-specific capture and playback DMA data config, which are of type `struct snd_dmaengine_dai_dma_data`) from within your CPU DAI driver's `.probe` callback, which will set both the `cpu_dai->playback_dma_data` and `cpu_dai->capture_dma_data` fields. The `snd_soc_dai_init_dma_data()` method simply sets the DMA settings (for either capture, playback, or both) for the DAI given as a parameter.

- `compat_request_channel`: This is used to request DMA channels for platforms that do not use the device tree. If set, `.compat_filter_fn` will be ignored.

- `compat_filter_fn`: It is used as the filter function when requesting a DMA channel for platforms that do not use the device tree. The filter parameter will be the DAI's DMA data.

- `dma_dev`: This allows requesting DMA channels for a device other than the device that is registering the PCM driver. If set, DMA channels will be requested on this device rather than the DAI device.

- `chan_names`: This is the array of names to use when requesting capture/playback DMA channels. This is useful when the default `"tx"` and `"rx"` channel names don't apply, for example, if an HW module supports multiple channels, each having different DMA channel names.

- `pcm_hardware`: This describes the PCM hardware capabilities. If not set, rely on the core to fill in the right flags derived from the DMA engine information. This field is of type `struct snd_pcm_hardware` and will be described in the next section.

- `prealloc_buffer_size`: This is the size of the preallocated audio buffer.

The PCM DMA config may not be supplied to the registration API (it might be NULL), and the registration would be `ret = devm_snd_dmaengine_pcm_register(&pdev->dev, NULL, 0)`. In this case, you should provide capture and playback DAI DMA channel configs via `snd_soc_dai_init_dma_data()`, as described earlier. By using this method, other elements will be derived from the system core. For example, to request a DMA channel, the PCM DMA engine core will rely on the device tree, assuming that the capture and playback DMA channel names are respectively `"rx"` and `"tx"`, unless the flag `SND_DMAENGINE_PCM_FLAG_HALF_DUPLEX` is set in `flags`, in which case it will consider capture and playback as using the same DMA channel, named `rx-tx` in the device tree node.

DMA channel settings will be derived from the system DMA engine too. The following is what `snd_soc_dai_init_dma_data()` looks like:

```
static inline void snd_soc_dai_init_dma_data(
                                        struct snd_soc_dai *dai,
                                        void *playback,
                                        void *capture)
{
    dai->playback_dma_data = playback;
    dai->capture_dma_data = capture;
}
```

Though `snd_soc_dai_init_dma_data()` accepts both capture and playback as void types, the values passed should actually be of type `struct snd_dmaengine_dai_dma_data`, defined in `include/sound/dmaengine_pcm.h` as follows:

```
struct snd_dmaengine_dai_dma_data {
    dma_addr_t addr;
    enum dma_slave_buswidth addr_width;
    u32 maxburst;
    unsigned int slave_id;
    void *filter_data;
    const char *chan_name;
    unsigned int fifo_size;
    unsigned int flags;
};
```

This structure represents DMA channel data (or config or whatever you prefer) for a DAI channel. You should refer to the header where it is defined for the meaning of its fields. Additionally, you can have a look at other drivers for more details on how to set up this data structure.

PCM hardware configuration

When the DMA settings are not automatically fed from the system by the PCM DMA engine core, the platform PCM driver may need to provide PCM hardware settings, which describe how hardware lays out the PCM data. Those settings are provided through the `snd_dmaengine_pcm_config.pcm_hardware` field, which is of type `struct snd_pcm_hardware`, defined as follows:

```
struct snd_pcm_hardware {
    unsigned int info;
    u64 formats;
    unsigned int rates;
    unsigned int rate_min;
    unsigned int rate_max;
    unsigned int channels_min;
    unsigned int channels_max;
    size_t buffer_bytes_max;
    size_t period_bytes_min;
    size_t period_bytes_max;
```

```
    unsigned int periods_min;
    unsigned int periods_max;
    size_t fifo_size;
};
```

This structure describes the hardware limitations of the platform itself (or should I say, it sets the allowed parameters), such as the number of channels/sampling rate/data format that can be supported, the range of period size supported by DMA, the range of period counts, and so on. In the preceding data structure, range values, period min, and period max depend on the capabilities of the DMA controller, the DAI hardware, and the codec. The following are the detailed meanings of each field:

- `info` contains the type and capabilities of this PCM. The possible values are bit flags that are all defined in `include/uapi/sound/asound.h` (this means user code should include `<sound/asound.h>`) as `SNDRV_PCM_INFO_XXX`. For example, `SNDRV_PCM_INFO_MMAP` would mean the hardware supports the `mmap()` system call. Here, at least, you have to specify whether the `mmap` system call is supported and which interleaved format is supported. When the `mmap()` system call is supported, add the `SNDRV_PCM_INFO_MMAP` flag here. When the hardware supports the interleaved or the non-interleaved formats, the `SNDRV_PCM_INFO_INTERLEAVED` or `SNDRV_PCM_INFO_NONINTERLEAVED` flag must be set, respectively. If both are supported, you can set both, too.

- The `formats` field contains the bit flags of supported formats (`SNDRV_PCM_FMTBIT_XXX`). If the hardware supports more than one format, you should use all OR'ed bits.

- The `rates` field contains the bit flags of supported rates (`SNDRV_PCM_RATE_XXX`).

- `rate_min` and `rate_max` define the minimum and maximum sample rate. This should correspond somehow to rate bits.

- `channel_min` and `channel_max` define, as you might have already guessed, the minimum and the maximum number of channels.

- `buffer_bytes_max` defines the maximum buffer size in bytes. There is no `buffer_bytes_min` field since it can be calculated from the minimum period size and the minimum number of periods. Meanwhile, `period_bytes_min` and `period_bytes_max` define the minimum and maximum size of the period in bytes.

- `periods_max` and `periods_min` define the maximum and the minimum number of periods in the buffer.

The other fields need the concept of a period to be introduced. The period defines the size at which a PCM interrupt is generated. The concept of a period is very important. A period basically describes an interrupt. It sums up the "chunk" size that the hardware supplies data in:

- `period_bytes_min` is the minimum transfer size of the DMA written to as the number of bytes processed between interrupts. For example, if the DMA can transmit a minimum of 2,048 bytes, it should be written as `2048`.

- `period_bytes_max` is the maximum transfer size of the DMA aka the maximum number of bytes processed between interrupts. For example, if the DMA can transmit a maximum of 4,096 bytes, it should be written as `4096`.

The following is an example of such PCM constraints from the STM32 I2S DMA driver, defined in `sound/soc/stm/stm32_i2s.c`:

```
static const struct snd_pcm_hardware stm32_i2s_pcm_hw = {
    .info = SNDRV_PCM_INFO_INTERLEAVED | SNDRV_PCM_INFO_MMAP,
    .buffer_bytes_max = 8 * PAGE_SIZE,
    .period_bytes_max = 2048,
    .periods_min = 2,
    .periods_max = 8,
};
```

Once set up, this structure should end up in the `snd_dmaengine_pcm_config.pcm_hardware` field prior to the `struct snd_dmaengine_pcm_config` object given to `devm_snd_dmaengine_pcm_register()`.

The following is a playback flow, showing the involved components and PCM data flow:

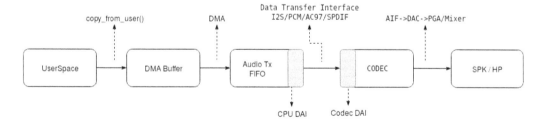

Figure 5.4 – ASoC audio playback flow

The preceding figure shows the audio playback flow and blocks involved in each step. We can see audio data is copied from the user to DMA buffers, followed by DMA transactions to move the data into the platform audio Tx FIFO, which, thanks to its link with the codec (via their respective DAIs), sends this data to the codec in charge of playing the audio through the speaker. The capture operation is the opposite flow with the speaker replaced by a microphone.

This brings us to the end of dealing with the platform class driver. We have seen the data structures and concepts it shares with the codec class driver. Note that both codec and platform drivers need to be linked together in order to build the real audio path from a system point of view. According to the ASoC architecture, this has to be done in another class driver, the so-called machine driver, which is the topic of the next chapter.

Summary

In this chapter, we analyzed the ASoC architecture. On this basis, we dealt with both the codec driver and the platform driver. By learning about these topics, we went through several concepts, such as controls and widgets. We have seen how the ASoC framework differs from the classic PC ALSA system, mostly by targeting code reusability and implementing power management.

Last but not least, we have seen that platform and codec drivers do not work standalone. They need to be bound together by the machine driver, which is responsible for registering the final audio device, and this is the main topic in the next chapter.

6
ALSA SoC Framework – Delving into the Machine Class Drivers

While starting our ALSA SoC framework series, we noticed that neither platform nor codec class drivers are intended to work on their own. The ASoC architecture is designed in such a way that platform and codec class drivers must be bound together in order to build the audio device. This binding can be done either from a so-called machine driver or from within the device tree, each of which being machine specific. It then goes without saying that the machine driver targets a specific system, and it may change from one board to another. In this chapter, we highlight the dark side of AsoC machine class drivers and discuss specific cases we may encounter when we need to write a machine class driver.

In this chapter, we will present the Linux ASoC driver architecture and implementation. This chapter will be split into different parts, which are as follows:

- Introduction to machine class drivers
- Machine routing considerations
- Clocking and formatting considerations
- Sound card registration
- Leveraging the simple-card machine driver

Technical requirements

You need the following for this chapter:

- Strong knowledge of the concept of device trees
- Familiarity with both platform and codec class drivers (discussed in *Chapter 5, ALSA SoC Framework – Leveraging Codec and Platform Class Drivers*)
- Linux kernel v4.19.X sources, available at `https://git.kernel.org/pub/scm/linux/kernel/git/stable/linux.git/refs/tags`

Introduction to machine class drivers

Codec and platform drivers cannot work alone. Machine drivers are responsible for binding them together in order to finish audio information processing. The machine driver class acts as the glue that describes and ties the other component drivers together to form an ALSA sound card device. It manages any machine-specific controls and machine-level audio events (such as turning on an amp at the start of playback). The machine drivers describe and bind the CPU **Digital Audio Interfaces** (**DAIs**) and codec drivers together to create the DAI links and the ALSA sound card. The machine driver connects the codec drivers by linking the DAIs exposed by each module (the CPU and codec) described in *Chapter 5, ALSA SoC Framework – Leveraging Codec and Platform Class Drivers*. It defines the `struct snd_soc_dai_link` structure and instantiates the sound card, `struct snd_soc_card`.

Platform and codec drivers are generally reusable, but machine drivers are not because they have specific hardware features that are non-reusable most of time. The so-called hardware characteristics refer to the link between DAIs; opening the amplifier through a GPIO; detecting the plug-in through a GPIO; using a clock such as MCLK/External OSC as the reference clock source of I2; the codec module, and so on. In general, machine driver responsibilities include the following:

- Populating the `struct snd_soc_dai_link` structure with appropriate CPU and codec DAIs
- Physical codec clock settings (if any) and codec initialization master/slave configurations (if any)
- Defining DAPM widgets to route through the physical codec internals and complete the DAPM path as needed
- Propagating the runtime sampling frequency to the individual codec drivers as needed

To put it together, we have the following flow:

1. The codec driver registers a component driver, a DAI driver, and their operation functions.
2. The platform driver registers a component driver, the PCM driver, the CPU DAI driver, and their operation functions and sets playback and capture operations as applicable.
3. The machine layer creates the DAI link between the codec and CPU and registers the sound card and PCM devices.

Now that we have seen the development flow of a machine class driver, let's start with the first step, which consists of populating the DAI link.

The DAI link

A DAI link is the logical representation of the link between the CPU and the codec DAIs. It is represented from within the kernel using `struct snd_soc_dai_link`, defined as follows:

```
struct snd_soc_dai_link {
    const char *name;
    const char *stream_name;
    const char *cpu_name;
```

```
    struct device_node *cpu_of_node;
    const char *cpu_dai_name;
    const char *codec_name;
    struct device_node *codec_of_node;
    const char *codec_dai_name;

    struct snd_soc_dai_link_component *codecs;
    unsigned int num_codecs;

    const char *platform_name;
    struct device_node *platform_of_node;
    int id;

    const struct snd_soc_pcm_stream *params;
    unsigned int num_params;
    unsigned int dai_fmt;
    enum snd_soc_dpcm_trigger trigger[2];

/* codec/machine specific init - e.g. add machine controls */
    int (*init)(struct snd_soc_pcm_runtime *rtd);
    /* machine stream operations */
    const struct snd_soc_ops *ops;

    /* For unidirectional dai links */
    unsigned int playback_only:1;
    unsigned int capture_only:1;

    /* Keep DAI active over suspend */
    unsigned int ignore_suspend:1;
[...]
    /* DPCM capture and Playback support */
    unsigned int dpcm_capture:1;
    unsigned int dpcm_playback:1;
```

```
    struct list_head list; /* DAI link list of the soc card */
};
```

> **Important note**
>
> The full `snd_soc_dai_link` data structure definition can be found at `https://elixir.bootlin.com/linux/v4.19/source/include/sound/soc.h#L880`.

This link is set up from within the machine driver. It should specify the `cpu_dai`, the `codec_dai`, and the platform that is used. Once set up, DAI links are fed to `struct snd_soc_card`, which represents a sound card. The following list describes the elements in the structure:

- name: This is chosen arbitrarily. It can be anything.

- `codec_dai_name`: This must match the `snd_soc_dai_driver.name` field from within the codec chip driver. Codecs may have one or more DAIs. Refer to the codec driver to identify the DAI names.

- `cpu_dai_name`: This must match the `snd_soc_dai_driver.name` field from within the CPU DAI driver.

- `stream_name`: This is the stream name of this link.

- `init`: This is the DAI link initialization callback. It is typically used to add DAI link-specific widgets or other types of one-time settings.

- `dai_fmt`: This should be set with the supported format and clock configuration, which should be coherent for both CPU and CODEC DAI drivers. Possible bit flags for this field are introduced later.

- ops: This field is of the `struct snd_soc_ops` type. It should be set with machine-level PCM operations of the DAI link: `startup`, `hw_params`, `prepare`, `trigger`, `hw_free`, `shutdown`. This field is described in detail later.

- `codec_name`: If set, this should be the name of the codec driver, such as `platform_driver.driver.name` or `i2c_driver.driver.name`.

- `codec_of_node`: The device tree node associated with the codec.

- `cpu_name`: If set, this should be the name of the CPU DAI driver CPU.

- `cpu_of_node`: This is the device tree node associated with the CPU DAI.

- `platform_name` or `platform_of_node`: This is the name or DT node reference to the platform node, which provides DMA capabilities.

- `playback_only` and `capture_only` are to be used in case of unidirectional links, such as SPDIF. If this is an output only link (playback only), then `playback_only` and `capture_only` must be set to `true` and `false` respectively. With an input-only link, the opposite values should be used.

In most cases, `.cpu_of_node` and `.platform_of_node` are the same, since the CPU DAI driver and the DMA PCM driver are implemented by the same device. That being said, you must specify the link's codec either by name or by `of_node`, but not both. You must do the same for the CPU and platform. However, at least one of the CPU DAI name or the CPU device name/node must be specified. This could be summarized as follows:

```
if (link->platform_name && link->platform_of_node)
    ==> Error
if (link->cpu_name && link->cpu_of_node)
    ==> Eror
if (!link->cpu_dai_name && !(link->cpu_name ||
                             link->cpu_of_node))
    ==> Error
```

There is a key point it is worth noting here. How do we reference the platform or CPU node in the DAI link? We will answer this question later. Let's first consider the following two device nodes. The first one (`ssi1`) is the SSI `cpu-dai` node for the i.mx6 SoC. The second node (`sgtl5000`) represents the sgtl5000 codec chip:

```
ssi1: ssi@2028000 {
    #sound-dai-cells = <0>;
    compatible = "fsl,imx6q-ssi", "fsl,imx51-ssi";
    reg = <0x02028000 0x4000>;
    interrupts = <0 46 IRQ_TYPE_LEVEL_HIGH>;
    clocks = <&clks IMX6QDL_CLK_SSI1_IPG>,
             <&clks IMX6QDL_CLK_SSI1>;
    clock-names = "ipg", "baud";
    dmas = <&sdma 37 1 0>, <&sdma 38 1 0>;
    dma-names = "rx", "tx";
    fsl,fifo-depth = <15>;
    status = "disabled";
```

```
};

&i2c0{
    sgtl5000: codec@0a {
        compatible = "fsl,sgtl5000";
        #sound-dai-cells = <0>;
        reg = <0x0a>;
        clocks = <&audio_clock>;
        VDDA-supply = <&reg_3p3v>;
        VDDIO-supply = <&reg_3p3v>;
        VDDD-supply = <&reg_1p5v>;
    };
};
```

> **Important note**
>
> In the SSI node, you can see the `dma-names = "rx", "tx";` property,
> which is the expected DMA channel names requested by the pcmdmaengine
> framework. This may also be an indication that the CPU DAI and platform
> PCM are represented by the same node.

We will consider a system where an i.MX6 SoC is connected to an sgtl5000 audio codec.
It is common for machine drivers to grab either CPU or CODEC device tree nodes by
referencing those nodes (their `phandle` actually) as its properties. This way, you can just
use one of the `OF` helpers (such as `of_parse_phandle()`) to grab a reference on these
nodes. The following is an example of a machine node that references both the codec and
the platform by an `OF` node:

```
sound {
    compatible = "fsl,imx51-babbage-sgtl5000",
                 "fsl,imx-audio-sgtl5000";
    model = "imx51-babbage-sgtl5000";
    ssi-controller = <&ssi1>;
    audio-codec = <&sgtl5000>;
    [...]
};
```

In the preceding machine node, the codec and CPUE are passed by reference (their `phandle`) via the `audio-codec` and `ssi-controller` properties. These property names are not standardized as long as the machine driver is written by you (this is not true if you use the `simple-card` machine driver, for example, which expects some predefined names). In the machine driver, you'll see something like this:

```
static int imx_sgtl5000_probe(struct platform_device *pdev)
{
    struct device_node *np = pdev->dev.of_node;
    struct device_node *ssi_np, *codec_np;
    struct imx_sgtl5000_data *data = NULL;
    int int_port, ext_port; int ret;
[...]
    ssi_np = of_parse_phandle(pdev->dev.of_node,
                              "ssi-controller", 0);
    codec_np = of_parse_phandle(pdev->dev.of_node,
                                "audio-codec", 0);
    if (!ssi_np || !codec_np) {
        dev_err(&pdev->dev, "phandle missing or invalid\n");
        ret = -EINVAL;
        goto fail;
    }

    data = devm_kzalloc(&pdev->dev, sizeof(*data), GFP_KERNEL);
    if (!data) {
        ret = -ENOMEM;
        goto fail;
    }

    data->dai.name = "HiFi";
    data->dai.stream_name = "HiFi";
    data->dai.codec_dai_name = "sgtl5000";
    data->dai.codec_of_node = codec_np;
    data->dai.cpu_of_node = ssi_np;
    data->dai.platform_of_node = ssi_np;
    data->dai.init = &imx_sgtl5000_dai_init;
```

```
        data->card.dev = &pdev->dev;
        [...]
};
```

The preceding excerpt used `of_parse_phandle()` to obtain node references. This is an excerpt from the `imx_sgtl5000` machine, which is `sound/soc/fsl/ imx-sgtl5000.c` in the kernel sources. Now that we are familiar with the way the DAI link should be handled, we can proceed to audio routing from within the machine driver in order to define the path the audio data should follow.

Machine routing consideration

The machine driver can alter (or should I say append) the routes defined from within the codec. It has the last word on which codec pins must be used, for example.

Codec pins

Codec pins are meant to be connected to the board connectors. The available codec pins are defined in the codec driver using the `SND_SOC_DAPM_INPUT` and `SND_SOC_DAPM_ OUTPUT` macros. These macros can be searched with the `grep` command in the codec driver in order to find the available PIN.

For example, the `sgtl5000` codec driver defines the following output and input:

```
static const struct snd_soc_dapm_widget sgtl5000_dapm_widgets[]
= {
    SND_SOC_DAPM_INPUT("LINE_IN"),
    SND_SOC_DAPM_INPUT("MIC_IN"),

    SND_SOC_DAPM_OUTPUT("HP_OUT"),
    SND_SOC_DAPM_OUTPUT("LINE_OUT"),
    SND_SOC_DAPM_SUPPLY("Mic Bias", SGTL5000_CHIP_MIC_CTRL, 8,
                        0,
                        mic_bias_event,
                        SND_SOC_DAPM_POST_PMU |
                        SND_SOC_DAPM_PRE_PMD),
    [...]
};
```

In the next sections, we will see how those pins are connected to the board.

Board connectors

The board connectors are defined in the machine driver in the `struct snd_soc_dapm_widget` part of the registered `struct snd_soc_card`. Most of the time, these board connectors are virtual. They are just logical stickers that are connected with codec pins (which are real this time). The following lists the connectors defined by the `imx-sgtl5000` machine driver, `sound/soc/fsl/imx-sgtl5000.c` (whose documentation is `Documentation/devicetree/bindings/sound/imx-audio-sgtl5000.txt`), which has been given as an example so far:

```
static const struct snd_soc_dapm_widget
imx_sgtl5000_dapm_widgets[] = {
    SND_SOC_DAPM_MIC("Mic Jack", NULL),
    SND_SOC_DAPM_LINE("Line In Jack", NULL),
    SND_SOC_DAPM_HP("Headphone Jack", NULL),
    SND_SOC_DAPM_SPK("Line Out Jack", NULL),
    SND_SOC_DAPM_SPK("Ext Spk", NULL),
};
```

The next section will connect this connector to the codec pins.

Machine routing

The final machine routing can be either static (that is, populated from within the machine driver itself) or populated from within the device tree. Moreover, the machine driver can optionally extend the codec power map and become an audio power map of the audio subsystem by connecting to the supply widget that has been defined in the codec driver with either `SND_SOC_DAPM_SUPPLY` or `SND_SOC_DAPM_REGULATOR_SUPPLY`.

Device tree routing

Let's take the node of our machine as an example, which connects an i.MX6 SoC to an sgtl5000 codec (this excerpt can be found in the machine documentation):

```
sound {
    compatible = "fsl,imx51-babbage-sgtl5000",
                 "fsl,imx-audio-sgtl5000";
    model = "imx51-babbage-sgtl5000";
    ssi-controller = <&ssi1>;
    audio-codec = <&sgtl5000>;
    audio-routing = "MIC_IN", "Mic Jack",
```

```
                        "Mic Jack", "Mic Bias",
                        "Headphone Jack", "HP_OUT";
[...]
};
```

Routing from the device tree expects the audio map to be given in a certain format. That is, entries are parsed as pairs of strings, the first being the connection's sink, the second being the connection's source. Most of the time, these connections are materialized as codec pins and board connector mappings. Valid names for sources and sinks depend on the hardware binding, which is as follows:

- **The codec**: This should have defined the pins whose names are used here.

- **The machine**: This should have defined the connectors or jacks whose names are used here.

In the preceding excerpt, what do you notice there? We can see MIC_IN, HP_OUT, and "Mic Bias", which are codec pins (coming from the codec driver), and "Mic Jack" and "Headphone Jack", which have been defined in the machine driver as board connectors.

In order to use the route defined in the DT, the machine driver must call snd_soc_of_parse_audio_routing(), which has the following prototype:

```
int snd_soc_of_parse_card_name(struct snd_soc_card *card,
                                const char *prop);
```

In the preceding prototype, card represents the sound card for which the routes are parsed, and prop is the name of the property that contains the routes in the device tree node. This function returns 0 on success and a negative error code on error.

Static routing

Static routing consists of defining a DAPM route map from the machine driver and assigning it to the sound card directly as follows:

```
static const struct snd_soc_dapm_widget rk_dapm_widgets[] = {
    SND_SOC_DAPM_HP("Headphone", NULL),
    SND_SOC_DAPM_MIC("Headset Mic", NULL),
    SND_SOC_DAPM_MIC("Int Mic", NULL),
    SND_SOC_DAPM_SPK("Speaker", NULL),
};
```

```
/* Connection to the codec pin */
static const struct snd_soc_dapm_route rk_audio_map[] = {
    {"IN34", NULL, "Headset Mic"},
    {"Headset Mic", NULL, "MICBIAS"},
    {"DMICL", NULL, "Int Mic"},
    {"Headphone", NULL, "HPL"},
    {"Headphone", NULL, "HPR"},
    {"Speaker", NULL, "SPKL"},
    {"Speaker", NULL, "SPKR"},
};

static struct snd_soc_card snd_soc_card_rk = {
    .name = "ROCKCHIP-I2S",
    .owner = THIS_MODULE,
[...]
    .dapm_widgets = rk_dapm_widgets,
    .num_dapm_widgets = ARRAY_SIZE(rk_dapm_widgets),
    .dapm_routes = rk_audio_map,
    .num_dapm_routes = ARRAY_SIZE(rk_audio_map),
    .controls = rk_mc_controls,
    .num_controls = ARRAY_SIZE(rk_mc_controls),
};
```

The preceding snippet is an excerpt from sound/soc/rockchip/rockchip_
rt5645.c. By using it this way, it is not necessary to use snd_soc_of_parse_
audio_routing(). However, a con of using this method is that it is not possible to
change the route without recompiling the kernel. Next, we will be looking at clocking and
formatting considerations.

Clocking and formatting considerations

Before delving deeper into this section, let's spend some time on the snd_soc_dai_
link->ops field. This field is of type struct snd_soc_ops, defined as follows:

```
struct snd_soc_ops {
    int (*startup)(struct snd_pcm_substream *);
    void (*shutdown)(struct snd_pcm_substream *);
    int (*hw_params)(struct snd_pcm_substream *,
```

```
                        struct snd_pcm_hw_params *);
    int (*hw_free)(struct snd_pcm_substream *);
    int (*prepare)(struct snd_pcm_substream *);
    int (*trigger)(struct snd_pcm_substream *, int);
};
```

These callback fields in this structure should remind you of those defined in the
`snd_soc_dai_driver->ops` field, which is of type `struct snd_soc_dai_ops`.
From within the DAI link, these callbacks represent the machine-level PCM operations
of the DAI link, while in `struct snd_soc_dai_driver`, they are either codec-DAI-
specific or CPU-DAI-specific.

`startup()` is invoked by ALSA when a PCM substream is opened (when someone has
opened the capture/playback device), while `hw_params()` is called when setting up the
audio stream. The machine driver may configure DAI link data format from within both
of these callbacks. `hw_params()` offers the advantage of receiving stream parameters
(*channel count*, *format*, *sample rate*, and so forth).

The data format configuration should be consistent between the CPU DAI and the codec.
The ASoC core provides helper functions to change those configurations. They are
as follows:

```
int snd_soc_dai_set_fmt(struct snd_soc_dai *dai,
                        unsigned int fmt)
int snd_soc_dai_set_pll(struct snd_soc_dai *dai, int pll_id,
                        int source, unsigned int freq_in,
                        unsigned int freq_out)
int snd_soc_dai_set_sysclk(struct snd_soc_dai *dai, int clk_id,
                        unsigned int freq, int dir)
int snd_soc_dai_set_clkdiv(struct snd_soc_dai *dai,
                        int div_id, int div)
```

In the preceding helper list, `snd_soc_dai_set_fmt` sets the DAI format for things
such as the clock master/slave relationship, audio format, and signal inversion; `snd_soc_
dai_set_pll` configures the clock PLL; `snd_soc_dai_set_sysclk` configures the
clock source; and `snd_soc_dai_set_clkdiv` configures the clock divider. Each of
these helpers will call the appropriate callback in the underlying DAI's driver ops. For
example, calling `snd_soc_dai_set_fmt()` with the CPU DAI will invoke this CPU
DAI's `dai->driver->ops->set_fmt` callback.

The following is the actual list of formats/flags that can be assigned either to DAIs or the `dai_link.format` field:

- **Format**: Configured through `snd_soc_dai_set_fmt()`:

  A) **Clock master/slave**:

     a) `SND_SOC_DAIFMT_CBM_CFM`: The CPU is the slave for the bit clock and frame sync. This also means the codec is the master for both.

     b) `SND_SOC_DAIFMT_CBS_CFS`. The CPU is the master for the bit clock and frame sync. This also means the codec is the slave for both.

     c) `SND_SOC_DAIFMT_CBM_CFS`. The CPU is the slave for the bit clock and the master for frame sync. This also means the codec is the master for the former and the slave for the latter.

  B) **Audio format**:

     a) `SND_SOC_DAIFMT_DSP_A`: Frame syncing is 1 bit-clock wide, 1-bit delay.

     b) `SND_SOC_DAIFMT_DSP_B`: Frame syncing is 1 bit-clock wide, 0-bit delay. This format can be used for the TDM protocol.

     c) `SND_SOC_DAIFMT_I2S`: Frame syncing is 1 audio word wide, 1-bit delay, I2S mode.

     d) `SND_SOC_DAIFMT_RIGHT_J`: Right justified mode.

     e) `SND_SOC_DAIFMT_LEFT_J`: Left justified mode.

     f) `SND_SOC_DAIFMT_DSP_A`: Frame syncing is 1 bit-clock wide,1-bit delay.

     g) `SND_SOC_DAIFMT_AC97`: AC97 mode.

     h) `SND_SOC_DAIFMT_PDM`: Pulse-density modulation.

     i) `SND_SOC_DAIFMT_DSP_B`: Frame sync is 1 bit-clock wide, 1-bit delay.

  C) **Signal inversion**:

     a) `SND_SOC_DAIFMT_NB_NF`: Normal bit clock, normal frame sync. The CPU transmitter shifts data out on the falling edge of the bit clock, the receiver samples data on the rising edge. The CPU frame sync generator starts the frame on the rising edge of the frame sync. This parameter is recommended for I2S on the CPU side.

b) `SND_SOC_DAIFMT_NB_IF`: Normal bit clock, inverted frame sync. The CPU transmitter shifts data out on the falling edge of the bit clock, and the receiver samples data on the rising edge. The CPU frame sync generator starts the frame on the falling edge of the frame sync.

c) `SND_SOC_DAIFMT_IB_NF`: Inverted bit clock, normal frame sync. The CPU transmitter shifts data out on the rising edge of the bit clock, and the receiver samples data on the falling edge. The CPU frame sync generator starts the frame on the rising edge of the frame sync.

d) `SND_SOC_DAIFMT_IB_IF`: Inverted bit clock, inverted frame sync. The CPU transmitter shifts data out on the rising edge of the bit clock, and the receiver samples data on the falling edge. The CPU frame sync generator starts the frame on the falling edge of the frame sync. This configuration can be used for PCM mode (such as Bluetooth or modem-based audio chips).

- **Clock source**: Configured through `snd_soc_dai_set_sysclk()`. The following are the direction parameters letting ALSA know which clock is used:

  a) `SND_SOC_CLOCK_IN`: This means an internal clock is used for sysclock.

  b) `SND_SOC_CLOCK_OUT`: This means an external clock is used for sysclock.

- **Clock divider**: Configured through `snd_soc_dai_set_clkdiv()`.

The preceding flags are the possible values that can be set in the `dai_link->dai_fmt` field or assigned to either codec or CPU DAIs from within the machine driver. The following is a typical `hw_param()` implementation:

```
static int foo_hw_params(struct snd_pcm_substream *substream,
                         struct snd_pcm_hw_params *params)
{
    struct snd_soc_pcm_runtime *rtd = substream->private_data;
    struct snd_soc_dai *codec_dai = rtd->codec_dai;
    struct snd_soc_dai *cpu_dai = rtd->cpu_dai;
    unsigned int pll_out = 24000000;
    int ret = 0;

    /* set the cpu DAI configuration */
    ret = snd_soc_dai_set_fmt(cpu_dai, SND_SOC_DAIFMT_I2S |
                              SND_SOC_DAIFMT_NB_NF |
                              SND_SOC_DAIFMT_CBM_CFM);
    if (ret < 0)
```

```
        return ret;

    /* set codec DAI configuration */
    ret = snd_soc_dai_set_fmt(codec_dai, SND_SOC_DAIFMT_I2S |
                              SND_SOC_DAIFMT_NB_NF |
                              SND_SOC_DAIFMT_CBM_CFM);
    if (ret < 0)
        return ret;

    /* set the codec PLL */
    ret = snd_soc_dai_set_pll(codec_dai, WM8994_FLL1, 0,
                          pll_out, params_rate(params) * 256);
    if (ret < 0)
        return ret;

    /* set the codec system clock */
    ret = snd_soc_dai_set_sysclk(codec_dai, WM8994_SYSCLK_FLL1,
                params_rate(params) * 256, SND_SOC_CLOCK_IN);
    if (ret < 0)
        return ret;

    return 0;
}
```

In the preceding implementation of the `foo_hw_params()` function, we can see how both codec and platform DAIs are configured, with both format and clock settings. Now we come to the last step of machine driver implementation, which consists of registering the audio sound card, which is the device through which audio operations on the system are performed.

Sound card registration

A sound card is represented in the kernel as an instance of `struct snd_soc_card`, defined as follows:

```
struct snd_soc_card {
    const char *name;
```

```
struct module *owner;
[...]
/* callbacks */
int (*set_bias_level)(struct snd_soc_card *,
                      struct snd_soc_dapm_context *dapm,
                      enum snd_soc_bias_level level);
int (*set_bias_level_post)(struct snd_soc_card *,
                           struct snd_soc_dapm_context *dapm,
                           enum snd_soc_bias_level level);
[...]
/* CPU <--> Codec DAI links    */
struct snd_soc_dai_link *dai_link;
int num_links;
const struct snd_kcontrol_new *controls;
int num_controls;

const struct snd_soc_dapm_widget *dapm_widgets;
int num_dapm_widgets;
const struct snd_soc_dapm_route *dapm_routes;
int num_dapm_routes;
const struct snd_soc_dapm_widget *of_dapm_widgets;
int num_of_dapm_widgets;
const struct snd_soc_dapm_route *of_dapm_routes;
int num_of_dapm_routes;
[...]
};
```

For the sake of readability, only the relevant field has been listed, and the full definition can be found at https://elixir.bootlin.com/linux/v4.19/source/include/sound/soc.h#L1010. That being said, the following list describes the fields we have listed:

- name is the name of the sound card.

- owner is the module owner for this sound card.

- `dai_link` is the array of DAI links this sound card is made of, and `num_links` specifies the number of entries in the array.

- `controls` is an array that contains the controls that are statically defined and set by the machine driver, and `num_controls` specifies the number of entries in the array.

- `dapm_widgets` is an array that contains the DAPM widgets that are statically defined and set by the machine driver, and `num_dapm_widgets` specifies the number of entries in the array.

- `damp_routes` is an array that contains the DAPM routes that are statically defined and set by the machine driver, and `num_dapm_routes` specifies the number of entries in the array.

- `of_dapm_widgets` represents the DAPM widgets fed from the DT (via `snd_soc_of_parse_audio_simple_widgets()`), and `num_of_dapm_widgets` is the actual number of widget entries.

- `of_dapm_routes` represents the DAPM routes fed from the DT (via `snd_soc_of_parse_audio_routing()`), and `num_of_dapm_routes` is the actual number of route entries.

After the sound card structure has been set up, it can be registered by the machine using the `devm_snd_soc_register_card()` method, whose prototype is as follows:

```
int devm_snd_soc_register_card(struct device *dev,
                               struct snd_soc_card *card);
```

In the preceding prototype, `dev` represents the underlying device used to manage the card, and `card` is the actual sound card data structure that was set up previously. This function returns 0 on success. However, when this function is called, every component driver and DAI driver will be probed. As a result, the `component_driver->probe()` and `dai_driver->probe()` methods will be invoked for both the CPU and CODEC. Additionally, a new PCM device will be created for each successfully probed DAI link.

The following excerpts (from a Rockchip machine ASoC driver for boards using a MAX90809 CODEC, implemented in `sound/soc/rockchip/rockchip_max98090.c` in kernel sources) will show the entire sound card creation, from widgets to routes, through DAI link configurations. Let's start by defining a widget and control for this machine, as well as the callback, which is used to configure the CPU and codec DAIs:

```
static const struct snd_soc_dapm_widget rk_dapm_widgets[] = {
    [...]
};
```

```
static const struct snd_soc_dapm_route rk_audio_map[] = {
    [...]
};


static const struct snd_kcontrol_new rk_mc_controls[] = {
    SOC_DAPM_PIN_SWITCH("Headphone"),
    SOC_DAPM_PIN_SWITCH("Headset Mic"),
    SOC_DAPM_PIN_SWITCH("Int Mic"),
    SOC_DAPM_PIN_SWITCH("Speaker"),
};


static const struct snd_soc_ops rk_aif1_ops = {
    .hw_params = rk_aif1_hw_params,
};


static struct snd_soc_dai_link rk_dailink = {
    .name = "max98090",
    .stream_name = "Audio",
    .codec_dai_name = "HiFi",
    .ops = &rk_aif1_ops,
    /* set max98090 as slave */
    .dai_fmt = SND_SOC_DAIFMT_I2S | SND_SOC_DAIFMT_NB_NF |
               SND_SOC_DAIFMT_CBS_CFS,
};
```

In the preceding excerpt, `rk_aif1_hw_params` can be seen in the original code implementation file. Now comes the data structure, which is used to build the sound card, defined as follows:

```
static struct snd_soc_card snd_soc_card_rk = {
    .name = "ROCKCHIP-I2S",
    .owner = THIS_MODULE,
    .dai_link = &rk_dailink,
    .num_links = 1,
    .dapm_widgets = rk_dapm_widgets,
    .num_dapm_widgets = ARRAY_SIZE(rk_dapm_widgets),
```

```
    .dapm_routes = rk_audio_map,
    .num_dapm_routes = ARRAY_SIZE(rk_audio_map),
    .controls = rk_mc_controls,
    .num_controls = ARRAY_SIZE(rk_mc_controls),
};
```

This sound card is finally created in the driver `probe` method as follows:

```
static int snd_rk_mc_probe(struct platform_device *pdev)
{
    int ret = 0;
    struct snd_soc_card *card = &snd_soc_card_rk;
    struct device_node *np = pdev->dev.of_node;
[...]
    card->dev = &pdev->dev;
    /* Assign codec, cpu and platform node */
    rk_dailink.codec_of_node = of_parse_phandle(np,
                                "rockchip,audio-codec", 0);
    rk_dailink.cpu_of_node = of_parse_phandle(np,
                                "rockchip,i2s-controller", 0);
    rk_dailink.platform_of_node = rk_dailink.cpu_of_node;
[...]
    ret = snd_soc_of_parse_card_name(card, "rockchip,model");
    ret = devm_snd_soc_register_card(&pdev->dev, card);
[...]
}
```

Once again, the three preceding code blocks are excerpts from `sound/soc/rockchip/rockchip_max98090.c`. So far, we have learned the main purpose of machine drivers, which is to bind Codec and CPU drivers together and to define the audio path. That being said, there are cases when we might need even less code. Such cases concern boards where neither the CPU nor the Codecs need special hacks before being bound together. In this case, the ASoC framework provides the **simple-card machine driver**, introduced in the next section.

Leveraging the simple-card machine driver

There are cases when your board does not require any hacks from the Codec nor the CPU DAI. The ASoC core provides the `simple-audio` machine driver, which can be used to describe a whole sound card from the DT. The following is an excerpt of such a node:

```
sound {
    compatible ="simple-audio-card";
    simple-audio-card,name ="VF610-Tower-Sound-Card";
    simple-audio-card,format ="left_j";
    simple-audio-card,bitclock-master = <&dailink0_master>;
    simple-audio-card,frame-master = <&dailink0_master>;
    simple-audio-card,widgets ="Microphone","Microphone Jack",
                               "Headphone","Headphone Jack",
                               "Speaker","External Speaker";
    simple-audio-card,routing = "MIC_IN","Microphone Jack",
                               "Headphone Jack","HP_OUT",
                               "External Speaker","LINE_OUT";

    simple-audio-card,cpu {
        sound-dai = <&sh_fsi20>;
    };
    dailink0_master: simple-audio-card,codec {
        sound-dai = <&ak4648>;
        clocks = <&osc>;
    };
};
```

This is fully documented in `Documentation/devicetree/bindings/sound/simple-card.txt`. In the preceding excerpt, we can see machine widgets and route maps being specified, as well as both the codec and the CPU nodes, which are referenced. Now that we are familiar with the simple-card machine driver, we can leverage it and try as much as possible not to write our own machine driver. Having said that, there are situations where the codec device can't be dissociated, and this changes the way the machine should be written. Such audio devices are called codec-less sound cards, and we discuss them in the next section.

Codec-less sound cards

There may be situations where digital audio data is sampled from an external system, such as when using the SPDIF interface, and the data is therefore preformatted. In this case, the sound card registration is the same, but the ASoC core needs to be aware of this particular case.

With the output, the DAI link object's `.capture_only` field should be `false`, while `.playback_only` should be `true`. The reverse should be done with the input. Additionally, the machine driver must set the DAI link's `codec_dai_name` and `codec_name` to `"snd-soc-dummy-dai"` and `"snd-soc-dummy"` respectively. This is, for example, the case for the `imx-spdif` machine driver (`sound/soc/fsl/imx-spdif.c`), which contains the following excerpt:

```
data->dai.name = "S/PDIF PCM";
data->dai.stream_name = "S/PDIF PCM";
data->dai.codecs->dai_name = "snd-soc-dummy-dai";
data->dai.codecs->name = "snd-soc-dummy";
data->dai.cpus->of_node = spdif_np;
data->dai.platforms->of_node = spdif_np;
data->dai.playback_only = true;
data->dai.capture_only = true;


if (of_property_read_bool(np, "spdif-out"))
    data->dai.capture_only = false;
if (of_property_read_bool(np, "spdif-in"))
    data->dai.playback_only = false;
if (data->dai.playback_only && data->dai.capture_only) {
    dev_err(&pdev->dev, "no enabled S/PDIF DAI link\n");
    goto end;
}
```

You can find the binding documentation of this driver in `Documentation/devicetree/bindings/sound/imx-audio-spdif.txt`. At the end of machine class driver study, we are done with the whole ASoC class driver development. In this machine class driver, in addition to bound CPU and Codec in the code, as well as providing a setup callback, we have seen how to avoid writing code by using the simple-card machine driver and implementing the rest in the device tree.

Summary

In this chapter, we have gone through the architecture of ASoC machine class drivers, which represents the last element in this ASoC series. We have learned how to bind platform and subdevice drivers, but also how to define routes for audio data.

In the next chapter, we will cover another Linux media subsystem, that is, V4L2, which is used to deal with video devices.

7
Demystifying V4L2 and Video Capture Device Drivers

Video has long been inherent in embedded systems. Given that Linux is the favorite kernel used in such systems, it goes without saying that it natively embeds its support for video. This is the so-called **V4L2**, which stands for **Video 4 (for) Linux 2**. Yes! *2* because there was a first version, *V4L*. V4L2 augments V4L with memory management features and other elements that make this framework as generic as possible. Through this framework, the Linux kernel is able to deal with camera devices and the bridge to which they are connected, as well as the associated DMA engines. These are not the only elements supported by V4L2. We will begin with an introduction to framework architecture, learning how it is organized, and walk through the main data structures it comprises. Then, we will learn how to design and write the bridge device driver, the one responsible for DMA operations, and finally, we will delve into sub-device drivers. That said, in this chapter, the following topics will be covered:

- Framework architecture and the main data structures
- Bridge video device drivers

- The concept of sub-devices
- V4L2 control infrastructure

Technical requirements

The following are prerequisites for this chapter:

- Advanced computer architecture knowledge and C programming skills

- Linux kernel v4.19.X sources, available at `https://git.kernel.org/pub/scm/linux/kernel/git/stable/linux.git/refs/tags`

Framework architecture and the main data structures

Video devices are becoming increasingly complex. In such devices, hardware often comprises several integrated IPs that need to cooperate with one another in a controlled manner, and this leads to complex V4L2 drivers. This requires figuring out the architecture prior to delving into the code and this is precisely the requirement that this section addresses.

It is known that drivers normally mirror the hardware model in programming. In the V4L2 context, the diverse IP components are modeled as software blocks called sub-devices. V4L2 sub-devices are usually kernel-only objects. Moreover, if the V4L2 driver implements the media device API (which we will discuss in the next chapter, *Chapter 8*, *Integrating with V4L2 Async and Media Controller Frameworks*), those sub-devices will automatically inherit from media entities, allowing applications to enumerate the sub-devices and to discover the hardware topology using the media framework's entities, pads, and link-related enumeration APIs.

Notwithstanding making sub-devices discoverable, drivers can likewise decide to make them configurable by applications in a straightforward manner. When both the sub-device driver and the V4L2 device driver uphold this, sub-devices will feature a character device node on which **ioctls** (**input/output controls**) can be invoked in order to query, read, and write sub-device capabilities (including controls), or to even negotiate image formats on individual sub-device pads.

At the driver level, V4L2 does a lot of work for the driver developer so that they just have to implement the hardware-related code and register the relevant device. Before going further, we must introduce several important structures that constitute the core of V4L2:

- `struct v4l2_device`: A hardware device may contain multiple child devices, such as a TV card in addition to a capture device, and possibly a VBI device or FM tuner. `v4l2_device` is the root node of all of these devices and is responsible for managing all child devices.

- `struct video_device`: The main purpose of this structure is to provide the well-known `/dev/videoX` or `/dev/v4l-subdevX` device nodes. This structure mainly abstracts the capture interface, also known as the **bridge interface** (bridge because it carries data from its data lines to the kernel memory). This will always be either part of the SoC or connected to high-speed buses such as PCI. Though sub-devices also inherit from this structure, they are not used in the same way as the bridge, but are limited to exposing their `/dev/v4l-subdevX` nodes and their file operations. From within the sub-device driver, only the core accesses this structure in the underlying sub-device.

- `struct vb2_queue`: For me, this is the main data structure in the video driver, as it is used in the real logic of data streaming and the center part of the DMA operations, along with `struct vb2_v4l2_buffer`.

- `struct v4l2_subdev`: This is the sub-device responsible for implementing specific functions and abstracting a specific function in the video system of the SoC.

`struct video_device` can be regarded as the base class for all devices and sub-devices. When we write our own drivers, access to this data structure may be direct (if we are dealing with a bridge driver) or indirect (if we are dealing with a sub-device, because sub-device APIs abstract and hide the underlying `struct video_device` embedded into each sub-device data structure).

Now we are aware of the data structures this framework is made of. Moreover, we introduced their relationships and their respective purposes. It is now time for us to go deeper into the details by introducing how to initialize and register a V4L2 device with the system.

Initializing and registering a V4L2 device

Prior to being used or part of the system, the V4L2 device must be initialized and registered, and this is the main topic of this section. Once the framework architecture description is complete, we can start going through the code. In this kernel, a V4L2 device is an instance of the `struct v4l2_device` structure. This is the highest data structure in the media framework, maintaining a list of sub-devices the media pipe is comprised of and acting as the parent of the bridge device. V4L2 drivers should include `<media/v4l2-device.h>`, which will bring in the following definition of a `struct v4l2_device`:

```
struct v4l2_device {
    struct device *dev;
    struct media_device *mdev;
    struct list_head subdevs;
    spinlock_t lock;
    char name[V4L2_DEVICE_NAME_SIZE];
    void (*notify)(struct v4l2_subdev *sd,
                   unsigned int notification, void *arg);
    struct v4l2_ctrl_handler *ctrl_handler;
    struct v4l2_prio_state prio;
    struct kref ref;
    void (*release)(struct v4l2_device *v4l2_dev);
};
```

Unlike other video-related data structures that we will introduce in the following sections, there are only a few fields in this structure. Their meanings are as follows:

- `dev` is a pointer to the parent `struct device` for this V4L2 device. This will be automatically set upon registration, and `dev->driver_data` will point to this v4l2 struct.

- `mdev` is a pointer to a `struct media_device` object to which this V4L2 device belongs. This field deals with the media controller framework and will be introduced in the related section. This may be NULL if integration with the media controller framework is not required.

- `subdevs` is the list of sub-devices for this V4L2 device.

- `lock` is the lock protecting access to this structure.

- name is a unique name for this V4L2 device. By default, it is derived from the driver name plus the bus ID.

- notify is a pointer to a notification callback, called by a sub-device to inform this V4L2 device of some events.

- ctrl_handler is the control handler associated with this device. It keeps track of all of the controls this V4L2 device has. This may be NULL if there are no controls.

- prio is the device's priority state.

- ref is internally used by the core for reference counting.

- release is the callback to be called when the last user of this structure goes off.

This top-level structure is initialized and registered with the core by the same function, v4l2_device_register(), whose prototype is the following:

```
int v4l2_device_register(struct device *dev,
                         struct v4l2_device *v4l2_dev);
```

The first dev argument is normally the struct device pointer of the bridge bus's related device-data structure. That is pci_dev, usb_device, or platform_device.

If the dev->driver_data field is NULL, this function will make it point to the actual v4l2_dev object being registered. Moreover, if v4l2_dev->name is empty, then it will be set to a value resulting from the concatenation of dev driver name + dev device name.

However, if the dev parameter is NULL, then you must set v4l2_dev->name before calling v4l2_device_register(). On the other hand, a previously registered V4L2 device can be unregistered using v4l2_device_unregister() as follows:

```
v4l2_device_unregister(struct v4l2_device *v4l2_dev);
```

Upon a call to this function, all sub-devices will be unregistered as well. This is all about the V4L2 device. However, you should keep in mind that it is the top-level structure, maintaining a list of sub-devices of the media device and acting as the parent of the bridge device.

Now that we are done with the main V4L2 device (the one that encompasses the other device-related data structures) initialization and registration, we can introduce specific device drivers, starting with the bridge driver, which is platform-specific.

Introducing video device drivers – the bridge driver

The bridge driver controls the platform /USB/PCI/... hardware that is responsible for the DMA transfers. This is the driver that handles data streaming from the device. One of the main data structures the bridge driver directly deals with is struct video_device. This structure embeds the entire element needed to perform video streaming, and one of its first interactions with the user space is to create device files in the /dev/ directory.

The struct video_device structure is defined in include/media/v4l2-dev.h, which means the driver code must contain #include <media/v4l2-dev.h>. The following is what this structure looks like from the header file where it is defined:

```
struct video_device
{
#if defined(CONFIG_MEDIA_CONTROLLER)
    struct media_entity entity;
    struct media_intf_devnode *intf_devnode;
    struct media_pipeline pipe;
#endif
    const struct v4l2_file_operations *fops;
    u32 device_caps;
    struct device dev; struct cdev *cdev;
    struct v4l2_device *v4l2_dev;
    struct device *dev_parent;
    struct v4l2_ctrl_handler *ctrl_handler;
    struct vb2_queue *queue;
    struct v4l2_prio_state *prio;
    char name[32];
    enum vfl_devnode_type vfl_type;
    enum vfl_devnode_direction vfl_dir;
    int minor;
    u16 num;
    unsigned long flags; int index;

    spinlock_t fh_lock;
    struct list_head fh_list;
```

```
    void (*release)(struct video_device *vdev);
    const struct v4l2_ioctl_ops *ioctl_ops;
    DECLARE_BITMAP(valid_ioctls, BASE_VIDIOC_PRIVATE);

    struct mutex *lock;
};
```

Not only does the bridge driver play with this structure – this structure is the main v4l2 structure when it comes to representing V4L2-compatible devices, including sub-devices. However, depending on the nature of the driver (be it a bridge driver or sub-device driver), some elements may vary or may be NULL. The following are descriptions of each element in the structure:

- entity, intf_node, and pipe are part of the integration with the media framework, as we will see in the section of the same name. The former abstracts the video device (which becomes an entity) from within the media framework, while intf_node represents the media interface device node, and pipe represents the streaming pipe to which the entity belongs.

- fops represents the file operations for the video device's file node. The V4L2 core overrides the virtual device file operation with some extra logic required by the subsystem.

- cdev is the character device structure, abstracting the underlying /dev/videoX file node. vdev->cdev->ops is set with v4l2_fops (defined in drivers/media/v4l2-core/v4l2-dev.c) by the V4L2 core. v4l2_fops is actually a generic (in term of ops implemented) and V4L2-oriented (in terms of what these ops do) file op assigned to each /dev/videoX char device and wraps the video device-specific ops defined in vdev->fops. At their return paths, each callback in v4l2_fops will call its counterpart in vdev->fops. v4l2_fops callbacks perform a sanity check prior to invoking the real ops in vdev->fops. For example, on a mmap() system call issued by the user space on a /dev/videoX file, v4l2_fops->mmap will be invoked first, which will make sure that vdev->fops->mmap is set prior to calling it and printing a debug message if needed.

- ctrl_handler: The default value is vdev->v4l2_dev->ctrl_handler.

- queue is the buffer management queue associated with this device node. This is one of the data structures only the bridge driver can play with. This may be NULL, especially when it comes to non-bridge video drivers (sub-devices, for example).

- `prio` is a pointer to `&struct v4l2_prio_state` with the device's priority state. If this state is `NULL`, then `v4l2_dev->prio` will be used.

- `name` is the name of the video device.

- `vfl_type` is the V4L device type. Possible values are defined by enum `vfl_devnode_type`, containing the following:

  - `VFL_TYPE_GRABBER`: For video input/output devices

  - `VFL_TYPE_VBI`: For vertical blank data (undecoded)

  - `VFL_TYPE_RADIO`: For radio cards

  - `VFL_TYPE_SUBDEV`: For V4L2 sub-devices

  - `VFL_TYPE_SDR`: Software-defined radio

  - `VFL_TYPE_TOUCH`: For touch sensors

- `vfl_dir` is a V4L receiver, transmitter, or memory-to-memory (denoted m2m or mem2mem) device. Possible values are defined by enum `vfl_devnode_direction`, containing the following:

  - `VFL_DIR_RX`: For capture devices

  - `VFL_DIR_TX`: For output devices

  - `VFL_DIR_M2M`: should be mem2mem devices (read mem-to-mem, and also known as memory-to-memory devices). A mem2mem device is a device that uses memory buffers passed by user space applications for both the source and destination. This is distinct from current and existing drivers that use memory buffers for only one of those at a time. Such a device would be of both the **OUTPUT** and **CAPTURE** types in terms of V4L2. Although no such devices are present in the V4L2 framework, a demand for such a model exists, for example, for 'resizer devices' or for the V4L2 loopback driver.

- `v4l2_dev` is the `v4l2_device` parent device of this video device.

- `dev_parent` is the device parent for this video device. If not set, the core will set it with `vdev->v4l2_dev->dev`.

- `ioctl_ops` is a pointer to `&struct v4l2_ioctl_ops`, which defines a set of ioctl callbacks.

- `release` is a callback called by the core when the last user of the video device exits. This must be non-`NULL`.

- lock is a mutex serializing access to this device. It is the principal serialization lock by means of which all ioctls are serialized. It is common for bridge drivers to set this field with the same mutex as the *queue->lock*, which is the lock for serializing access to the queue (serializing streaming). However, if *queue->lock* is set, then the streaming ioctls are serialized by that separate lock.

- num is the actual device node index assigned by the core. It corresponds to the *X* in /dev/videoX.

- flags are video device flags. You should use bit operations to set/clear/test flags. They contain a set of &enum v4l2_video_device_flags flags.

- fh_list is a list of struct v4l2_fh, which describes a V4L2 file handler, enabling tracking of the number of opened file handles for this video device. fh_lock is the lock associated with this list.

- class corresponds to the sysfs class. It is assigned by the core. This class entry corresponds to the /sys/video4linux/ sysfs directory.

Initializing and registering the video device

Prior to its registration, the video device can be allocated either dynamically using video_device_alloc() (which simply invokes kzalloc()), or statically embedded into a dynamically allocated structure, which is the device state structure most of time.

The video device is dynamically allocated using video_device_alloc(), as in the following example:

```
struct video_device * vdev;

vdev = video_device_alloc();
if (!vdev)
     return ERR_PTR(-ENOMEM);
vdev->release = video_device_release;
```

In the preceding excerpt, the last line provides the release method for the video device since the .release field must be non-NULL. The video_device_release() callback is provided by the kernel. It just calls kfree() to free the allocated memory.

When it is embedded into a device state structure, the code becomes as follows:

```
struct my_struct {
    [...]
    struct video_device vdev;
```

```
};

[...]
struct my_struct *my_dev;
struct video_device *vdev;

my_dev =  kzalloc(sizeof(struct my_struct), GFP_KERNEL);
if (!my_dev)
    return ERR_PTR(-ENOMEM);
vdev = &my_vdev->vdev;

/* Now work with vdev as our video_device struct */
vdev->release = video_device_release_empty;
[...]
```

Here, the video device must not be released alone as it is part of a bigger picture. When the video device is embedded into another structure, as in the preceding example, it does not require anything to be deallocated. At this point, since the release callback must be non-NULL, we can assign an empty function, such as `video_device_release_empty()`, also provided by the kernel.

We are done with allocation. At this point, we can use `video_register_device()` in order to register the video device. The following is the prototype of this function:

```
int video_register_device(struct video_device *vdev,
                          enum vfl_devnode_type type, int nr)
```

In the preceding prototype, `type` specifies the type of bridge device being registered. It will be assigned to the `vdev->vfl_type` field. In the remainder of the chapter, we will consider it set to `VFL_TYPE_GRABBER` since we are dealing with the video capture interface. `nr` is the desired device node number (*0 == /dev/video0, 1 == /dev/video1, ...*). However, setting its value to `-1` will instruct the kernel to pick the first free index and use it. Specifying a fixed index may be useful to build fancy *udev* rules since the device node name is known in advance. In order for the registration to succeed, the following requirements must be met:

- First, you *MUST* set the `vdev->release` function as it can't be empty. If you don't need it, you can pass the V4L2 core's empty release method.

- Second, you *MUST* set the `vdev->v4l2_dev` pointer; it should point to the V4L2 parent of the video device.

- Finally, but not mandatorily, you should set vdev->fops and vdev->ioctl_ops.

video_register_device() returns 0 when successful. However, it might fail if there is no free minor, if the device node number could be found, or if the registration of the device node failed. In either error case, it returns a negative error number. Each registered video device creates a directory entry in /sys/class/video4linux with some attributes inside.

> **Important note**
>
> Minor numbers are allocated dynamically unless the kernel is compiled with the kernel option CONFIG_VIDEO_FIXED_MINOR_RANGES. In that case, minor numbers are allocated in ranges depending on the device node type (video, radio, and so on), with a total limit for VIDEO_NUM_DEVICES, which is set to 256.

The vdev->release() callback will never be called if registration fails. In this case, you need to call video_device_release() to free the allocated video_device struct if it has been allocated dynamically, or free your own struct if the video_device was embedded in it.

On the unloading path of the driver, or when the video nodes are no longer needed, you should call video_unregister_device() on the video device in order to unregister it so that its nodes can be removed:

```
void video_unregister_device(struct video_device *vdev)
```

After the preceding call, the device sysfs entries will be removed, causing *udev* to remove nodes in /dev/.

So far, we have only discussed the simplest part of the registration process, but there are some complex fields in the video device that need to be initialized prior to registration. Those fields extend the driver capabilities by providing the video device file operations, a coherent set of ioctl callbacks, and, most importantly, the media's queue and memory management interface. We will discuss these in the forthcoming sections.

Video device file operations

The video device (by means of its driver) is meant to be exposed to the user space as a special file in the /dev/ directory, which the user space can use to interact with the underlying device: streaming the data. In order for the video device to be able to address user space queries (by means of system calls), a set of standard callbacks has to be implemented from within the driver. These callbacks form what are known today as **file operations**. The file operation structure for video devices is of the struct v4l2_file_operations type, defined in include/media/v4l2-dev.h as follows:

```
struct v4l2_file_operations {
    struct module *owner;
    ssize_t (*read) (struct file *file, char user *buf,
                        size_t, loff_t *ppos);
    ssize_t (*write) (struct file *file, const char user *buf,
                        size_t, loff_t *ppos);
    poll_t (*poll) (struct file *file,
                        struct poll_table_struct *);
    long (*unlocked_ioctl) (struct file *file,
                            unsigned int cmd, unsigned long arg);
#ifdef CONFIG_COMPAT
    long (*compat_ioctl32) (struct file *file,
                            unsigned int cmd, unsigned long arg);
#endif
    unsigned long (*get_unmapped_area) (struct file *file,
                                unsigned long, unsigned long,
                                unsigned long, unsigned long);
    int (*mmap) (struct file *file,
                    struct vm_area_struct *vma);
    int (*open) (struct file *file);
    int (*release) (struct file *file);
};
```

These can be regarded as top-level callbacks as they are actually called (following a number of sanity checks, of course) by another low-level device file op associated with the vdev->cdev field this time, and which is set with vdev->cdev->ops = &v4l2_fops; upon file node creation. This allows the kernel to implement an extra logic and enforce sanity:

- owner is the pointer to the module. Most of the time, it is THIS_MODULE.

- open should contain operations needed to implement the open() system call. Most of the time, this could be set to v4l2_fh_open, which is a V4L2 helper that simply allocates and initializes a v4l2_fh struct and adds it to the vdev->fh_list list. However, if your device requires some extra initialization, perform your initialization inside, and then call v4l2_fh_open(struct file * filp). In any case, you *MUST* deal with v4l2_fh_open.

- release should contain operations needed to implement the close() system call. This callback must deal with v4l2_fh_release. It can be set to either of the following:

  – vb2_fop_release, which is a videobuf2-V4L2 release helper that will clean up any ongoing streaming. This helper will call v4l2_fh_release.

  – Your custom callback, undoing what has been done in .open, and which must call v4l2_fh_release either directly or indirectly (using the _vb2_fop_release() helper, for example, in order for the V4L2 core to handle the cleanup of any ongoing streaming).

- read should contain operations needed to implement the read() system call. Most of the time, the videobuf2-V4L2 helper vb2_fop_read is enough.

- write is not needed in our case as it is for an OUTPUT type device. However, using vb2_fop_write here does the job.

- unlocked_ioctl must be set to video_ioctl2 if you use v4l2_ioctl_ops. The next section explains this in detail. This V4L2 core helper is a wrapper around __video_do_ioctl(), which handles the real logic, and which routes each ioctl to the appropriate callback in vdev->ioctl_ops, which is where individual ioctl handlers are defined.

- mmap should contain operations needed to implement the mmap() system call. Most of the time, the videobuf2-V4L2 helper vb2_fop_mmap is enough, unless additional elements are required prior to performing mapping. Video buffers in the kernel (allocated in response to the VIDIOC_REQBUFS ioctl) have to be mapped individually prior to being accessed in the user space. This is the purpose of this .mmap callback, which just has to map one, and only one, video buffer to the user space. Information needed to map a buffer to a user space is queried to the kernel using the VIDIOC_QUERYBUF ioctl. Given the vma parameter, you can grab a pointer to the corresponding video buffer as follows:

```
struct vb2_queue *q = container_of_myqueue_wrapper();
unsigned long off = vma->vm_pgoff << PAGE_SHIFT;
struct vb2_buffer *vb;
unsigned int buffer = 0, plane = 0;


for (i = 0; i < q->num_buffers; i++) {
    struct vb2_buffer *buf = q->bufs[i];


    /* The below assume we are on a single-planar system,
     * else we would have loop over each plane
     */
    if (buf->planes[0].m.offset == off)
        break;
    return i;
}
videobuf_queue_unlock(myqueue);
```

- poll should contain operations needed to implement the poll() system call. Most of the time, the videobuf2-V4L2 helper vb2_fop_call is enough. If this helper doesn't know how to lock (neither queue->lock nor vdev->lock are set), then you shouldn't be using it, but you should write your own, which can rely on the vb2_poll() helper that does not handle locking.

In either of these callbacks, you can use the v4l2_fh_is_singular_file() helper in order to check whether the given file is the only file handle opened for the associated video_device. Its alternative is v4l2_fh_is_singular(), which relies on v4l2_fh this time:

```
int v4l2_fh_is_singular_file(struct file *filp)
int v4l2_fh_is_singular(struct v4l2_fh *fh);
```

To summarize, the following is what a capture video device driver's file operation may look like:

```c
static int foo_vdev_open(struct file *file)
{
    struct mydev_state_struct *foo_dev = video_drvdata(file);
    int ret;
[...]
    if (!v4l2_fh_is_singular_file(file))
        goto fh_rel;
[...]
fh_rel:
    if (ret)
        v4l2_fh_release(file);


    return ret;
}

static int foo_vdev_release(struct file *file)
{
    struct mydev_state_struct *foo_dev = video_drvdata(file);
    bool fh_singular;
    int ret;
[...]
    fh_singular = v4l2_fh_is_singular_file(file);
    ret = _vb2_fop_release(file, NULL);

    if (fh_singular)
        /* do something */
        [...]
    return ret;
}

static const struct v4l2_file_operations foo_fops = {
    .owner = THIS_MODULE,
    .open = foo_vdev_open,
    .release = foo_vdev_release,
```

```
    .unlocked_ioctl = video_ioctl2,
    .poll = vb2_fop_poll,
    .mmap = vb2_fop_mmap,
    .read = vb2_fop_read,
};
```

You can observe that in the preceding block, we used only standard core helpers in our file operations.

> **Important note**
>
> Mem2mem devices may use their related v4l2-mem2mem-based helpers. Have a look at `drivers/media/v4l2-core/v4l2-mem2mem.c`.

V4L2 ioctl handling

Let's talk a bit more about the `v4l2_file_operations.unlocked_ioctl` callback. As we have seen in the previous section, it should be set to `video_ioctl2`. `video_ioctl2` takes care of argument copying between the kernel and user space and performs some sanity checks (for example, whether the ioctl command is valid) prior to dispatching each individual `ioctl()` call to the driver, which ends up in a callback entry in the `video_device->ioctl_ops` field, which is of the `struct v4l2_ioctl_ops` type.

The `struct v4l2_ioctl_ops` structure contains callbacks for every possible ioctl in the V4L2 framework. However, you should only set these depending on the type of your device and the capability of the driver. Each callback in the structure maps an ioctl, and the structure is defined as follows:

```
struct v4l2_ioctl_ops {
    /* VIDIOC_QUERYCAP handler */
    int (*vidioc_querycap)(struct file *file, void *fh,
                             struct v4l2_capability *cap);
    /* Buffer handlers */
    int (*vidioc_reqbufs)(struct file *file, void *fh,
                             struct v4l2_requestbuffers *b);
    int (*vidioc_querybuf)(struct file *file, void *fh,
                             struct v4l2_buffer *b);
    int (*vidioc_qbuf)(struct file *file, void *fh,
                          struct v4l2_buffer *b);
    int (*vidioc_expbuf)(struct file *file, void *fh,
```

```
                           struct v4l2_exportbuffer *e);
   int (*vidioc_dqbuf)(struct file *file, void *fh,
                           struct v4l2_buffer *b);

   int (*vidioc_create_bufs)(struct file *file, void *fh,
                               struct v4l2_create_buffers *b);
   int (*vidioc_prepare_buf)(struct file *file, void *fh,
                               struct v4l2_buffer *b);

   int (*vidioc_overlay)(struct file *file, void *fh,
                           unsigned int i);
[...]
};
```

This structure has more than 120 entries describing operations for each and every possible V4L2 ioctl, whatever the device type is. In the preceding excerpt, only those that may be of interest to us are listed. We will not introduce callbacks into this structure. However, when you reach *Chapter 9, Leveraging V4L2 API from the User Space*, I encourage you to come back to this structure and things will be clearer.

That said, because you have provided a callback, it remains accessible. There are situations where you may want a callback that you had specified in v4l2_ioctl_ops to be ignored. This tends to be needed if, based on external factors (for example, which card is being used), you want to turn off certain features in v4l2_ioctl_ops without having to make a new struct. In order for the core to be aware of that and to ignore the callback, you should call v4l2_disable_ioctl() on the ioctl commands in question before video_register_device() is called:

```
 v4l2_disable_ioctl (vdev, cmd)
```

The following is an example: v4l2_disable_ioctl(&tea->vd, VIDIOC_S_HW_FREQ_SEEK);. The previous call will mark the VIDIOC_S_HW_FREQ_SEEK ioctl to be ignored on the tea->vd video device.

The videobuf2 interface and APIs

The videobuf2 framework is used to connect the V4L2 driver layer to the user space layer, providing a channel for data exchange that can allocate and manage video frame data. The videobuf2 memory management backend is fully modular. This allows custom memory management routines for devices and platforms with non-standard memory management requirements to be plugged in, without changing the high-level buffer management functions and API. The framework provides the following:

- Implementation of streaming I/O V4L2 ioctls and file operations
- High-level video buffer, video queue, and state management functions
- Video buffer memory allocation and management

Videobuf2 (or just vb2) facilitates driver development, reduces the code size of drivers, and aids in the proper and consistent implementation of the V4L2 API in drivers. V4L2 drivers are then charged with the task of acquiring video data from a sensor (usually via some sort of DMA controller) and feeding to the buffer managed by the vb2 framework.

This framework implements many ioctl functions, including buffer allocation, enqueue, dequeue, and data flow control. It then deprecates any vendor-specific solutions, reducing significantly the media framework code size and easing efforts required to write V4L2 device drivers.

> **Important note**
>
> Every videobuf2 helper, API, and data structure is prefixed with vb2_, while the version 1 (videobuf, defined in `drivers/media/v4l2-core/videobuf-core.c`) counterpart used the `videobuf_` prefix.

This framework includes a number of concepts that may be familiar to some of you, but that need to be discussed in detail nevertheless.

Concept of buffers

A buffer is the unit of data exchanged in a single shot between vb2 and the user space. From the point of view of user space code, a V4L2 buffer represents the data corresponding to a video frame (in the case of a capture device, for example). Streaming entails exchanging buffers between the kernel and user spaces. vb2 uses the `struct vb2_buffer` data structure to describe a video buffer. This structure is defined in `include/media/videobuf2-core.h` as follows:

```
struct vb2_buffer {
    struct vb2_queue *vb2_queue;
    unsigned int index;
    unsigned int type;
    unsigned int memory;
    unsigned int num_planes;
    u64 timestamp;
    /* private: internal use only
     *
     * state: current buffer state; do not change
     * queued_entry: entry on the queued buffers list, which
     * holds all buffers queued from userspace
     * done_entry: entry on the list that stores all buffers
     * ready to be dequeued to userspace
     * vb2_plane: per-plane information; do not change
     */
    enum vb2_buffer_state state;

    struct vb2_plane planes[VB2_MAX_PLANES];
    struct list_head queued_entry;
    struct list_head done_entry;
[...]
};
```

In the preceding data structure, those fields of no interest to us have been removed. The remaining fields are defined as follows:

- vb2_queue is the vb2 queue to which this buffer belongs. This will lead us to the next section, where we introduce the concept of queues according to videobuf2.

- index is the ID for this buffer.

- `type` is the type of buffer. It is set by `vb2` at the time of allocation. It matches the type of queue it belongs to: `vb->type = q->type`.

- `memory` is the type of memory model used to make the buffers visible on user spaces. The value of this field is of the `enum vb2_memory` type, which matches its V4L2 user space counterpart, `enum v4l2_memory`. This field is set by `vb2` at the time of buffer allocation and reports the `vb2` equivalent of the user space value assigned to the `.memory` field of `v4l2_requestbuffers` given to `vIDIOC_REQBUFS`. Possible values include the following:

    – `VB2_MEMORY_MMAP`: Its equivalent that is assigned in user space is `V4L2_MEMORY_MMAP`, which indicates that the buffer is used for memory mapping I/O.

    – `VB2_MEMORY_USERPTR`: Its equivalent that is assigned in user space is `V4L2_MEMORY_USERPTR`, indicating that the user allocates buffers in the user space, and passes a pointer via the `buf.m.userptr` member of `v4l2_buffer`. The purpose of `USERPTR` in V4L2 is to allow users to pass buffers allocated in user space directly by `malloc()` or statically.

    – `VB2_MEMORY_DMABUF`. Its equivalent that is assigned in user space is `V4L2_MEMORY_DMABUF`, indicating that the memory is allocated by the driver and exported as a DMABUF file handler. This DMABUF file handler may be imported in another driver.

- `state` is of the `enum vb2_buffer_state` type and represents the current state of this video buffer. Drivers can use the `void vb2_buffer_done(struct vb2_buffer *vb, enum vb2_buffer_state state)` API in order to change this state. Possible state values include the following:

    – `VB2_BUF_STATE_DEQUEUED` means the buffer is under user space control. It is set by the videobuf2 core in the execution path of the `VIDIOC_REQBUFS` ioctl.

    – `VB2_BUF_STATE_PREPARING` means the buffer is being prepared in videobuf2. This flag is set by the videobuf2 core in the execution path of the `VIDIOC_PREPARE_BUF` ioctl for drivers supporting it.

    – `VB2_BUF_STATE_QUEUED` means the buffer is queued in videobuf, but not yet in the driver. This is set by the videobuf2 core in the execution path of the `VIDIOC_QBUF` ioctl. However, the driver must set the state of all buffers to `VB2_BUF_STATE_QUEUED` if it fails to start streaming. This is the equivalent of returning the buffer back to videobuf2.

– VB2_BUF_STATE_ACTIVE means the buffer is actually queued in the driver and possibly used in a hardware operation (DMA, for example). There is no need for the driver to set this flag as it is set by the core right before calling the buffer .buf_queue callback.

– VB2_BUF_STATE_DONE means the driver should set this flag on the success path of the DMA operation on this buffer in order to pass the buffer to vb2. This means to the videobuf2 core that the buffer is returned from the driver to videobuf, but is not yet dequeued to the user space.

– VB2_BUF_STATE_ERROR is the same as the above, but the operation on the buffer has ended with an error, which will be reported to the user space when it is dequeued.

If the concept of buffer skills appears complex to you after this, then I encourage you to first read *Chapter 9*, *Leveraging V4L2 API from the User Space*, prior to coming back here.

The concept of planes

There are devices that require data for each input or output video frame to be placed in discontiguous memory buffers. In such cases, one video frame has to be addressed using more than one memory address, in other words, one pointer per "plane." A plane is a sub-buffer of the current frame (or a chunk of the frame).

Thus, on a single-planar system, a plane represents a whole video frame, whereas it represents only a chunk of the video frame in a multi-planar system. Because memory is discontiguous, multi-planar devices use Scatter/Gather DMA.

The concept of queue

The queue is the central element of streaming and is the DMA engine-related part of the bridge driver. In fact, it is the element through which the driver introduces itself to videobuf2. It helps us to implement the data flow management module in the driver. A queue is represented through the following structure:

```
struct vb2_queue {
    unsigned int type;
    unsigned int io_modes;
    struct device *dev;
    struct mutex *lock;
    const struct vb2_ops *ops;
    const struct vb2_mem_ops *mem_ops;
    const struct vb2_buf_ops *buf_ops;
```

```
    u32 min_buffers_needed;
    gfp_t gfp_flags;
    void *drv_priv;
    struct vb2_buffer *bufs[VB2_MAX_FRAME];
    unsigned int num_buffers;


    /* Lots of private and debug stuff omitted */
    [...]
};
```

The structure should be zeroed, and the preceding fields filled in. The following are the meanings of each element in the structure:

- `type` is the buffer type. This should be set with one of the values present in `enum v4l2_buf_type`, defined in `include/uapi/linux/videodev2.h`. This must be `V4L2_BUF_TYPE_VIDEO_CAPTURE` in our case.

- `io_modes` is a bitmask describing what types of buffers can be handled. Possible values include the following:

  – `VB2_MMAP`: Buffers allocated within the kernel and accessed via `mmap()`; vmalloc'ed and contiguous DMA buffers will usually be of this type.

  – `VB2_USERPTR`: This is for buffers allocated in user space. Normally, only devices that can do Scatter/Gather I/O can deal with user space buffers. However, contiguous I/O to huge pages is not supported. Interestingly, videobuf2 supports contiguous buffers allocated by the user space. The only way to get those, though, is to use some sort of special mechanism, such as the out-of-tree Android `pmem` driver.

  – `VB2_READ`, `VB2_WRITE`: These are user space buffers provided via the `read()` and `write()` system calls.

- `lock` is the mutex for serialization locks for the streaming ioctls. It is common to set this lock with the same mutex as `video_device->lock`, which is the main serialization lock. However, if some of the non-streaming ioctls were to take a long time to execute, then you might want to have a different lock here to prevent `VIDIOC_DQBUF` from being blocked while waiting for another action to finish.

- `ops` represents the driver-specific callbacks to set up this queue and control streaming operations. It is of the `struct vb2_ops` type. We will examine this structure in detail in the next section.

- The mem_ops field is where the driver tells videobuf2 what kind of buffers it is actually using; it should be set to one of vb2_vmalloc_memops, vb2_dma_contig_memops, or vb2_dma_sg_memops. These are the three basic types of buffer allocation videobuf2 implements:

  – The first one is the **vmalloc buffers** allocator, by means of which the memory for buffers is allocated with vmalloc() and is thus virtually contiguous in the kernel space and is not guaranteed to be physically contiguous.

  – The second one is the **contiguous DMA buffers** allocator, by means of which the memory is physically contiguous in memory, usually because the hardware cannot perform DMA to any other type of buffer. This allocator is backed by coherent DMA allocation.

  – The final one is the **S/G DMA buffers** allocator, where buffers are scattered through the memory. This is the way to go if the hardware can perform Scatter/Gather DMA. Obviously, this involves streaming DMA.

  Depending on the type of memory allocator used, the driver should include one of the following three headers:

  ```
  /* => vb_queue->mem_ops = &vb2_vmalloc_memops;*/
  #include <media/videobuf2-vmalloc.h>


  /* => vb_queue->mem_ops = &vb2_dma_contig_memops; */
  #include <media/videobuf2-dma-contig.h>


  /* => vb_queue->mem_ops = &vb2_dma_sg_memops; */
  #include <media/videobuf2-dma-sg.h>
  ```

  So far, there have not been situations where none of the existing allocators can do the job for a device. However, if this arises, the driver author could create a custom set of operations through vb2_mem_ops in order to meet that need. There are no limits.

- You may not care about buf_ops as it is provided by the vb2 core if not set. However, it contains callbacks to deliver buffer information between the user space and kernel space.

- min_buffers_needed is the minimum number of buffers needed before you can start streaming. If that is non-zero, then vb2_queue->ops->start_streaming won't be called until at least that many buffers have been queued up by the user space. In other words, it represents the number of available buffers the DMA engine needs to have before it can be started.

- **bufs** is an array of pointers to buffers in this queue. Its maximum is VB2_MAX_
  FRAME, which corresponds to the maximum number of buffers allowed per queue
  by the vb2 core. It is set to 32, which is already a quite considerable value.

- **num_buffers** is the number of allocated/used buffers in the queue.

Driver-specific streaming callbacks

A bridge driver needs to expose a collection of functions for managing buffer queues,
including queue and buffer initialization. These functions will handle buffer allocation,
queueing, and streaming-related requests coming from the user space. This can be done
by setting up an instance of struct vb2_ops, defined as follows:

```
struct vb2_ops {
    int (*queue_setup)(struct vb2_queue *q,
                        unsigned int *num_buffers,
                        unsigned int *num_planes,

                        unsigned int sizes[],
                        struct device *alloc_devs[]);


    void (*wait_prepare)(struct vb2_queue *q);
    void (*wait_finish)(struct vb2_queue *q);
    int (*buf_init)(struct vb2_buffer *vb);
    int (*buf_prepare)(struct vb2_buffer *vb);
    void (*buf_finish)(struct vb2_buffer *vb);
    void (*buf_cleanup)(struct vb2_buffer *vb);
    int (*start_streaming)(struct vb2_queue *q,
                            unsigned int count);
    void (*stop_streaming)(struct vb2_queue *q);
    void (*buf_queue)(struct vb2_buffer *vb);
};
```

The following are the purpose of each callback in this structure:

- `queue_setup`: This callback function is called by the driver's `v4l2_ioctl_ops.vidioc_reqbufs()` method (in response to `VIDIOC_REQBUFS` and `VIDIOC_CREATE_BUFS` ioctls) to adjust the buffer count and size. This callback's goal is to inform videobuf2-core of how many buffers and planes per buffer it requires, as well as the size and allocator context for each plane. In other words, the chosen vb2 memory allocator calls this method for negotiating with the driver about the number of buffers and planes per buffer to be used during streaming. 3 is a good choice for the minimum number of buffers since most DMA engines need at least 2 buffers in the queue. The parameters of this callback are defined as follows:

  – `q` is the `vb2_queue` pointer.

  – `num_buffers` is the pointer to the number of buffers requested by the application. The driver should then set the granted number of buffers allocated in this `*num_buffers` field. Since this callback can be called twice during the negotiation process, you should check `queue->num_buffers` to have an idea of the number of buffers already allocated prior to setting this.

  – `num_planes` contains the number of distinct video planes needed to hold a frame. This should be set by the driver.

  – `sizes` contains the size (in bytes) of each plane. For a single-planar system, only `size[0]` should be set.

  – `alloc_devs` is an optional per-plane allocator-specific device array. Consider it as a pointer to the allocation context.

The following is an example of the `queue_setup` callback:

```
/* Setup vb_queue minimum buffer requirements */
static int rcar_drif_queue_setup(struct vb2_queue *vq,
                                 unsigned int *num_buffers,
                                 unsigned int *num_planes,
                                 unsigned int sizes[],
                                 struct device *alloc_devs[])
{
    struct rcar_drif_sdr *sdr = vb2_get_drv_priv(vq);

    /* Need at least 16 buffers */
    if (vq->num_buffers + *num_buffers < 16)
        *num_buffers = 16 - vq->num_buffers;
```

```
    *num_planes = 1;
    sizes[0] = PAGE_ALIGN(sdr->fmt->buffersize);
    rdrif_dbg(sdr, "num_bufs %d sizes[0] %d\n",
              *num_buffers, sizes[0]);


    return 0;
}
```

- buf_init is called once on a buffer after memory has been allocated for it, or after a new USERPTR buffer is queued. This can be used, for example, to pin pages, verify contiguity, and set up IOMMU mappings.

- buf_prepare is called on the execution path of the VIDIOC_QBUF ioctl. It should prepare the buffer for queueing to the DMA engine. The buffer is prepared and the user space virtual address or user address is converted into a physical address.

- buf_finish is called on each DQBUF ioctl. It can be used, for example, for cache syncing and copying back from bounce buffers.

- buf_cleanup is called before freeing/releasing memory. It can be used for unmapping memory and suchlike.

- buf_queue: The videobuf2 core sets the VB2_BUF_STATE_ACTIVE flag in the buffer right before invoking this callback. However, it is invoked on behalf of the VIDIOC_QBUF ioctl. The user space queues buffers one by one, one after the other. Moreover, buffers may be queued faster than the bridge device grabs data from the capture device to the buffer. In the meantime, VIDIOC_QBUF may be called several times before issuing VIDIOC_DQBUF. It is recommended for the driver to maintain a list of buffers queued for DMA, so that in the event of any DMA completion, the filled buffer is moved off the list, given to the vb2 core at the same time by filling its timestamp and adding the buffer to the videobuf2's done buffers list, and DMA pointers are updated if necessary. Roughly speaking, this callback function should add a buffer to the driver DMA queue and start DMA on that buffer.
In the meantime, it is common for drivers to reimplement their own buffer data structure, built on top of the generic vb2_v4l2_buffer structure, but adding a list in order to address the queueing issue we just described. The following is an example of such a custom buffer data structure:

```
struct dcmi_buf {
    struct vb2_v4l2_buffer vb;
    dma_addr_t paddr; /* the bus address of this buffer */
```

```
        size_t size;
        struct list_head list; /* list entry for tracking
        buffers */
    };
```

- `start_streaming` starts the DMA engine for streaming. Prior to starting streaming, you must first check whether the minimum number of buffers have been queued. If not, you should return `-ENOBUFS` and the vb2 framework will call this function again the next time a buffer has been queued until enough buffers are available to actually start the DMA engine. You should also enable streaming on the sub-devices if the following is supported: `v4l2_subdev_call(subdev, video, s_stream, 1)`. You should get the next frame from the buffer queue and start DMA on it. Typically, interrupts happen after a new frame has been captured. It is the job of the handler to remove the new frame from the internal buffers (using the `list_del()`) list and give it back to the vb2 framework (by means of `vb2_buffer_done()`), updating the sequence counter field and timestamp at the same time.

- `stop_streaming` stops all pending DMA operations, stops the DMA engine, and releases DMA channel resources.
  You should also disable streaming on the sub-devices if the following is supported: `v4l2_subdev_call(subdev, video, s_stream, 0)`. Disable interruptions if necessary. Since the driver maintains a list of buffers queued for DMA, all buffers queued in that list must be returned to vb2 in the ERROR state.

Initializing and releasing the vb2 queue

In order for the driver to complete the queue initialization, it should call the `vb2_queue_init()` function, given the queue as an argument. However, the `vb2_queue` structure should first be allocated by the driver. Additionally, the driver must have cleared its content and set initial values for some requisite entries before calling this function. Those required values are `q->ops`, `q->mem_ops`, `q->type`, and `q->io_modes`. Otherwise, the queue initialization will fail, as shown in the following `vb2_core_queue_init()` function, which is invoked and whose return value is checked from within `vb2_queue_init()`:

```
int vb2_core_queue_init(struct vb2_queue *q)
{
    /*
     * Sanity check
     */
    if (WARN_ON(!q) || WARN_ON(!q->ops) ||
```

```
          WARN_ON(!q->mem_ops) ||
          WARN_ON(!q->type) || WARN_ON(!q->io_modes) ||
          WARN_ON(!q->ops->queue_setup) ||
          WARN_ON(!q->ops->buf_queue))
     return -EINVAL;

   INIT_LIST_HEAD(&q->queued_list);
   INIT_LIST_HEAD(&q->done_list);
   spin_lock_init(&q->done_lock);
   mutex_init(&q->mmap_lock);
   init_waitqueue_head(&q->done_wq);

   q->memory = VB2_MEMORY_UNKNOWN;

   if (q->buf_struct_size == 0)
       q->buf_struct_size = sizeof(struct vb2_buffer);
   if (q->bidirectional)
       q->dma_dir = DMA_BIDIRECTIONAL;
   else
       q->dma_dir = q->is_output ? DMA_TO_DEVICE :
       DMA_FROM_DEVICE;

   return 0;
}
```

The preceding excerpt shows the body of vb2_core_queue_init() in the kernel. This internal API is a pure basic initialization method that simply does some sanity checks and initializes basic data structures (lists, mutexes, and spinlocks).

The concept of sub-devices

In the early days of the V4L2 subsystem, there were just two main data structures:

- struct video_device: This is the structure through which /dev/<type>X appears.

- struct vb2_queue: This is responsible for buffer management.

This was enough in an era when there were not that many IP blocks embedded with the video bridge. Nowadays, image blocks in SoCs embed so many IP blocks, each of which plays a specific role by offloading specific tasks, such as image resizing, image converting, and video deinterlacing functionalities. In order to use a modular approach for addressing this diversity, the concept of the sub-device has been introduced. This brings a modular approach to the software modeling of the hardware, allowing to abstract each hardware component as a software block.

With this approach, each IP block (except the bridge device) participating in the processing pipe is seen as a sub-device, even the camera sensor itself. Whereas the bridge video device node has the `/dev/videoX` pattern, sub-devices on their side use the `/dev/v4l-subdevX` pattern (assuming they have the appropriate flag set prior to having their nodes created).

> **Important note**
>
> For a better understanding of the difference between the bridge device and the sub-devices, you can consider the bridge device as the final element in the processing pipeline, sometimes the one responsible for the DMA transactions. One example is the Atmel-**ISC** (**Image Sensor Controller**), extracted from its driver in `drivers/media/platform/atmel/atmel-isc.c`: Sensor-->PFE-->WB-->CFA-->CC-->GAM-->CSC-->CBC-->SUB-->RLP-->DMA. You are encouraged to have a look in this driver for the meaning of each element.

From a coding point of view, the driver should include `<media/v4l-subdev.h>`, which defines the `struct v4l2_subdev` structure, which is the abstraction data structure used to instantiate a sub-device in the kernel. This structure is defined as follows:

```
struct v4l2_subdev {
#if defined(CONFIG_MEDIA_CONTROLLER)
    struct media_entity entity;
#endif
    struct list_head list;
    struct module *owner;
    bool owner_v4l2_dev;
    u32 flags;
    struct v4l2_device *v4l2_dev;
    const struct v4l2_subdev_ops *ops;
    [...]
    struct v4l2_ctrl_handler *ctrl_handler;
```

```
        char name[V4L2_SUBDEV_NAME_SIZE];
        u32 grp_id; void *dev_priv;
        void *host_priv;
        struct video_device *devnode;
        struct device *dev;
        struct fwnode_handle *fwnode;
        struct device_node *of_node;
        struct list_head async_list;
        struct v4l2_async_subdev *asd;
        struct v4l2_async_notifier *notifier;
        struct v4l2_async_notifier *subdev_notifier;
        struct v4l2_subdev_platform_data *pdata;
};
```

The `entity` field of this structure will be discussed in the next chapter, *Chapter 8, Integrating with V4L2 Async and Media Controller Frameworks*. In the meantime, there are fields of no interest to us that have been removed.

However, the other fields in the structure are defined as follows:

- `list` is of the `list_head` type, and is used by the core to insert the current sub-device in the list of sub-devices maintained by `v4l2_device` to which it belongs.

- `owner` is set by the core and represents the module owning this structure.

- `flags` represents the sub-device flags that the driver can set and that can have the following values:

  – `V4L2_SUBDEV_FL_IS_I2C`: You should set this flag if this sub-device is actually an I2C device.

  – `V4L2_SUBDEV_FL_IS_SPI` should be set if this sub-device is an SPI device.

  – `V4L2_SUBDEV_FL_HAS_DEVNODE` should be set if the sub-device needs a device node (the famous `/dev/v4l-subdevX` entry). An API using this flag is `v4l2_device_register_subdev_nodes()`, which is discussed later and called by the bridge in order to create the sub-device node entries.

  – `V4L2_SUBDEV_FL_HAS_EVENTS` means this sub-device generates events.

- `v4l2_dev` is set by the core on the sub-device registration and is a pointer to the `struct 4l2_device` to which this sub-device belongs.

- `ops` is optional. This is a pointer to `struct v4l2_subdev_ops`, which represents a set of operations and which should be set by the driver to provide the callbacks the core can rely upon for this sub-device.

- `ctrl_handler` is a pointer to `struct v4l2_ctrl_handler`. It represents the list of controls provided by this sub-device, as we will see in the *V4L2 controls infrastructure* section.

- `name` is a unique name for the sub-device. It should be set by the driver after the sub-device has been initialized. For the I2C variant's initialization, the default name assigned by the core is (`"%s %d-%04x"`, `driver->name`, `i2c_adapter_id(client->adapter)`, `client->addr`). When including the support of **media controller**, this name is used as the media entity name.

- `grp_id` is driver-specific and provided by the core when in asynchronous mode, and is used to group similar sub-devices.

- `dev_priv` is the pointer to the device's private data, if any.

- `host_priv` is a pointer to private data used by the device where the sub-device is attached.

- `devnode` is the device node for this sub-device, set by the core upon a call to `v4l2_device_register_subdev_nodes()`, not to be confused with the bridge device built on top of the same structure. You should keep in mind that every v4l2 element (be they sub-devices or bridges) is a video device.

- `dev` is the pointer to the physical device, if any. The driver can set this value using `void v4l2_set_subdevdata(struct v4l2_subdev *sd, void *p)` or can get it using `void *v4l2_get_subdevdata(const struct v4l2_subdev *sd)`.

- `fwnode` is the firmware node object handle for this sub-device. In older kernel versions, this member used to be `struct device_node *of_node` and pointed to the **device-tree (DT)** node of the sub-device. However, kernel developers found that it is better to use the generic `struct fwnode_handle`, as it allows switching to/from the device-tree node/acpi device according to which it is used on the platform. In other words, it is either `dev->of_node->fwnode` or `dev->fwnode`, whichever is non-NULL.

The `async_list`, `asd`, `subdev_notifier`, and `notifier` elements are part of the v4l2-async framework, as we will see in the next section. However, brief descriptions of these elements are provided here:

- `async_list`: When registered with the async core, this member is used by the core to link this sub-device to a global `subdev_list` (which is a list of orphan sub-devices that do not belong to any notifier, meaning this sub-device has been registered prior to its parent, the bridge) or to its parent bridge's `notifier->done` list. We discuss this in detail later in the next chapter, *Chapter 8, Integrating with V4L2 Async and Media Controller Frameworks*.

- `asd`: This field is of the `struct v4l2_async_subdev` type and abstracts this sub-device in the async core.

- `subdev_notifier`: This is the notifier implicitly registered by this sub-device in case it needs to be notified of the probing of some other sub-devices. It is commonly used on systems where the streaming pipeline involves several sub-devices, where the sub-device N needs to be notified of the probing of the sub-device N-1.

- `notifier`: This is set by the async core and corresponds to the notifier with which its underlying `.asd` async sub-device is matched.

- `pdata`: This is a common part of the sub-device platform data.

Sub-device initialization

Each sub-device driver must have a `struct v4l2_subdev` structure, either standalone or embedded in the larger and device-specific structure. The second case is recommended as it allows the device state to be tracked. The following is an example of a typical device-specific structure:

```
struct mychip_struct {
    struct v4l2_subdev sd;
[...]
    /* device speific fields*/
[...]
};
```

Prior to being accessed, a V4L2 sub-device need to be initialized using the `v4l2_subdev_init()` API. However, when it comes to sub-devices with an I2C- or SPI-based control interface (typically camera sensors), the kernel provides `v4l2_spi_subdev_init()` and `v4l2_i2c_subdev_init()` variants:

```
void v4l2_subdev_init(struct v4l2_subdev *sd,
                        const struct v4l2_subdev_ops *ops)

void v4l2_i2c_subdev_init(struct v4l2_subdev *sd,
                        struct i2c_client *client,
                        const struct v4l2_subdev_ops *ops)

void v4l2_spi_subdev_init(struct v4l2_subdev *sd,
                        struct spi_device *spi,
                        const struct v4l2_subdev_ops *ops)
```

All of these APIs take a pointer to the `struct v4l2_subdev` structure as a first argument. Registering our sub-device using our device-specific data structure would then appear as follows:

```
v4l2_i2c_subdev_init(&mychip_struct->sd, client, subdev_ops);
/*or*/
v4l2_subdev_init(&mychip_struct->sd, subdev_ops);
```

The spi/i2c variants wrap the `v4l2_subdev_init()` function. Additionally, they require the underlying low-level, bus-specific structure as a second argument. Moreover, these bus-specific variants will store the sub-device object (given as the first argument) as low-level, bus-specific device data and vice versa by storing the low-level, bus-specific structure as the sub-device's private data. This way, `i2c_client` (or `spi_device`) and `v4l2_subdev` point to one another, meaning that by having a pointer to the I2C client, for example, you can call `i2c_set_clientdata()` (such as `struct v4l2_subdev *sd = i2c_get_clientdata(client);`) in order to grab the pointer to our internal sub-device object, and use the `container_of` macro (such as `struct mychip_struct *foo = container_of(sd, struct mychip_struct, sd);`) in order to grab the pointer to the chip-specific structure. On the other hand, having a pointer to the sub-device object, you can use `v4l2_get_subdevdata()` in order to grab the underlying bus-specific structure.

Least but not last, these bus-specific variants will mangle the sub-device name, as explained when introducing the `struct v4l2_subdev` data structure. An excerpt of `v4l2_i2c_subdev_init()` can provide a better understanding of this:

```
void v4l2_i2c_subdev_init(struct v4l2_subdev *sd,
                          struct i2c_client *client,
                          const struct v4l2_subdev_ops *ops)
{
    v4l2_subdev_init(sd, ops);
    sd->flags |= V4L2_SUBDEV_FL_IS_I2C;

    /* the owner is the same as the i2c_client's driver owner */
    sd->owner = client->dev.driver->owner;
    sd->dev = &client->dev;

    /* i2c_client and v4l2_subdev point to one another */
    v4l2_set_subdevdata(sd, client);
    i2c_set_clientdata(client, sd);

    /* initialize name */
    snprintf(sd->name, sizeof(sd->name),
            "%s %d-%04x", client->dev.driver->name,
            i2c_adapter_id(client->adapter), client->addr);
}
```

In each of the preceding three initialization APIs, `ops` is the last argument and is a pointer to a `struct v4l2_subdev_ops` representing operations exposed/supported by the sub-device. However, let's discuss this in the next section.

Sub-device operations

Sub-devices are devices that are somehow connected to the main bridge device. In the whole media device, each IP (sub-device) has its set of functionalities. These functionalities have to be exposed to the core by means of callbacks well defined by kernel developers for commonly used functionalities. This is the purpose of `struct v4l2_subdev_ops`.

However, some sub-devices can perform so many different and unrelated things that even `struct v4l2_subdev_ops` has been split into small and categorized coherent sub-structure ops, each gathering related functionalities so that `struct v4l2_subdev_ops` becomes the top-level ops structure, described as follows:

```
struct v4l2_subdev_ops {
    const struct v4l2_subdev_core_ops        *core;
    const struct v4l2_subdev_tuner_ops       *tuner;
    const struct v4l2_subdev_audio_ops       *audio;
    const struct v4l2_subdev_video_ops       *video;
    const struct v4l2_subdev_vbi_ops         *vbi;
    const struct v4l2_subdev_ir_ops          *ir;
    const struct v4l2_subdev_sensor_ops      *sensor;
    const struct v4l2_subdev_pad_ops         *pad;
};
```

> **Important note**
>
> Operations should only be provided for sub-devices exposed to the user space by an underlying char device file node. When registered, this device file node will have the same file operations as discussed earlier, that is, `v4l2_fops`. However, as we have seen earlier, these low-level ops only wrap (deal with) `video_device->fops`. Therefore, in order to reach `v4l2_subdev_ops`, the core uses `subdev->video_device->fops` as an intermediate and assigns it another file ops upon initialization (`subdev->vdev->fops = &v4l2_subdev_fops;`), which will wrap and call the real subdev ops. The call chain here is
>
> `v4l2_fops ==> v4l2_subdev_fops ==> our_custom_subdev_ops.`

You can see that the preceding top-level ops structure is made of pointers to category ops structures, which are as follows:

- `core` of the `v4l2_subdev_core_ops` type: This is the core ops category, providing generic callbacks, such as logging and debugging. It also allows the provision of additional and custom ioctls (especially useful if the ioctl does not fit in any category).

- `video` of the `v4l2_subdev_video_ops` type: `.s_stream` is called when streaming starts. It writes different configuration values to a camera's registers based on the chosen frame size and format.

- `pad` of the `v4l2_subdev_pad_ops` type: For cameras that support multiple frame sizes and image sample formats, these operations allow users to choose from the available options.

- `tuner`, `audio`, `vbi`, and `ir` are beyond the scope of this book.

- `sensor` of the `v4l2_subdev_sensor_ops` type: This covers camera sensor operations, typically for known buggy sensors that need some frames or lines to be skipped because they are corrupted.

Each callback in each category structure corresponds to an ioctl. The routing is actually done at a low level by `subdev_do_ioctl()`, defined in `drivers/media/v4l2-core/v4l2-subdev.c`, and which is indirectly called by `subdev_ioctl()`, which corresponds to `v4l2_subdev_fops.unlocked_ioctl`. The real call chain should be `v4l2_fops ==> v4l2_subdev_fops.unlocked_ioctl ==> our_custom_subdev_ops`.

The nature of this top-level `struct v4l2_subdev_ops` structure just confirms how wide the range of devices is that may be supported by V4L2. Ops categories that are of no interest to the sub-device driver can be left `NULL`. Do also note that the `.core` ops are common to all sub-devs. This does not mean it is mandatory; it merely means that any sub-device driver of whatever category is free to implement the `.core` ops as its callbacks are category-independent.

struct v4l2_subdev_core_ops

This structure implements generic callbacks and has the following definition:

```
struct v4l2_subdev_core_ops {
    int (*log_status)(struct v4l2_subdev *sd);
    int (*load_fw)(struct v4l2_subdev *sd);
    long (*ioctl)(struct v4l2_subdev *sd, unsigned int cmd,
```

```
                            void *arg);
[...]
#ifdef CONFIG_COMPAT
    long (*compat_ioctl32)(struct v4l2_subdev *sd,
                           unsigned int cmd,
                           unsigned long arg);
#endif
#ifdef CONFIG_VIDEO_ADV_DEBUG
    int (*g_register)(struct v4l2_subdev *sd,
                      struct v4l2_dbg_register *reg);
    int (*s_register)(struct v4l2_subdev *sd,
                      const struct v4l2_dbg_register *reg);
#endif
    int (*s_power)(struct v4l2_subdev *sd, int on);
    int (*interrupt_service_routine)(struct v4l2_subdev *sd,
                             u32 status,
                             bool *handled);
    int (*subscribe_event)(struct v4l2_subdev *sd,
                      struct v4l2_fh *fh,
                      struct v4l2_event_subscription *sub);
    int (*unsubscribe_event)(struct v4l2_subdev *sd,
                      struct v4l2_fh *fh,
                      struct v4l2_event_subscription *sub);
};
```

In the preceding structure, fields of no interest to us have been removed. Those remaining are defined as follows:

- `.log_status` is for logging purposes. You should use the `v4l2_info()` macro for this.

- `.s_power` puts the sub-device (the camera, for example) in power-saving mode (`on==0`) or normal operation mode (`on==1`).

- The `.load_fw` operation has to be called to load the sub-device's firmware.

- `.ioctl` should be defined if the sub-device provides extra ioctl commands.

- `.g_register` and `.s_register` are to be used for advanced debugging only and require the kernel config option `CONFIG_VIDEO_ADV_DEBUG` to be set. These operations allow the reading and writing of hardware registers in response to the `VIDIOC_DBG_G_REGISTER` and `VIDIOC_DBG_S_REGISTER` ioctls. The reg parameters (of the type `v4l2_dbg_register`, defined in `include/uapi/linux/videodev2.h`) are filled and given by the application.

- `.interrupt_service_routine` is called by the bridge from within its IRQ handler (it should use `v4l2_subdev_call` for this) when an interrupt status has been raised due to this sub-device, in order for the sub-device to handle the details. `handled` is an output parameter provided by the bridge driver, but has to be filled by the sub-device driver in order to inform (as *true or false*) on the result of its processing. We are in the IRQ context, so must not sleep. Sub-devices behind I2C/SPI buses should probably schedule their work in a threaded context.

- `.subscribe_event` and `.unsubscribe_event` are used to subscribe or unsubscribe to control change events. Please have a look at other V4L2 drivers implementing this to see how to implement yours.

struct v4l2_subdev_video_ops or struct v4l2_subdev_pad_ops

People often need to decide whether to implement `struct v4l2_subdev_video_ops` or `struct v4l2_subdev_pad_ops`, because some of the callbacks are redundant in both of these structures. The thing is, the callbacks of the `struct v4l2_subdev_video_ops` structure are used when the V4L2 device was opened in video mode, which includes TVs, camera sensors, and framebuffer. So far, so good. The concept of **pad** is tightly tied to the media controller framework. This means that as long as integration with the media controller framework is not needed, `struct v4l2_subdev_pad_ops` is not needed either. However, the media controller framework abstracts the sub-device by means of an entity object (we will see this later), which connects to other elements via PAD. In this case, it makes sense to use PAD-related functionalities instead of sub-device-related ones, hence, using `struct v4l2_subdev_pad_ops` instead of `struct v4l2_subdev_video_ops`.

Since we have not introduced the media framework yet, we are only interested in the `struct v4l2_subdev_video_ops` structure, which is defined as follows:

```
struct v4l2_subdev_video_ops {
    int (*querystd)(struct v4l2_subdev *sd, v4l2_std_id *std);
[...]
    int (*s_stream)(struct v4l2_subdev *sd, int enable);
    int (*g_frame_interval)(struct v4l2_subdev *sd,
```

```
                    struct v4l2_subdev_frame_interval *interval);
    int (*s_frame_interval)(struct v4l2_subdev *sd,
                    struct v4l2_subdev_frame_interval *interval);
[...]
};
```

In the preceding excerpt, for the sake of readability, I removed the TV and video output-related callbacks as well as those not related to camera devices, which are also somehow useless for us. For the commonly used ones, they are defined as follows:

- querystd: This is the callback for the VIDIOC_QUERYSTD() ioctl handler code.

- s_stream: This is used to notify the driver that a video stream will start or has stopped, depending on the value of the enable parameter.

- g_frame_interval: This is the callback for the VIDIOC_SUBDEV_G_FRAME_INTERVAL() ioctl handler code.

- s_frame_interval: This is the callback for the VIDIOC_SUBDEV_S_FRAME_INTERVAL() ioctl handler code.

struct v4l2_subdev_sensor_ops

There are sensors that produce initial garbage frames when they start streaming. Such sensors possibly require some time in order to ensure the stability of some of their properties. This structure makes it possible to inform the core of the number of frames to skip in order to avoid garbage. Moreover, some sensors may always produce images with a certain number of corrupted lines at the top, or embed their metadata in these lines. In both cases, the resulting frames they produce are always corrupted. This structure also allows us to specify the number of lines to skip on each frame before it is grabbed.

The following is the definition of the v4l2_subdev_sensor_ops structure:

```
struct v4l2_subdev_sensor_ops {
    int (*g_skip_top_lines)(struct v4l2_subdev *sd,
                            u32 *lines);
    int (*g_skip_frames)(struct v4l2_subdev *sd, u32 *frames);
};
```

`g_skip_top_lines` is used to specify the number of lines to skip in each image of the sensors, while `g_skip_frames` allows us to specify the initial number of frames to skip in order to avoid garbage, as in the following example:

```
#define OV5670_NUM_OF_SKIP_FRAMES  2
static int ov5670_get_skip_frames(struct v4l2_subdev *sd,
                                   u32 *frames)
{
    *frames = OV5670_NUM_OF_SKIP_FRAMES;
    return 0;
}
```

The `lines` and `frames` parameters are output parameters. Each callback should return 0.

Calling sub-device operations

After all, if `subdev` callbacks are provided, then they are intended to be called. That said, invoking an ops callback is as simple as calling it directly, as follows:

```
err = subdev->ops->video->s_stream(subdev, 1);
```

However, there is a more convenient and safer way to achieve this, by using the `v4l2_subdev_call()` macro:

```
err = v4l2_subdev_call(subdev, video, s_stream, 1);
```

The macro, defined in `include/media/v4l2-subdev.h`, will perform the following actions:

- It will first check whether the sub-device is `NULL` and return `-ENODEV` otherwise.
- It will return `-ENOIOCTLCMD` if either the category (`subdev->video` in our example) or the callback itself (`subdev->video->s_stream` in our example) is `NULL`, or it will return the actual result of the `subdev->ops->video->s_stream` ops.

It is also possible to call all, or a subset, of the sub-devices:

```
v4l2_device_call_all(dev, 0, core, g_chip_ident, &chip);
```

Any sub-device that does not support this callback is skipped and error results are ignored. If you want to check for errors, use the following command:

```
err = v4l2_device_call_until_err(dev, 0, core,
                                  g_chip_ident, &chip);
```

Any error except -ENOIOCTLCMD will exit the loop with that error. If no errors (except -ENOIOCTLCMD) occurred, then 0 is returned.

Traditional sub-device (un)registration

There are two ways for a sub-device to be registered with the bridge, depending on the nature of the media device:

1. **Synchronous mode**: This is the traditional method. In this mode, the bridge driver has the responsibility of registering sub-devices. The sub-device driver is either implemented from within the bridge driver or you have to find a way for the bridge driver to grab the handles of sub-devices it is responsible for. This is usually achieved by means of platform data, or by the bridge driver exposing a set of APIs that will be used by the sub-device drivers, which would allow the bridge driver to be aware of these sub-devices (by tracking them in a private internal list, for example).

   With this method, the bridge driver must be aware of the sub-devices connected to it, and know exactly when to register them. This is typically the case for internal sub-devices, such as video data processing units within SoCs or complex PCI(e) boards, or camera sensors in USB cameras or connected to SoCs.

2. **Asynchronous mode**: This is where information about sub-devices is made available to the system independently of the bridge devices, which is typically the case on devicetree-based systems. This will be discussed in the next chapter, *Chapter 8, Integrating with V4L2 Async and Media Controller Frameworks*.

However, in order for a bridge driver to register a sub-device, it has to call `v4l2_device_register_subdev()`, while it has to call `v4l2_device_unregister_subdev()` to unregister this sub-device. In the meantime, after registering sub-devices with the core, it might be necessary to create their respective char file nodes, `/dev/v4l-subdevX`, only for sub-devices with the flag `V4L2_SUBDEV_FL_HAS_DEVNODE` set. You can use `v4l2_device_register_subdev_nodes()` to this end:

```
int v4l2_device_register_subdev(struct v4l2_device *v4l2_dev,
                                struct v4l2_subdev *sd)
void v4l2_device_unregister_subdev(struct v4l2_subdev *sd)
int v4l2_device_register_subdev_nodes(struct
                                v4l2_device *v4l2_dev)
```

`v4l2_device_register_subdev()` will insert `sd` into `v4l2_dev->subdevs`, which is the list of sub-devices maintained by this V4L2 device. This can fail if the `subdev` module disappeared before it could be registered. After this function has been called successfully, the `subdev->v4l2_dev` field points to the `v4l2_device`. This function returns `0` in the event of success or `v4l2_device_unregister_subdev()` will take `sd` off that list. Then, `v4l2_device_register_subdev_nodes()` walks through `v4l2_dev->subdevs` and creates a special char file node (`/dev/v4l-subdevX`) for each sub-device with the flag `V4L2_SUBDEV_FL_HAS_DEVNODE` set.

> **Important note**
>
> The `/dev/v4l-subdevX` device nodes allow direct control of the advanced and hardware-specific features of sub-devices.

Now that we have learned about sub-device initialization, operations, and registration, let's look at V4L2 controls in the next section.

V4L2 controls infrastructure

Some devices have controls that are settable by the user in order to modify some defined properties. Some of these controls may support a list of predefined values, a default value, an adjustment, and so on. The thing is, different devices may provide different controls with different values. Moreover, while some of these controls are standard, others may be vendor-specific. The main purpose of the control framework is to present controls to the user without assumptions relating to their purpose. In this section, we only address standard controls.

The control framework relies on two main objects, both defined in `include/media/v4l2- ctrls.h`, like the rest of the data structures and APIs provided by this framework. The first is `struct v4l2_ctrl`. This structure describes the control properties and keeps track of the control's value. The second and final one is `struct v4l2_ctrl_handler`, which keeps track of all the controls. Their detailed definitions are presented here:

```
struct v4l2_ctrl_handler {
    [...]
    struct mutex *lock;
    struct list_head ctrls;
    v4l2_ctrl_notify_fnc notify;
    void *notify_priv;
    [...]
};
```

In the preceding definition excerpt of `struct v4l2_ctrl_handler`, `ctrls` represents the list of controls owned by this handler. `notify` is a notify callback that is called whenever the control changes value. This callback is invoked with the handler's `lock` held. At the end, `notify_priv` is the context data given as the argument to notify. The next one is `struct v4l2_ctrl`, defined as follows:

```
struct v4l2_ctrl {
    struct list_head node;
    struct v4l2_ctrl_handler *handler;
    unsigned int is_private:1;
    [...]
    const struct v4l2_ctrl_ops *ops;
    u32 id;
    const char *name;
    enum v4l2_ctrl_type type;
    s64 minimum, maximum, default_value;
    u64 step;
    unsigned long flags; [...]
}
```

This structure represents the control on its own, with important members present. These are defined as follows:

- `node` is used to insert the control in the handler's control list.

- `handler` is the handler to which this control belongs.

- `ops` is of the `struct v4l2_ctrl_ops` type and represents the get/set operations for this control.

- `id` is the ID of this control.

- `name` is the name of the control.

- `minimum` and `maximum` are the minimum and maximum values accepted by the control, respectively.

- `default_value` is the default value of the control.

- `step` is the incrementation/decrementation step for this non-menu control.

- `flags` covers the control's flags. While the whole flag list is defined in `include/uapi/linux/videodev2.h`, some of the commonly used ones are as follows:

  - `V4L2_CTRL_FLAG_DISABLED`, which means the control is disabled

  - `V4L2_CTRL_FLAG_READ_ONLY`, for a read-only control

  - `V4L2_CTRL_FLAG_WRITE_ONLY`, for a write-only control

  - `V4L2_CTRL_FLAG_VOLATILE`, for a volatile control

- `is_private`, if set, will prevent this control from being added to any other handlers. It makes this control private to the initial handler where it is added. This can be used to prevent making a `subdev` control available in the V4L2 driver controls.

> **Important note**
>
> **Menu controls** are controls that do not require their values according to min/max/step, but instead allow a choice to be made between specific elements (in an `enum` usually) like a kind of menu, hence the name *menu control*.

V4L2 controls are identified by a unique ID. They are prefixed with `V4L2_CID_` and are all available in `include/uapi/linux/v4l2-controls.h`. The common standard controls supported in video capture devices are as follows (the following list is non-exhaustive):

```
#define V4L2_CID_BRIGHTNESS          (V4L2_CID_BASE+0)
#define V4L2_CID_CONTRAST            (V4L2_CID_BASE+1)
#define V4L2_CID_SATURATION          (V4L2_CID_BASE+2)
#define V4L2_CID_HUE    (V4L2_CID_BASE+3)
#define V4L2_CID_AUTO_WHITE_BALANCE       (V4L2_CID_BASE+12)
#define V4L2_CID_DO_WHITE_BALANCE   (V4L2_CID_BASE+13)
#define V4L2_CID_RED_BALANCE (V4L2_CID_BASE+14)
#define V4L2_CID_BLUE_BALANCE        (V4L2_CID_BASE+15)
#define V4L2_CID_GAMMA        (V4L2_CID_BASE+16)
#define V4L2_CID_EXPOSURE     (V4L2_CID_BASE+17)
#define V4L2_CID_AUTOGAIN     (V4L2_CID_BASE+18)
#define V4L2_CID_GAIN   (V4L2_CID_BASE+19)
#define V4L2_CID_HFLIP  (V4L2_CID_BASE+20)
#define V4L2_CID_VFLIP  (V4L2_CID_BASE+21)
[...]
#define V4L2_CID_VBLANK   (V4L2_CID_IMAGE_SOURCE_CLASS_BASE + 1)
#define V4L2_CID_HBLANK   (V4L2_CID_IMAGE_SOURCE_CLASS_BASE + 2)
#define V4L2_CID_LINK_FREQ (V4L2_CID_IMAGE_PROC_CLASS_BASE + 1)
```

The preceding list includes standard controls only. To support a custom control, you should add its ID based on the control's base class descriptor and make sure that the ID is not a duplication. To add control support to the driver, the control handler should first be initialized using the `v4l2_ctrl_handler_init()` macro. This macro accepts the handler to be initialized as well as the number of controls this handler can refer to, as shown in its following prototype:

```
v4l2_ctrl_handler_init(hdl, nr_of_controls_hint)
```

Once finished with the control handler, you can call `v4l2_ctrl_handler_free()` on this control handler in order to release its resources. Once the control handler is initialized, controls can be created and added to it. When it comes to standard V4L2 controls, you can use `v4l2_ctrl_new_std()` in order to allocate and initialize the new control:

```
struct v4l2_ctrl *v4l2_ctrl_new_std(
                        struct v4l2_ctrl_handler *hdl,
                        const struct v4l2_ctrl_ops *ops,
                        u32 id, s64 min, s64 max,
                        u64 step, s64 def);
```

This function will, in most fields, be based on the control ID. For custom controls, however (not discussed here), you should use the `v4l2_ctrl_new_custom()` helper instead. In the preceding prototype, the following elements are defined as follows:

- `hdl` represents the control handler initialized previously.

- `ops` is of the `struct v4l2_ctrl_ops` type and represents the control ops.

- `id` is the control ID, defined as `V4L2_CID_*`.

- `min` is the minimum value this control can accept. Depending on the control ID, this value can be mangled by the core.

- `max` is the maximum value this control can accept. Depending on the control ID, this value can be mangled by the core.

- `step` is the control's step value.

- `def` is the control's default value.

Controls are meant to be set/get. This is the purpose of the preceding ops argument. This means that prior to initializing a control, you should first define the ops that will be called when setting/getting this control's value. That said, the whole control list can be addressed by the same ops. In this case, the ops callback will have to `switch ... case` to handle different controls.

As we have seen earlier, control operations are of the `struct v4l2_ctrl_ops` type and are defined as follows:

```
struct v4l2_ctrl_ops {
    int (*g_volatile_ctrl)(struct v4l2_ctrl *ctrl);
    int (*try_ctrl)(struct v4l2_ctrl *ctrl);
    int (*s_ctrl)(struct v4l2_ctrl *ctrl);
};
```

The preceding structure is made of three callbacks, each with a specific purpose:

- `g_volatile_ctrl` gets the new value for the given control. Providing this callback only makes sense for volatile (those changed by the hardware itself, and that are read-only most of the time, such as the signal strength or autogain, for example) controls.

- `try_ctrl`, if set, is invoked to test whether the control's value to be applied is valid. Providing this callback only makes sense if the usual min/max/step checks are insufficient.

- `s_ctrl` is invoked to set the control's value.

Optionally, you can call `v4l2_ctrl_handler_setup()` on the control handler in order to set up this handler's controls to their default values. This helps to ensure that the hardware and the driver's internal data structures are in sync:

```
int v4l2_ctrl_handler_setup(struct v4l2_ctrl_handler *hdl);
```

This function iterates over all the controls in the given handler and calls the `s_ctrl` callback with each control's default value.

To summarize what we have seen throughout this V4L2 control interface section, let's now study in more detail an excerpt of the OV7740 camera sensor's driver (present in `drivers/media/i2c/ov7740.c`), especially the part dealing with V4L2 controls.

First, we have the implementation of the control `ops->sg_ctrl` callback:

```
static int ov7740_get_volatile_ctrl(struct v4l2_ctrl *ctrl)
{
    struct ov7740 *ov7740 = container_of(ctrl->handler,
    struct ov7740, ctrl_handler);
    int ret;
```

```
    switch (ctrl->id) {
    case V4L2_CID_AUTOGAIN:
        ret = ov7740_get_gain(ov7740, ctrl);
        break;
    default:
        ret = -EINVAL;
        break;
    }
    return ret;
}
```

The preceding callback only addresses the control ID of V4L2_CID_AUTOGAIN. It makes sense as the gain value may be changed by the hardware while in *auto* mode. This driver implements the ops->s_ctrl control as follows:

```
static int ov7740_set_ctrl(struct v4l2_ctrl *ctrl)
{
    struct ov7740 *ov7740 =
            container_of(ctrl->handler, struct ov7740,
                         ctrl_handler);
    struct i2c_client *client =
    v4l2_get_subdevdata(&ov7740->subdev);
    struct regmap *regmap = ov7740->regmap;
    int ret;
    u8 val = 0;
[...]
    switch (ctrl->id) {
    case V4L2_CID_AUTO_WHITE_BALANCE:
        ret = ov7740_set_white_balance(ov7740, ctrl->val);
break;
    case V4L2_CID_SATURATION:
        ret = ov7740_set_saturation(regmap, ctrl->val); break;
    case V4L2_CID_BRIGHTNESS:
        ret = ov7740_set_brightness(regmap, ctrl->val); break;
    case V4L2_CID_CONTRAST:
        ret = ov7740_set_contrast(regmap, ctrl->val); break;
    case V4L2_CID_VFLIP:
        ret = regmap_update_bits(regmap, REG_REG0C,
```

```
                                    REG0C_IMG_FLIP, val); break;
    case V4L2_CID_HFLIP:
        val = ctrl->val ? REG0C_IMG_MIRROR : 0x00;
        ret = regmap_update_bits(regmap, REG_REG0C,
                                    REG0C_IMG_MIRROR, val);
        break;
    case V4L2_CID_AUTOGAIN:
        if (!ctrl->val)
            return ov7740_set_gain(regmap, ov7740->gain->val);
        ret = ov7740_set_autogain(regmap, ctrl->val); break;
    case V4L2_CID_EXPOSURE_AUTO:
        if (ctrl->val == V4L2_EXPOSURE_MANUAL)
            return ov7740_set_exp(regmap, ov7740->exposure->val);
        ret = ov7740_set_autoexp(regmap, ctrl->val); break;
    default:
        ret = -EINVAL; break;
    }
[...]
    return ret;
}
```

The preceding code block also shows how easy it is to implement a menu control using the V4L2_CID_EXPOSURE_AUTO control as an example and whose possible values are enumerated in enum v4l2_exposure_auto_type. Finally, the control ops structure that will be given for control creation is defined as follows:

```
static const struct v4l2_ctrl_ops ov7740_ctrl_ops = {
    .g_volatile_ctrl = ov7740_get_volatile_ctrl,
    .s_ctrl = ov7740_set_ctrl,
};
```

Once defined, this control op can be used to initialize controls. The following is the ov7740_init_controls() method (invoked in the probe() function) excerpt, mangled and shrunk for the purposes of readability:

```
static int ov7740_init_controls(struct ov7740 *ov7740)
{
[...]
```

```
    struct v4l2_ctrl *auto_wb;
    struct v4l2_ctrl *gain;
    struct v4l2_ctrl *vflip;
    struct v4l2_ctrl *auto_exposure;
    struct v4l2_ctrl_handler *ctrl_hdlr

    v4l2_ctrl_handler_init(ctrl_hdlr, 12);
    auto_wb = v4l2_ctrl_new_std(ctrl_hdlr, &ov7740_ctrl_ops,
                                V4L2_CID_AUTO_WHITE_BALANCE,
                                0, 1, 1, 1);
    vflip = v4l2_ctrl_new_std(ctrl_hdlr, &ov7740_ctrl_ops,
                                V4L2_CID_VFLIP, 0, 1, 1, 0);

    gain = v4l2_ctrl_new_std(ctrl_hdlr, &ov7740_ctrl_ops,
                                V4L2_CID_GAIN, 0, 1023, 1, 500);

    /* let's mark this control as volatile*/
    gain->flags |= V4L2_CTRL_FLAG_VOLATILE;

    contrast = v4l2_ctrl_new_std(ctrl_hdlr, &ov7740_ctrl_ops,
                                V4L2_CID_CONTRAST, 0, 127,
                                1, 0x20);

    ov7740->auto_exposure =
                v4l2_ctrl_new_std_menu(ctrl_hdlr,
                                &ov7740_ctrl_ops,
                                V4L2_CID_EXPOSURE_AUTO,
                                V4L2_EXPOSURE_MANUAL,
                                0, V4L2_EXPOSURE_AUTO);
[...]
    ov7740->subdev.ctrl_handler = ctrl_hdlr;
    return 0;
}
```

You can see the control handler being assigned to the sub-device at the return path of the preceding function. Finally, somewhere in the code (the ov7740's driver does this from within the sub-device's `v4l2_subdev_video_ops.s_stream` callback), you should set all controls to their default values:

```
ret = v4l2_ctrl_handler_setup(ctrl_hdlr);
if (ret) {
    dev_err(&client->dev, "%s control init failed (%d)\n",
            __func__, ret);
    goto error;
}
```

There is more on V4L2 controls at `https://www.kernel.org/doc/html/v4.19/media/kapi/v4l2-controls.html`.

A word about control inheritance

It is common for sub-device drivers to implement controls already implemented by the bridge's V4L2 driver.

When `v4l2_device_register_subdev()` is invoked on `v4l2_subdev` and `v4l2_device` and the `ctrl_handler` fields of both are set, then the sub-device's controls will be added (by means of the `v4l2_ctrl_add_handler()` helper, which adds a given handler's control to another handler) to the `v4l2_device` controls. Sub-device controls that are already implemented by `v4l2_device` will be skipped. This means that a V4L2 driver can always override a `subdev` control.

That said, a control may perform low-level, hardware-specific operations on a given sub-device and the sub-device driver may not want this control to be available to the V4L2 driver (and so is not added to its control handler). In this case, the sub-device driver has to set the `is_private` member of the control to `1` (or `true`). This will make the control private to the sub-device.

> **Important note**
> Even though sub-device controls are added to the V4L2 device, they remain accessible through the control device node.

Summary

In this chapter, we dealt with V4L2 bridge device driver development, as well as the concept of sub-devices. We learned about the V4L2 architecture and are now familiar with its data structures. We studied the videobuf2 API and are now able to write platform bridge device drivers. Moreover, we should be able to implement sub-device operations, and to leverage the videobuf2 core.

This chapter can be regarded as the first part of a big picture, since the next chapter still addresses V4L2, but we will deal with the async core and integration with the media controller frameworks.

8
Integrating with V4L2 Async and Media Controller Frameworks

Over time, media support has become a must and a saling argument for **System on Chips** (**SoCs**), which keep becoming more and more complex. The complexity of those media IP cores is such that grabbing sensor data requires a whole pipeline (made of several sub-devices) to be set up by the software. The asynchronous nature of a device tree-based system means the setup and probing of those sub-devices are not straightforward. Thus entered the async framework, which addresses the unordered probing of sub-devices in order for the media device to be popped on time, when all of the media sub-devices are ready. Last but not least, because of the complexity of the media pipe, it became necessary to find a way to ease the configuration of the sub-devices it is made of. Thus came the media controller framework, which wraps the whole media pipe in a single element, the media device. It comes with some abstractions, one of which is that each sub-device is considered as an entity, with either a sink pad, a source pad, or both.

This chapter will focus on how both the async and media controller frameworks work and how they are designed, and we will go through their APIs to learn how to leverage them in **Video4Linux2 (V4L2)** device driver development.

In other words, in this chapter, we will cover the following topics:

- The V4L2 async interface and the concept of graph binding
- The V4L2 async and graph-oriented API
- The V4L2 async framework and APIs
- The Linux media controller framework

Technical requirements

In this chapter, you'll need the following elements:

- Advanced computer architecture knowledge and C programming skills
- Linux kernel v4.19.X sources, available at `https://git.kernel.org/pub/scm/linux/kernel/git/stable/linux.git/refs/tags`

The V4L2 async interface and the concept of graph binding

So far, with V4L2 driver development, we have not actually dealt with the probing order. That being said, we considered the synchronous approach, where bridge device drivers register devices for all sub-devices synchronously during their probing. However, this approach cannot be used with intrinsically asynchronous and unordered device registration systems, such as the **flattened device tree**. To address this, what we currently call the async interface has been introduced.

With this new approach, bridge drivers register lists of sub-device descriptors and notifier callbacks, and sub-device drivers register sub-devices that they are about to probe or have successfully probed. The async core will take care of matching sub-devices against hardware descriptors and calling bridge driver callbacks when matches are found. Another callback is called when the sub-device is unregistered. The async subsystem relies on device declaration in a special way, called **graph binding**, which we will deal with in the next section.

Graph binding

Embedded systems have a reduced set of devices, some of which are not discoverable. The device tree, however, came into the picture to allow describing the actual system (from a hardware point of view) to the kernel. Sometimes (if not always), these devices are somehow interconnected.

While the `phandle` properties pointing to other nodes could be used in the device tree to describe simple and direct connections, such as parent/child relationships, there was no way to model compound devices made of several interconnections. There were situations where the relationship modeling resulted in a quite complete graph – for example, the i.MX6 **Image Processing Unit (IPU)**, which is a logical device on its own, but made up of several physical IP blocks whose interconnections may result in a quite complex pipe.

This is where the so-called **Open Firmware (OF) graph** intervenes, along with its API and some new concepts, the concepts of **port** and **endpoint**:

- A **port** can be seen as an interface in a device (as in an IP block).

- An **endpoint** can be seen as a pad, as it describes one end of a connection to a remote port.

However, `phandle` properties are still used to refer to other nodes in the tree. More documentation on this can be found in `Documentation/devicetree/bindings/graph.txt`.

Port and endpoint representations

A port is an interface to a device. A device can have one or several ports. Ports are represented by port nodes contained in the node of the device they belong to. Each port node contains an endpoint subnode for each remote device port to which this port is connected. This means a single port can be connected to more than one port on the remote device(s) and that each link must be represented by an endpoint child node. Now, if a device node contains more than one port, if there is more than one endpoint at a port, or a port node needs to be connected to a selected hardware interface, a popular scheme using the `#address-cells`, `#size-cells`, and `reg` properties is used to number the nodes.

The following excerpt shows how to use the `#address-cells`, `#size-cells`, and `reg` properties to handle those cases:

```
device {
    ...
    #address-cells = <1>;
```

```
    #size-cells = <0>;

    port@0 {
        #address-cells = <1>;
        #size-cells = <0>;
        reg = <0>;

        endpoint@0 {
            reg = <0>;
            ...
        };
        endpoint@1 {
            reg = <1>;
            ...
        };
    };

    port@1 {
        reg = <1>;
        endpoint { ... };
    };
};
```

Complete documentation of this can be found in `Documentation/devicetree/bindings/graph.txt`. Now that we are done with port and endpoint representation, we need to learn how to link each with the other, as explained in the next section.

Endpoint linking

For two endpoints to be linked together, each of them should contain a `remote-endpoint phandle` property that points to the corresponding endpoint in the port of the remote device. In turn, the remote endpoint should contain a `remote-endpoint` property. Two endpoints with their `remote-endpoint` phandles pointing at each other form a link between the containing ports, as in the following example:

```
device-1 {
    port {
        device_1_output: endpoint {
```

```
                remote-endpoint = <&device_2_input>;
            };
        };
    };
    device-2 {
        port {
            device_2_input: endpoint {
                remote-endpoint = <&device_1_output>;
            };
        };
    }
```

Introducing the graph binding concept without talking at all about its API would be a waste of time. Let's jump to the API that comes along with this new binding method.

The V4L2 async and graph-oriented API

This section heading must not mislead you since graph binding is not just intended for the V4L2 subsystem. The Linux DRM subsystem also takes advantage of it. That being said, the async framework heavily relies on the device tree to describe either media devices along with their endpoints and connections, or links between those endpoints along with their bus configuration properties.

From the DT (of_graph_*) API to the generic fwnode graph API (fwnode_graph_*)

The fwnode graph API is a successful attempt at changing the device tree-only-based OF graph API to a generic API, merging both ACPI and device tree OF APIs together in order to have unified and generic APIs. This extends the concept of the graph with ACPI by using the same APIs. By having a look at the struct device_node and struct acpi_device structures, you can see the members they have in common: struct fwnode_handle fwnode:

```
struct device_node {
[...]
    struct fwnode_handle fwnode;
[...]
};
```

The preceding excerpt represents a device node from a device tree point of view, while the following is related to ACPI:

```
struct acpi_device    {
[...]
    struct fwnode_handle fwnode;
[...]
};
```

The `fwnode` member, which is of the `struct fwnode_handle` type, is a lower level and generic data structure abstracting either `device_node` or `acpi_device` as they both inherit from this data structure. This makes `struct fwnode_handle` a good client for graph API homogenization so that an endpoint (by means of its field of the `fwnode_handle` type) can refer to either an ACPI device or an OF-based device. This abstraction model is now used in graph APIs, allowing us to abstract an endpoint by a generic data structure (`struct fwnode_endpoint`, described as follows) embedding a pointer to `struct fwnode_handle`, which may refer to either an ACPI or OF node. In addition to the genericity, this allows the underlying sub-device to this endpoint to be either ACPI- or OF-based:

```
struct fwnode_endpoint {
    unsigned int port;
    unsigned int id;
    const struct fwnode_handle *local_fwnode;
};
```

This structure deprecates the old `struct of_endpoint` structure and the member of type `device_node*` leaves room for a member of the `fwnode_handle*` type. In the preceding structure, `local_fwnode` points to the related firmware node, `port` is the port number (that is, it corresponds to `0` in `port@0` or `1` in `port@1`), and `id` is the index of this endpoint from within the port (that is, it corresponds to the `0` in `endpoint@0` and to the `1` in `endpoint@1`).

The V4L2 framework uses this model for abstracting V4L2-related endpoints by means of `struct v4l2_fwnode_endpoint`, which is built on top of `fwnode_endpoint`, as follows:

```
struct v4l2_fwnode_endpoint {
    struct fwnode_endpoint base;
    /*
     * Fields below this line will be zeroed by
```

```
    * v4l2_fwnode_endpoint_parse()
    */
   enum v4l2_mbus_type bus_type;
   union {
       struct v4l2_fwnode_bus_parallel parallel;
       struct v4l2_fwnode_bus_mipi_csi1 mipi_csi1;
       struct v4l2_fwnode_bus_mipi_csi2 mipi_csi2;
   } bus;
   u64 *link_frequencies;
   unsigned int nr_of_link_frequencies;
};
```

This structure deprecates and replaces `struct v4l2_of_endpoint` since kernel v4.13, formerly used by V4L2 to represent endpoint nodes in the era of the **V4L2 OF API**. In the preceding data structure definition, `base` represents the `struct fwnode_endpoint` structure of the underlying ACPI or device node. Other fields are V4L2-related, as follows:

- `bus_type` is the type of media bus through which this sub-device streams data. The value of this member determines which underlying bus structure should be filled with the parsed bus properties from the `fwnode` endpoint (either device tree or ACPI). Possible values are listed in enum `v4l2_mbus_type`, as follows:

  ```
     enum v4l2_mbus_type {
         V4L2_MBUS_PARALLEL,
         V4L2_MBUS_BT656,
         V4L2_MBUS_CSI1,
         V4L2_MBUS_CCP2,
         V4L2_MBUS_CSI2,
     };
  ```

- bus is the structure representing the media bus itself. Possible values are already present in the union, and `bus_type` determines which one to consider. These bus structures are all defined in `include/media/v4l2-fwnode.h`.

- `link_frequencies` is the list of frequencies supported by this link.

- `nr_of_link_frequencies` is the number of elements in `link_frequencies`.

> **Important note**
>
> In kernel v4.19, the bus_type member is exclusively set according to the bus-type property in fwnode. The driver can check the read value and adapt its behavior. This means the V4L2 fwnode API will always base its parsing strategy on this fwnode property. However, as of kernel v5.0, drivers have to set this member to an expected bus type (prior to calling the parsing function), which will be compared to the value of the bus-type property read in fwnode and will raise an error if they don't match. If the bus type is not known or if the driver can deal with several bus types, the V4L2_MBUS_UNKNOWN value has to be used. This value is also part of enum v4l2_mbus_type, as of kernel v5.0.
>
> In the kernel code, you may find the enum v4l2_fwnode_bus_type enum type. This is a V4L2 fwnode local enum type that is the counterpart of the global enum v4l2_mbus_type enum type and whose values map each other. Their respective values are kept in sync as the code evolves.

The V4L2-related binding then requires additional properties. Part of these properties is used to build v4l2_fwnode_endpoint, while the other part is used to build the underlying bus (the media bus, actually) structure. All are described in a dedicated and video-related binding documentation, Documentation/devicetree/bindings/media/video- interfaces.txt, which I strongly recommend checking out.

The following is a typical binding between a bridge (isc) and a sensor sub-device (mt9v032):

```
&i2c1 {
    #address-cells = <1>;
    #size-cells = <0>;
    mt9v032@5c {
        compatible = "aptina,mt9v032";
        reg = <0x5c>;

        port {
            mt9v032_out: endpoint {
                remote-endpoint = <&isc_0>;
                link-frequencies =
                        /bits/ 64 <13000000 26600000 27000000>;
                hsync-active = <1>;
                vsync-active = <0>;
                pclk-sample = <1>;
```

```
            };
        };
    };
};

&isc {
    port {
        isc_0: endpoint@0 {
            remote-endpoint = <&mt9v032_out>;
            hsync-active = <1>;
            vsync-active = <0>;
            pclk-sample = <1>;
        };
    };
};
```

In the preceding binding, `hsync-active`, `vsync-active`, `link-frequencies`, and `pclk- sample` are all V4L2-specific properties and describe the media bus. Their values are not coherent here and do not really make sense but fit well for our learning purpose. This excerpt shows well the concepts of endpoint and remote endpoint; the use of `struct v4l2_fwnode_endpoint` is discussed in detail in the *The Linux media controller framework* section.

Important note

The part of V4L2 dealing with the `fwnode` API is called the **V4L2 fwnode API**. It is a replacement of the device tree-only API, the **V4L2 OF API**. The former has a set of APIs prefixed with `v4l2_fwnode_`, while the second's API set is prefixed with `v4l2_of_`. Do note that in OF-only-based APIs, an endpoint is represented by `struct of_endpoint`, and a V4L2-related endpoint is represented by `struct v4l2_of_endpoint`. There are APIs that allow switching from OF- to `fwnode`-based models and vice versa.

V4L2 `fwnode` and V4L2 OF are fully interoperable. For example, a sub-device driver using V4L2 `fwnode` will work with a media device driver using V4L2 OF without any effort, and vice versa! However, new drivers must use the `fwnode` API, including `#include <media/v4l2- fwnode. h>`, which should replace `#include <media/v4l2-of.h>` in the old driver when switching to the `fwnode` API.

That being said, `struct fwnode_endpoint`, which was discussed earlier, is just for showing the underlying mechanisms. We could have completely skipped it since only the core deals with this data structure. For a more generic approach, instead of using `struct device_node` to refer to the device's firmware node, you're better off using the new `struct fwnode_handle`. This definitely makes sure that DT and ACPI bindings are compatible/interoperable using the same code in the driver. The following is a short excerpt of how changes should look in new drivers:

```
-    struct device_node *of_node;
+    struct fwnode_handle *fwnode;


-    of_node = ddev->of_node;
+    fwnode = dev_fwnode(dev);
```

Some of the common `fwnode` node-related APIs are as follows:

```
[...]
struct fwnode_handle *fwnode_get_parent(
                    const struct fwnode_handle *fwnode);

struct fwnode_handle *fwnode_get_next_child_node(
                    const struct fwnode_handle *fwnode,
                    struct fwnode_handle *child);

struct fwnode_handle *fwnode_get_next_available_child_node(
                    const struct fwnode_handle *fwnode,
                    struct fwnode_handle *child);

#define fwnode_for_each_child_node(fwnode, child) \
    for (child = fwnode_get_next_child_node(fwnode, NULL); \
child; \
        child = fwnode_get_next_child_node(fwnode, child))

#define fwnode_for_each_available_child_node(fwnode, child) \
    for (child = fwnode_get_next_available_child_node(fwnode,
                                                    NULL); \
        child; \
    child = fwnode_get_next_available_child_node(fwnode, child))
```

```
struct fwnode_handle *fwnode_get_named_child_node(
                        const struct fwnode_handle *fwnode,
                        const char *childname);

struct fwnode_handle *fwnode_handle_get(struct
                                        fwnode_handle *fwnode);
void fwnode_handle_put(struct fwnode_handle *fwnode);
```

The aforementioned APIs have the following description:

- `fwnode_get_parent()` returns the parent handle of the node whose `fwnode` value is given in an argument, or `NULL` otherwise.

- `fwnode_get_next_child_node()` takes a parent node as its first argument and returns the next child (or `NULL` otherwise) after a given child (given as the second argument) in this parent. If `child` (the second argument) is `NULL`, then the first child of this parent will be returned.

- `fwnode_get_next_available_child_node()` is the same as `fwnode_get_next_child_node()` but makes sure that the device actually exists (has been probed successfully) prior to returning the `fwnode` handle.

- `fwnode_for_each_child_node()` iterates over the child in a given node (the first argument) and the second argument is used as an iterator.

- `fwnode_for_each_available_child_node` is the same as `fwnode_for_each_child_node()` but iterates only over nodes whose device is actually present on the system.

- `fwnode_get_named_child_node()` gets a child in a given node by its name.

- `fwnode_handle_get()` obtains a reference to a device node and `fwnode_handle_put()` drops this reference.

Some of the `fwnode`-related properties are as follows:

```
[...]
bool fwnode_device_is_available(const
                                struct fwnode_handle *fwnode);
bool fwnode_property_present(const
                                struct fwnode_handle *fwnode,
                                const char *propname);


int fwnode_property_read_string(const
```

```
                                    struct fwnode_handle *fwnode,
                              const char *propname,
                              const char **val);
int fwnode_property_match_string(const
                                    struct fwnode_handle *fwnode,
                              const char *propname,
                              const char *string);
```

Both property- and node-related `fwnode` APIs are available in `include/linux/property.h`. However, there are helpers that allow switching back and forth between OF, ACPI, and `fwnode`. The following is a short example:

```
/* to switch from fwnode to of */
struct device_node *of_node = to_of_node(fwnode);


/* to switch from of to fw */
struct fwnode_handle *fwnode = of_fwnode_handle(node)


/* to switch from fwnode to acpi handle, the below macro has
 * been introduced
 *
 * #define ACPI_HANDLE_FWNODE(fwnode)     \
 *        acpi_device_handle(to_acpi_device_node(fwnode))
 *
 * and to switch from acpi device to fwnode:
 *
 *    struct fwnode_handle *
 *          acpi_fwnode_handle(struct acpi_device *adev)
 *
 */
```

Finally, and most important for us, is the `fwnode` graph API. In the following code snippet, we enumerate the most important function of this API:

```
struct fwnode_handle
    *fwnode_graph_get_next_endpoint(const
                                struct fwnode_handle *fwnode,
                                struct fwnode_handle *prev);
struct fwnode_handle
    *fwnode_graph_get_port_parent(const
                                struct fwnode_handle *fwnode);

struct fwnode_handle
    *fwnode_graph_get_remote_port_parent(
                        const struct fwnode_handle *fwnode);

struct fwnode_handle
    *fwnode_graph_get_remote_port(const
                                struct fwnode_handle *fwnode);

struct fwnode_handle
    *fwnode_graph_get_remote_endpoint(
                        const struct fwnode_handle *fwnode);

#define fwnode_graph_for_each_endpoint(fwnode, child) \
    for (child = NULL; \
    (child = fwnode_graph_get_next_endpoint(fwnode, child)); )

int fwnode_graph_parse_endpoint(const
                                struct fwnode_handle *fwnode,
                                struct fwnode_endpoint *endpoint);
[...]
```

Though the preceding function names talk about themselves, the following are better descriptions of what they do:

- `fwnode_graph_get_next_endpoint()` returns the next endpoint (or `NULL` otherwise) in a given node (the first argument) after a previous endpoint (`prev`, the second argument). If `prev` is `NULL`, then the first endpoint is returned. This function obtains a reference to the returned endpoint that must be dropped after use. See `fwnode_handle_put()`.

- `fwnode_graph_get_port_parent()` returns the parent of the port node given in the argument.

- `fwnode_graph_get_remote_port_parent()` returns the firmware node of the remote device containing the endpoint whose firmware node is given through the `fwnode` argument.

- `fwnode_graph_get_remote_endpoint()` returns the firmware node of the remote endpoint corresponding to a local endpoint whose firmware node is given through the `fwnode` argument.

- `fwnode_graph_parse_endpoint()` parses common endpoint node properties in `fwnode` (the first argument) representing a graph endpoint node and stores the information in `endpoint` (the second and output argument). The V4L2 firmware node API heavily uses this.

The V4L2 firmware node (V4L2 fwnode) API

The main data structure in the V4L2 fwnode API is `struct v4l2_fwnode_endpoint`. This structure is nothing but `struct fwnode_handle` augmented with some V4L2-related properties. However, there is a V4L2-related fwnode graph function that it is worth talking about here: `v4l2_fwnode_endpoint_parse()`. This function's prototype is declared `include/media/v4l2-fwnode.h`, as follows:

```
int v4l2_fwnode_endpoint_parse(struct fwnode_handle *fwnode,
                               struct v4l2_fwnode_endpoint *vep);
```

Given `fwnode_handle` (the first argument in the preceding function) of an endpoint, you can use `v4l2_fwnode_endpoint_parse()` to parse all the fwnode node properties. This function also recognizes and takes care of the V4L2-specific properties, which are, if you remember, those documented in `Documentation/devicetree/bindings/media/video-interfaces.txt`. `v4l2_fwnode_endpoint_parse()` uses `fwnode_graph_parse_endpoint()` to parse common fwnode properties and uses V4L2-specific parser helpers to parse V4L2-related properties. It returns `0` on success or a negative error code on failure.

If we consider the `mt9v032` CMOS image sensor node in `dts`, we can have the following code in the `probe` method:

```
int err;
struct fwnode_handle *ep;
struct v4l2_fwnode_endpoint bus_cfg;

/* We grab the fwnode corresponding to the device */
struct fwnode_handle *fwnode = dev_fwnode(dev);

/* We grab its endpoint(s) node */
ep = fwnode_graph_get_next_endpoint(fwnode, NULL);

/* We parse the endpoint common properties as well as
 * v4l2 related properties
 */
err = v4l2_fwnode_endpoint_parse(ep, &bus_cfg);
if (err) {   /* handle error */ }

/* At this point we can access parameters such as bus_type,
 * bus.flags
 * (in case of mipi csi2 or parallel buses), V4L2_MBUS_*
 * which are the
 * media bus flags
 */

/* we drop the reference on the enpoint */
fwnode_handle_put(ep);
```

The preceding code shows how you use the fwnode API, as well as its V4L2 version, for accessing node and endpoint properties. There are, however, V4L2-specific properties being parsed upon the `v4l2_fwnode_endpoint_parse()` call. These properties describe the so-called **media bus** through which data is carried from one interface to another. We will discuss this in the next section.

V4L2 fwnode or media bus types

Most media devices support a particular media bus type. While endpoints are linked together, they are actually connected through buses, whose properties need to be described to the V4L2 framework. For V4L2 to be able to find this information, it is provided as properties in the device's fwnode (DT or ACPI). As these are specific properties, the V4L2 fwnode API is able to recognize and parse them. Each bus has its specificities and properties.

First of all, let's have a look at the currently supported buses, along with their data structures:

- **MIPI CSI-1**: This is MIPI Alliance's **Camera Serial Interface** (**CSI**) version 1. This bus is represented as instances of `struct v4l2_fwnode_bus_mipi_csi1`.

- **CCP2**: This stands for **Compact Camera Port 2**, made by the **Standard Mobile Imaging Architecture** (**SMIA**), which is an open standard for companies dealing with camera modules for use in mobile applications (such as SMIA CCP2). This bus is represented in this framework with instances of `struct v4l2_fwnode_bus_mipi_csi1` too.

- **Parallel bus**: This is the classic parallel interface, with `HSYNC` and `VSYNC` signals. The structure used to represent this bus is `struct v4l2_fwnode_bus_parallel`.

- **BT656**: This is for BT.1120 or any parallel bus that transmits the conventional video timing and synchronization signal (`HSYNC`, `VSYNC`, and `BLANK`) in the data. These buses have a reduced number of pins compared to the standard parallel bus. This framework uses `struct v4l2_fwnode_bus_parallel` to represent this bus.

- **MIPI CSI-2**: This is version 2 of MIPI Alliance's CSI interface. This bus is abstracted by the `struct v4l2_fwnode_bus_mipi_csi2` structure. However, this data structure does not differentiate between D-PHY and C-PHY. This lack of differentiation is addressed as of kernel v5.0.

As we will see later in the chapter, in the *The concept of a media bus* section, this concept of a bus can be used to detect compatibility between a local endpoint and its remote counterpart so that two sub-devices can't be linked together if they don't have the same bus properties, which makes complete sense.

Earlier, in the *The V4L2 fwnode API* section, we saw that `v4l2_fwnode_endpoint_parse()` is responsible for parsing the endpoint's fwnode and filling the appropriate bus structure. This function first calls `fwnode_graph_parse_endpoint()` in order to parse the common fwnode graph-related properties, and then checks the value of the `bus-type` property, as follows, in order to determine the appropriate `v4l2_fwnode_endpoint.bus` data type:

```
u32 bus_type = 0;
fwnode_property_read_u32(fwnode, "bus-type", &bus_type);
```

Depending on this value, a bus data structure will be chosen. The following are expected possible values from the `fwnode` device:

- 0: This means auto-detect. The core will try to guess the bus type according to the properties present in the fwnode (MIPI CSI-2 D-PHY, parallel, or BT656).
- 1: This means MIPI CSI-2 C-PHY.
- 2: This means MIPI CSI-1.
- 3: This means CCP2.

For the CPP2 bus, for example, the device's fwnode would contain the following line:

```
bus-type = <3>;
```

> **Important note**
>
> As of kernel v5.0, drivers can specify the expected bus type in the `bus_type` member of `v4l2_fwnode_endpoint` prior to giving it as a second argument to `v4l2_fwnode_endpoint_parse()`. This way, parsing will fail if the value returned by the preceding `fwnode_property_read_u32` does not match the expected one, except if the expected bus type was set to `V4L2_MBUS_UNKNOWN`.

BT656 and parallel buses

Those bus types are all represented by `struct v4l2_fwnode_bus_parallel`, as follows:

```
struct v4l2_fwnode_bus_parallel {
    unsigned int flags;
    unsigned char bus_width;
    unsigned char data_shift;
};
```

In the preceding data structure, `flags` represents the flags of the bus. Those flags will be set according to the properties present in the device's firmware node. `bus_width` represents the number of data lines actively used, not necessarily the total number of lines of the bus. `data_shift` is used to specify which data lines are really used by specifying the number of lines to skip prior to reaching the first active data line. The following are the binding properties of these media buses, which are used to set up `struct v4l2_fwnode_bus_parallel`:

- `hsync-active`: Active state of the HSYNC signal; 0/1 for LOW/HIGH, respectively. If this property's value is 0, then the V4L2_MBUS_HSYNC_ACTIVE_LOW flag is set in the `flags` member. Any other value will set the V4L2_MBUS_HSYNC_ACTIVE_HIGH flag instead.

- `vsync-active`: Active state of the VSYNC signal; 0/1 for LOW/HIGH, respectively. If this property's value is 0, then the V4L2_MBUS_VSYNC_ACTIVE_LOW flag is set in the `flags` member. Any other value will set the V4L2_MBUS_VSYNC_ACTIVE_HIGH flag instead.

- `field-even-active`: The field signal level during the even field data transmission. This is the same as the preceding, but the concerned flags are V4L2_MBUS_FIELD_EVEN_HIGH and V4L2_MBUS_FIELD_EVEN_LOW.

- `pclk-sample`: Sample data on the rising (1) or falling (0) edge of the pixel clock signal, V4L2_MBUS_PCLK_SAMPLE_RISING and V4L2_MBUS_PCLK_SAMPLE_FALLING.

- `data-active`: Similar to HSYNC and VSYNC, specifies data line polarity, V4L2_MBUS_DATA_ACTIVE_HIGH and V4L2_MBUS_DATA_ACTIVE_LOW.

- `slave-mode`: This is a Boolean property whose presence indicates that the link is run in slave mode, and the V4L2_MBUS_SLAVE flag is set. Otherwise, the V4L2_MBUS_MASTER flag will be set.

- `data-enable-active`: Similar to `HSYNC` and `VSYNC`, specifies the data-enable signal polarity.

- `bus-width`: This property concerns parallel buses only and represents the number of data lines actively used. The `V4L2_MBUS_DATA_ENABLE_HIGH` or `V4L2_MBUS_DATA_ENABLE_LOW` flags are set accordingly.

- `data-shift`: On parallel data buses where `bus-width` is used to specify the number of data lines, this property can be used to specify which data lines are really used; for example, `bus-width=<8>; data-shift=<2>;` means that lines 9:2 are used.

- `sync-on-green-active`: The active state of the **Sync-on-Green (SoG)** signal; 0/1 for LOW/HIGH, respectively. The `V4L2_MBUS_VIDEO_SOG_ACTIVE_HIGH` or `V4L2_MBUS_VIDEO_SOG_ACTIVE_LOW` flags are set accordingly.

The type of these buses is either `V4L2_MBUS_PARALLEL` or `V4L2_MBUS_BT656`. The underlying function responsible for parsing these buses is `v4l2_fwnode_endpoint_parse_parallel_bus()`.

MIPI CSI-2 bus

This is version 2 of MIPI Alliance's CSI bus. This bus involves two PHYs: either D- PHY or C-PHY. D-PHY has been around for a while and targets cameras, displays, and lower-speed applications. C-PHY is a newer and more complex PHY, where a clock is embedded into the data, rendering a separate clock lane unnecessary. It has fewer wires, a smaller number of lanes, and lower power consumption, and can achieve a higher data rate compared to D-PHY. C-PHY provides high throughput performance over bandwidth-limited channels.

Both C-PHY- and D-PHY-enabled buses are represented using one data structure, `struct v4l2_fwnode_bus_mipi_csi2`, as follows:

```
struct v4l2_fwnode_bus_mipi_csi2 {
    unsigned int flags;
    unsigned char data_lanes[V4L2_FWNODE_CSI2_MAX_DATA_LANES];
    unsigned char clock_lane;
    unsigned short num_data_lanes;
    bool lane_polarities[1 + V4L2_FWNODE_CSI2_MAX_DATA_LANES];
};
```

In the preceding block, `flags` represents the flags of the bus and will be set according to the properties present in the firmware node:

- `data-lanes` is an array of physical data lane indexes.

- `lane-polarities`: This property is valid for serial busses only. It is an array of polarities of the lanes, starting from the clock lane and followed by the data lanes, in the same order as in the `data-lanes` property. Valid values are 0 (normal) and 1 (inverted). The length of this array should be the combined length of the `data-lanes` and `clock-lanes` properties. Valid values are 0 (normal) and 1 (inverted). If the `lane-polarities` property is omitted, the value must be interpreted as 0 (normal).

- `clock-lanes` is the physical lane index of the clock lane. This is the clock lane position.

- `clock-noncontinuous`: If present, the `V4L2_MBUS_CSI2_` `NONCONTINUOUS_CLOCK` flag is set. Otherwise, `V4L2_MBUS_CSI2_` `CONTINUOUS_CLOCK` is set.

These buses have the `V4L2_MBUS_CSI2` type. Until Linux kernel v4.20, there were no differences between C-PHY- and D-PHY-enabled CSI buses. However as of Linux kernel v5.0, this difference has been introduced and `V4L2_MBUS_CSI2` has been replaced with either `V4L2_MBUS_CSI2_DPHY` or `V4L2_MBUS_CSI2_CPHY`, respectively, for D-PHY- or C-PHY-enabled buses.

The underlying function responsible for parsing these buses is `v4l2_fwnode_` `endpoint_parse_csi2_bus()`. An example is as follows:

```
[...]
    port {
        tc358743_out: endpoint {
            remote-endpoint = <&mipi_csi2_in>;
            clock-lanes = <0>;
            data-lanes = <1 2 3 4>;
            lane-polarities = <1 1 1 1 1>;
            clock-noncontinuous;
        };
    };
```

CPP2 and MIPI CSI-1 buses

These are older single data lane serial buses. Their type corresponds to either V4L2_ FWNODE_BUS_TYPE_CCP2 or V4L2_FWNODE_BUS_TYPE_CSI1. The kernel uses struct v4l2_fwnode_bus_mipi_csi1 to represent these buses:

```
struct v4l2_fwnode_bus_mipi_csi1 {
    bool clock_inv;
    bool strobe;
    bool lane_polarity[2];
    unsigned char data_lane;
    unsigned char clock_lane;
};
```

The following are the meanings of the elements in this structure:

- clock-inv: The polarity of the clock/strobe signal (false means not inverted, true means inverted). 0 means false, and other values mean true.

- strobe: False – data/clock, true – data/strobe.

- data-lanes: The number of the data lanes.

- clock-lanes: The number of the clock lanes.

- lane-polarities: This is the same as the preceding, but since CPP2 and MIPI CSI-1 are single data serial buses, the array can have only two entries: the polarities of the clock (index 0) and data lanes (index 1).

The preceding data structure is filled with v4l2_fwnode_endpoint_parse_csi1_ bus() after parsing the given node.

Bus guessing

Specifying the bus type to 0 (or V4L2_MBUS_UNKNOWN) will instruct the V4L2 core to try to guess the actual media bus according to the properties found in the firmware node. It will first consider whether the device is on a CSI-2 bus and try to parse the endpoint node accordingly, looking for CSI-2-related properties. Fortunately, CSI-2 and parallel buses have no properties in common. This way, if, and only if, no MIPI CSI-2-specific properties were found, the core will parse the parallel video bus properties. The core does not guess V4L2_MBUS_CCP2 nor V4L2_MBUS_CSI1. For these buses, the bus-type property must be specified.

V4L2 async

Because of the complexity of video-based hardware that sometimes integrates non-V4L2 devices (sub-devices, actually) sitting on different buses, the need has come for sub-devices to defer initialization until the bridge driver has been loaded, and on the other hand, the bridge driver needs to postpone initializing sub-devices until all required sub-devices have been loaded; that is, V4L2 async.

In asynchronous mode, sub-device probing can be invoked independently of bridge driver availability. The sub-device driver then has to verify whether all the requirements for a successful probing are satisfied. This can include a check for master clock availability, a GPIO, or anything else. If any of the conditions aren't satisfied, the sub-device driver might decide to return -EPROBE_DEFER to request further re-probing attempts. Once all the conditions are met, the sub-device will be registered with the V4L2 async core using the v4l2_async_register_subdev() function. The unregistration is performed using the v4l2_async_unregister_subdev() call.

We saw earlier where synchronous registration applies. It is a mode where the bridge driver is aware of the context of all the sub-devices it is responsible for. It has the responsibility of registering all the sub-devices using v4l2_device_register_subdev() on each of them during its probing, as is the case with the drivers/media/platform/exynos4-is/media-dev.c driver.

In the V4L2 async framework, the concept of a sub-device is abstracted. A sub-device is known in the async framework as an instance of a struct v4l2_async_subdev structure. Along with this structure, there is another struct v4l2_async_notifier structure. Both are defined in include/media/v4l2-async.h and somehow form the center part of the V4L2 async core. Prior to going further, we have to introduce the center part of the V4L2 async framework, struct v4l2_async_notifier, as follows:

```
struct v4l2_async_notifier {
    const struct v4l2_async_notifier_operations *ops;
    unsigned int num_subdevs;
    unsigned int max_subdevs;
    struct v4l2_async_subdev **subdevs;
    struct v4l2_device *v4l2_dev;
    struct v4l2_subdev *sd;
    struct v4l2_async_notifier *parent;
    struct list_head waiting;
    struct list_head done;
```

```
    struct list_head list;
};
```

The preceding structure is mostly used by the bridge drivers and the async core. In some cases, however, sub-device drivers may need to be notified by some other sub-devices. In either case, the uses and meanings of the members are the same:

- `ops` is a set of callbacks to be provided by the owner of this notifier that are invoked by the async core as and when sub-devices waiting in this notifier are probed.

- `v4l2_dev` is the V4L2 parent of the bridge driver that registered this notifier.

- `sd`, if this notifier has been registered by a sub-device, will point to this sub-device. We do not address this case here.

- `subdevs` is an array of sub-devices for which the registrar of this notifier (either the bridge driver or another sub-device driver) should be notified.

- `waiting` is a list of the sub-devices in this notifier waiting to be probed.

- `done` is a list of the sub-devices actually bound to this notifier.

- `num_subdevs` is the number of sub-devices in `**subdevs`.

- `list` is used by the async core during the registration of this notifier in order to link this notifier to the global list of notifiers, `notifier_list`.

Back to our `struct v4l2_async_subdev` structure, which is defined as follows:

```
struct v4l2_async_subdev {
    enum v4l2_async_match_type match_type;
    union {
        struct fwnode_handle *fwnode;
        const char *device_name;
        struct {
            int adapter_id;
            unsigned short address;
        } i2c;
        struct {
        bool (*match)(struct device *,
                      struct v4l2_async_subdev *);
            void *priv;
        } custom;
    } match;
```

```
    /* v4l2-async core private: not to be used by drivers */
    struct list_head list;
};
```

The preceding data structure is a sub-device in the eyes of the V4L2 async framework. Only the bridge driver (which allocates the async sub-device) and the async core can play with this structure. The sub-device driver is not aware of this at all. The meanings of its members are as follows:

- `match_type` is of the enum `v4l2_async_match_type` type. A match is a comparison of some criteria (occurring **strictly** between a sub-device of the `struct v4l2_subdev` type and an async sub-device of the `struct v4l2_async_subdev` type). Since each `struct v4l2_async_subdev` structure must be associated with its `struct v4l2_subdev` structure, this field specifies the algorithm used by the async core to match both. This field is set by the driver (which is also responsible for allocating asynchronous sub-devices). Possible values are as follows:

  --`V4L2_ASYNC_MATCH_DEVNAME`, which instructs the async core to use the device name for the matching. In this case, the bridge driver must set the `v4l2_async_subdev.match.device_name` field so that it can match the sub-device device name (that is, `dev_name(v4l2_subdev->dev)`) when that sub-device will be probed.

  --`V4L2_ASYNC_MATCH_FWNODE`, which means the async core should use the firmware node for the match. In this case, the bridge driver must set `v4l2_async_subdev.match.fwnode` with the firmware node handle corresponding to the sub-device's device node so that they can match.

  --`V4L2_ASYNC_MATCH_I2C` is to be used to perform the match by checking for the I2C adapter ID and address. Using this, the bridge driver must set both `v4l2_async_subdev.match.i2c.adapter_id` and `v4l2_async_subdev.match.i2c.address`. These values will be compared with the address and the adapter number of the `i2c_client` object associated with `v4l2_subdev.dev`.

  --`V4L2_ASYNC_MATCH_CUSTOM` is the last possibility and means the async core should use the matching callback set by the bridge driver in `v4l2_async_subdev.match.custom.match`. If this flag is set and there is no custom matching callback provided, any matching attempt will immediately return true.

- `list` is used to add this async sub-device waiting to be probed in the waiting list of a notifier.

Sub-device registration does not depend on the bridge availability anymore and only consists of calling the `v4l2_async_unregister_subdev()` method. However, prior to registering itself, the bridge driver will have to do the following:

1. Allocate a notifier for later use. It is better to embed this notifier in a larger device state data structure. This notifier object is of the `struct v4l2_async_notifier` type.

2. Parse its port node(s) and create an async sub-device (`struct v4l2_async_subdev`) for each sensor (or IP block) specified there and that it needs for its operations:

   a) This parsing is done using the `fwnode` graph API (old drivers still use the `of_graph` API), such as the following:

   `--fwnode_graph_get_next_endpoint()` (or `of_graph_get_next_endpoint()` in old drivers) to grab the `fw_handle` (or the `of_node` in old drivers) of an endpoint from within the bridge's port subnode.

   `--fwnode_graph_get_remote_port_parent()` (or `of_graph_get_remote_port_parent()` in old drivers) to grab the `fw_handle` (or the device's `of_node` in old drivers) corresponding to the parent of the remote port of the current endpoint.

   Optionally (in old drivers using the OF API), `of_fwnode_handle()` is used in order to convert the `of_node` grabbed in the previous state into an `fw_handle`.

   b) Set up the current async sub-device according to the matching logic that should be used. It should set the `v4l2_async_subdev.match_type` and `v4l2_async_subdev.match` members.

   c) Add this async sub-device to the list of async sub-devices of the notifier. As of version 4.20 of the kernel, there is a helper, `v4l2_async_notifier_add_subdev()`, allowing you to do this.

3. Register the notifier object (this notifier will be stored in the global `notifier_list` list defined in `drivers/media/v4l2-core/v4l2-async.c`) using the `v4l2_async_notifier_register(&big_struct->v4l2_dev, &big_struct->notifier)` call. To unregister the notifier, the driver has to call `v4l2_async_notifier_unregister(&big_struct->notifier)`.

When the bridge driver invokes `v4l2_async_notifier_register()`, the async core iterates over async sub-devices in the `notifier->subdevs` array. For each async sub-device inside, the core checks whether this `asd->match_type` value is `V4L2_ASYNC_MATCH_FWNODE`. If applicable, the async core makes sure `asd` is not present in the `notifier->waiting` list or in the `notifier->done` list by comparing fwnodes. This provides assurance that `asd` was not already set up for `fwnode` and it does not already exist in the given notifier. If `asd` is not already known, it is added to `notifier->waiting`. After this, the async core will test all async sub-devices in the `notifier->waiting` list for a match with all sub-devices present in `subdev_list`, which is the list of "kind-of" orphan sub-devices, those that were registered prior to their bridge driver (thus prior to their notifier). The async core uses the `asd->match` value of each current `asd` for this. If a match occurs (the `asd->match` callback returns true), the current async sub-device (from `notifier->waiting`) and the current sub-device (from `subdev_list`) will be bound, the async sub-device will be removed from the `notifier->waiting` list, the sub-device will be registered with the V4L2 core using `v4l2_device_register_subdev()`, and the sub-device will be moved from the global `subdev_list` list to the `notifier->done` list.

Finally, the actual notifier being registered will be added to the global list of notifiers, `notifier_list`, so that it can be used later for matching attempts whenever a new sub-device is registered with the async core.

> **Important note**
>
> What the async core does when the sub-device driver invokes `v4l2_async_register_subdev()` can be guessed from the preceding matching and bounding logic descriptions. Effectively, upon this call, the async core will attempt to match the current sub-device with all the async sub-devices waiting in each notifier present in the `notifier_list` global list. If no match occurs, it means this sub-device's bridge has not been probed yet, and the sub-device is added to the global list of sub-devices, `subdev_list`. If a match occurs, the sub-device will not be added to this list at all.
>
> Do also keep in mind that a match test is a comparison of some criteria, occurring strictly between a sub-device of the `struct v4l2_subdev` type and an async sub-device of the `struct v4l2_async_subdev` type.

In the preceding paaragraphs, we said the async sub-device and the sub-device are bound. But what does this mean? Here is where the `notifier->ops` member comes into the picture. It is of the `struct v4l2_async_notifier_operations` type and is defined as follows:

```
struct v4l2_async_notifier_operations {
    int (*bound)(struct v4l2_async_notifier *notifier,
                    struct v4l2_subdev *subdev,
                    struct v4l2_async_subdev *asd);
    int (*complete)(struct v4l2_async_notifier *notifier);
    void (*unbind)(struct v4l2_async_notifier *notifier,
                    struct v4l2_subdev *subdev,
                    struct v4l2_async_subdev *asd);
};
```

The following are the meanings of each callback in this structure despite the fact that all three callbacks are optional:

- bound: If set, this callback will be invoked by the async core in response to a successful sub-device probing by its (sub-device) driver. This also implies that an async sub-device has successfully matched this sub-device. This callback takes as an argument the notifier that originated the match, as well as the sub-device (`subdev`) and the async sub-device (`asd`) that matched. Most drivers simply print debug messages here. However, you can perform additional setup on the sub-device here – that is, `v4l2_subdev_call()`. If everything seems OK, it should return a positive value; otherwise, the sub-device is unregistered.

- unbind is invoked when a sub-device is removed from the system. In addition to printing debug messages here, the bridge driver must unregister the video device if the unbound sub-device was a requirement for it to work normally – that is, `video_unregister_device()`.

- complete is invoked when there are no more async sub-devices waiting in the notifier. The async core can detect when the `notifier->waiting` list is empty (which would mean sub-devices have been probed successfully and are all moved into the `notifier->done` list). The complete callback is only executed for the root notifier. Sub-devices that registered notifiers will not have their `.complete` callback invoked. The root notifier is usually the one registered by the bridge device.

There is no doubt, then, that, prior to registering the notifier object, the bridge driver must set the notifier's `ops` member. The most important callback for us is `.complete`.

While you can call `v4l2_device_register()` from within the bridge driver's `probe` function, it is a common practice to register the actual video device from within the `notifier.complete` callback, as all sub-devices would be registered, and the presence of `/dev/videoX` would mean it is really usable. The `.complete` callback is also suitable for both registering the actual video device's subnode and registering the media device by means of `v4l2_device_register_subdev_nodes()` and `media_device_register()`.

Note that `v4l2_device_register_subdev_nodes()` will create a device node (`/dev/v4l2-subdevX`, actually) for every `subdev` object marked with the `V4L2_SUBDEV_FL_HAS_DEVNODE` flag.

Async bridge and sub-device probing example

We will go through this section with a simple use case. Consider the following config:

- One bridge device (our CSI controller) – let's say the `omap` ISP, with `foo` as its name.

- One off-chip sub-device, the camera sensor, with `bar` as its name.

   Both are connected this way: `CSI <-- Camera Sensor`.

In the `bar` driver, we could register an async sub-device as follows:

```
static int bar_probe(struct device *dev)
{
    int ret;
    ret = v4l2_async_register_subdev(subdev);
    if (ret) {
        dev_err(dev, "ouch\n");
        return -ENODEV;
    }
    return 0;
}
```

The `probe` function of the `foo` driver could be as follows:

```
/* struct foo_device */
struct foo_device {
    struct media_device mdev;
    struct v4l2_device v4l2_dev;
    struct video_device *vdev;
    struct v4l2_async_notifier notifier;
    struct *subdevs[FOO_MAX_SUBDEVS];
};


/* foo_probe() */
static int foo_probe(struct device *dev)
{
    struct foo_device *foo = kmalloc(sizeof(*foo));
    media_device_init(&bar->mdev);
    foo->dev = dev;
    foo->notifier.subdevs = kcalloc(FOO_MAX_SUBDEVS,
                            sizeof(struct v4l2_async_subdev));
    foo_parse_nodes(foo);
    foo->notifier.bound = foo_bound;
    foo->notifier.complete = foo_complete;
    return
        v4l2_async_notifier_register(&foo->v4l2_dev,
                                     &foo->notifier);
}
```

In the following code, we implement the `foo` fwnode (or `of_node`) parser helper, `foo_parse_nodes()`:

```
struct foo_async {
    struct v4l2_async_subdev asd;
    struct v4l2_subdev *sd;
};


/* Either */
static void foo_parse_nodes(struct device *dev,
                            struct v4l2_async_notifier *n)
```

```
{
    struct device_node *node = NULL;
    while ((node = of_graph_get_next_endpoint(dev->of_node,
                                         node))) {
        struct foo_async *fa = kmalloc(sizeof(*fa));
        n->subdevs[n->num_subdevs++] = &fa->asd;
        fa->asd.match.of.node =
        of_graph_get_remote_port_parent(node);
        fa->asd.match_type = V4L2_ASYNC_MATCH_OF;
    }
}
/* Or */
static void foo_parse_nodes(struct device *dev,
                            struct v4l2_async_notifier *n)
{
    struct fwnode_handle *fwnode = dev_fwnode(dev);
    struct fwnode_handle *ep = NULL;
    while ((ep = fwnode_graph_get_next_endpoint(ep, fwnode))) {
        struct foo_async *fa = kmalloc(sizeof(*fa));
        n->subdevs[n->num_subdevs++] = &fa->asd;
        fa->asd.match.fwnode =
                fwnode_graph_get_remote_port_parent(ep);
        fa->asd.match_type = V4L2_ASYNC_MATCH_FWNODE;
    }
}
```

In the preceding code, both of_graph_get_next_endpoint() and
fwnode_graph_get_next_endpoint() have been used in order to show
how to play with the two. That being said, you're better off using the fwnode version,
as it is much more generic.

In the meantime, we need to write foo's notifier operations, which could look as follows:

```
/* foo_bound() and foo_complete() */
static int foo_bound(struct v4l2_async_notifier *n,
                struct v4l2_subdev *sd,
                struct v4l2_async_subdev *asd)
{
    struct foo_async *fa = container_of(asd, struct bar_async,
```

```
                                                    asd);
    /* One can use subdev_call here */
    [...]
    fa->sd = sd;
}


static int foo_complete(struct v4l2_async_notifier *n)
{
    struct foo_device *foo =
            container_of(n, struct foo_async, notifier);
    struct v4l2_device *v4l2_dev = &isp->v4l2_dev;

    /* Create /dev/sub-devX if applicable */
    v4l2_device_register_subdev_nodes(&foo->v4l2_dev);

    /* setup the video device: fops, queue, ioctls ... */
[...]
    /* Register the video device */
        ret = video_register_device(foo->vdev,
                                    VFL_TYPE_GRABBER, -1);

    /* Register with the media controller framework */
    return media_device_register(&bar->mdev);
}
```

In the device tree, the V4L2 bridge device can be declared as follows:

```
csi1: csi@1cb4000 {
    compatible = "allwinner,sun8i-v3s-csi";
    reg = <0x01cb4000 0x1000>;
    interrupts = <GIC_SPI 84 IRQ_TYPE_LEVEL_HIGH>;
    /* we omit clock and others */
[...]

    port {
        csi1_ep: endpoint {
            remote-endpoint = <&ov7740_ep>;
```

```
                  /* We omit v4l2 related properties */
    [...]
          };
       };
    };
```

The camera node from within the I2C controller node can be declared as follows:

```
&i2c1 {
    #address-cells = <1>;
    #size-cells = <0>;

    ov7740: camera@21 {
        compatible = "ovti,ov7740";
        reg = <0x21>;
        /* We omit clock or pincontrol or everything else */

        [...]
        port {
            ov7740_ep: endpoint {
                remote-endpoint = <&csi1_ep>;
                /* We omit v4l2 related properties */
                [...]
            };
        };
    };
};
```

Now we are familiar with the V4L2 async framework and we have seen how the asynchronous sub-device registration eases both the probe and the code. We ended with a concrete example that highlights each aspect we have discussed. Now we can move forward and integrate with the media controller framework, which is the last improvement we can add to our V4L2 drivers.

The Linux media controller framework

Media devices turn out to be very complex, involving several IP blocks of the SoC and thus requiring video stream (re)routing.

Now, let's consider a case where we have a much more sophisticated SoC made of two more on-chip sub-devices – let's say a resizer and an image converter, called `baz` and `biz`.

In the previous example in the *V4L2 async* section, the setup was made up of one bridge device and one sub-device (the fact that it is off-chip does not matter), the camera sensor. This was quite straightforward. Luckily, things worked. But what if now we have to route the stream through the image converter or the image resizer, or even through both IPs? Or, say we have to switch from one to the other (dynamically)?

We could achieve this either via `sysfs` or `ioctls`, but this would have the following problems:

- It would be too ugly (no doubt) and probably buggy.

- It would be too hard (a lot of work).

- It would be deeply SoC vendor-dependent, with possibly a lot of code duplication, no unified user space API and ABI, and no consistency between drivers.

- It would be not a very credible solution.

Many SoCs can reroute internal video streams – for example, capturing them from a sensor and doing memory-to-memory resizing, or sending the sensor output directly to the resizer. Since the V4L2 API did not support these advanced devices, SoC manufacturers made their own custom drivers. However, V4L2 is undisputably the Linux API for capturing images and is sometimes used for specific display devices (these are mem2mem devices).

It is becoming clear that we need another subsystem and framework that covers the limits of V4L2. This is how the Linux media controller framework was born.

The media controller abstraction model

Discovering a device's internal topology and configuring it at runtime is one of the goals of the media framework. To achieve this, it comes with a layer of abstraction. With the media controller framework, hardware devices are represented through an oriented graph made of **entities** whose **pads** are connected via **links**. This set of elements put together forms the so-called **media device**. A source pad can only generate data.

The preceding short description deserves some attention. There are three highlighted words that are of high interest: entity, pad, and link:

- **Entities** are represented by a `struct media_entity` instance, defined in `include/media/media-entity.h`. The structure is usually embedded into a higher-level structure, such as a `v4l2_subdev` or `video_device` instance, although drivers can allocate entities directly.

- **Pads** are the entity's interface to the outside world. These are input- and output-connectable points of a media entity. However, a pad can be either an input (sink pad) or an output (source pad), not both. Data streams from one entity's source pad to another entity's sink pad. Typically, a device such as a sensor or a video decoder would have only an output pad since it only feeds video into the system, and a `/dev/videoX` pad would be modeled as an input pad since it is the end of the stream.

- **Links**: These links can be set, fetched, and enumerated through the media device. The application, for a driver to properly work, is responsible for setting up the links properly so that the driver understands the source and destination of the video data.

All the entities on the system, along with their pads and the connection links between them, give the **media device** shown in the following diagram:

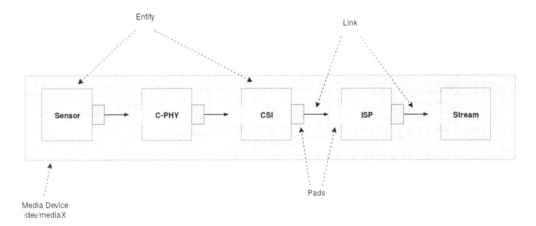

Figure 8.1 – Media controller abstraction model

In the preceding diagram, **Stream** would be the equivalent of a `/dev/videoX` char device as it is the end of the stream.

V4L2 device abstraction

At a higher level, the media controller uses struct media_device to abstract struct v4l2_device in the V4L2 framework. That being said, struct media_device is to the media controller what struct v4l2_device is to V4L2, englobing other lower-level structures. Back to struct v4l2_device, the mdev member is used by the media controller framework to abstract this structure. The following is an excerpt:

```
struct v4l2_device {
[...]
    struct media_device *mdev;
[...]
};
```

However, from a media controller point of view, V4L2 video devices and sub-devices are all seen as media entities, represented in this framework as instances of struct media_entity. It is then obvious for the video device and sub-device data structures to embed a member of this type, as shown in the following excerpt:

```
struct video_device
{
#if defined(CONFIG_MEDIA_CONTROLLER)
    struct media_entity entity;
    struct media_intf_devnode *intf_devnode;
    struct media_pipeline pipe;
#endif
[...]
};

struct v4l2_subdev {
#if defined(CONFIG_MEDIA_CONTROLLER)
    struct media_entity entity;
#endif
[...]
};
```

The video device has additional members, intf_devnode and pipe. The former, of the struct media_intf_devnode type, represents the media controller interface to the video device node. This structure gives the media controller access to information of the underlying video device node, such as its major and minor numbers. The other additional member, pipe, which is of the struct media_pipeline type, stores information related to the streaming pipeline of this video device.

Media controller data structures

The media controller framework is based on a few data structures, among which is the struct media_device structure, which is on top of the hierarchy and defined as follows:

```
struct media_device {
    /* dev->driver_data points to this struct. */
    struct device *dev;
    struct media_devnode *devnode;

    char model[32];
    char driver_name[32];
[...]
    char serial[40];
    u32 hw_revision;
    u64 topology_version;
    struct list_head entities;
    struct list_head pads;
    struct list_head links;

    struct list_head entity_notify;
    struct mutex graph_mutex;
[...]
    const struct media_device_ops *ops;
};
```

This structure represents a high-level media device. It allows easy access to entities and provides basic media device-level support:

- `dev` is the parent device for this media device (usually a `&pci_dev`, `&usb_interface`, or `&platform_device` instance).

- `devnode` is the media device node, abstracting the underlying `/dev/mediaX`.

- `driver_name` is an optional but recommended field, representing the media device driver name. If not set, it defaults to `dev->driver->name`.

- `model` is the model name of this media device. It doesn't have to be unique.

- `serial` is an optional member that should be set with the device serial number. `hw_revision` is the hardware device revision for this media device.

- `topology_version`: Monotonic counter for storing the version of the graph topology. Should be incremented each time the topology changes.

- `entities` is the list of registered entities.

- `pads` is the list of pads registered with this media device.

- `links` is the list of links registered with this media device.

- `entity_notify` is the notify callback list invoked when a new entity is registered with this media device. Drivers may register this callback to take action via `media_device_unregister_entity_notify()` and unregister it using `media_device_register_entity_notify()`. All the registered `media_entity_notify` callbacks are invoked when a new entity is registered.

- `graph_mutex`: Protects access to `struct media_device` data. It should, for example, be held when using `media_graph_*` family functions.

- `ops` is of the `struct media_device_ops` type and represents the operation handler callbacks for this media device.

In addition to being manipulated by the media controller framework, `struct media_device` is essentially used in the bridge driver, where it is initialized and registered. That being said, the media device on its own is made up of several entities. This concept of entities allows the media controller to be the central authority when it comes to modern and complex V4L2 drivers that may also support framebuffers, ALSA, I2C, LIRC, and/or DVB devices at the same time and is used to inform user space of what is what.

A media entity is represented as an instance of `struct media_entity`, defined in `include/media/media-entity.h` as follows:

```
struct media_entity {
    struct media_gobj graph_obj;
    const char *name;
    enum media_entity_type obj_type;
    u32 function;
    unsigned long flags;

    u16 num_pads;
    u16 num_links;
    u16 num_backlinks;
    int internal_idx;

    struct media_pad *pads;
    struct list_head links;
    const struct media_entity_operations *ops;
    int stream_count;
    int use_count;
    struct media_pipeline *pipe;
[...]
};
```

This is the second data structure in the media framework in terms of hierarchy. The preceding definition has been shrunk to the minimum that we are interested in. The following are the meanings of the members in this structure:

- `name` is the name of this entity. It should be meaningful enough as it is used as it is in user space with the `media-ctl` tool.

- `type` is most of the time set by the core depending on the type of V4L2 video data structure this struct is embedded in. It is the type of the object that implements `media_entity` – for example, set with `MEDIA_ENTITY_TYPE_V4L2_SUBDEV` at the sub-device initialization by the core. This allows runtime type identification of media entities and safe casting to the correct object type using the `container_of` macro, for instance. Possible values are as follows:

  --`MEDIA_ENTITY_TYPE_BASE`: This means the entity is not embedded in another.

  --`MEDIA_ENTITY_TYPE_VIDEO_DEVICE`: This indicates the entity is embedded in a `struct video_device` instance.

  --`MEDIA_ENTITY_TYPE_V4L2_SUBDEV`: This means the entity is embedded in a `struct v4l2_subdev` instance.

- `function` represents the entity's main function. This must be set by the driver according to the value defined in `include/uapi/linux/media.h`. The following are commonly used values while dealing with video devices:

  --`MEDIA_ENT_F_IO_V4L`: This flag means the entity is a data streaming input and/or output entity.

  --`MEDIA_ENT_F_CAM_SENSOR`: This flag means this entity is a camera video sensor entity.

  --`MEDIA_ENT_F_PROC_VIDEO_SCALER`: Means this entity can perform video scaling. These entities have at least one sink pad, from which they receive frame(s) (on the active one) and one source pad where they output the scaled frame(s).

--MEDIA_ENT_F_PROC_VIDEO_ENCODER: Means this entity is capable of compressing video. These entities must have one sink pad and at least one source pad.

--MEDIA_ENT_F_VID_MUX: This is to be used for a video multiplexer. This entity has at least two sink pads and one source pad and must pass the video frame(s) received from the active sink pad to the source pad.

--MEDIA_ENT_F_VID_IF_BRIDGE: Video interface bridge. A video interface bridge entity should have at least one sink pad and one source pad. It receives video frames on its sink pad from an input video bus of one type (HDMI, eDP, MIPI CSI-2, and so on) and outputs them on its source pad to an output video bus of another type (eDP, MIPI CSI-2, parallel, and so on).

- flags is set by the driver. It represents the flags for this entity. Possible values are the MEDIA_ENT_FL_* flag family defined in include/uapi/linux/media.h. The following link may be of help to you to understand the possible values: https://linuxtv.org/downloads/v4l-dvb-apis/userspace-api/mediactl/media-types.html.

- function represents this entity's function and by default is MEDIA_ENT_F_V4L2_SUBDEV_UNKNOWN. Possible values are the MEDIA_ENT_F_* function family defined in include/uapi/linux/media.h. For example, a camera sensor sub-device driver must contain sd->entity.function = MEDIA_ENT_F_CAM_SENSOR;. You can follow this link to find detailed information on what may be suitable for your media entity: https://linuxtv.org/downloads/v4l-dvb-apis/uapi/mediactl/media-types.html.

- num_pads is the total number of pads of this entity (sink and source).

- num_links is the total number of links of this entity (forward, back, enabled, and disabled)

- num_backlinks is the numbers of backlinks of this entity. Backlinks are used to help graph traversal and are not reported to user space.

- internal_idx: A unique entity number assigned by the media controller core when the entity is registered.

- pads is the array of pads of this entity. Its size is defined by num_pads.

- links is the list of data links of this entity. See media_add_link().

- ops is of the media_entity_operations type and represents operations for this entity. This structure will be discussed later.

- stream_count: Stream count for the entity.

- `use_count`: The use count for the entity. Used for power management purposes.
- `pipe` is the media pipeline that this entity belongs to.

Naturally, the next data structure that seems obvious for us to introduce is the `struct media_pad` structure, which represents a pad in this framework. A pad is a connection endpoint through which an entity can interact with other entities. Data (not restricted to video) produced by an entity flows from the entity's output to one or more entity inputs. Pads should not be confused with the physical pins at chip boundaries. `struct media_pad` is defined as follows:

```
struct media_pad {
[...]
    struct media_entity *entity;
    u16 index;
    unsigned long flags;
};
```

Pads are identified by their entity and their 0-based `index` in the entity's pads array. In the `flags` field, either `MEDIA_PAD_FL_SINK` (which indicates that the pad supports sinking data) or `MEDIA_PAD_FL_SOURCE` (which indicates that the pad supports sourcing data) can be set, but not both at the same time, since a pad can't both sink and source.

Pads are meant to be bound together to allow data flow paths. Two pads, either from the same entity or from different entities, are bound together by means of point-to-point-oriented connections called links. Links are represented in the media framework as instances of `struct media_link`, defined as follows:

```
struct media_link {
    struct media_gobj graph_obj;
    struct list_head list;
[...]
    struct media_pad *source;
    struct media_pad *sink;
[...]
    struct media_link *reverse;
    unsigned long flags;
    bool is_backlink;
};
```

In the preceding code block, only a few fields have been listed for the sake of readability. The following are the meanings of those fields:

- `list`: Used to associate this link with the entity or interface owning the link.

- `source`: Where this link originates from.

- `sink`: The link target.

- `flags`: Represents the link flags, as defined in `uapi/media.h` (with the `MEDIA_LNK_FL_*` pattern). The following are the possible values:

  --`MEDIA_LNK_FL_ENABLED`: This flag means the link is enabled and is ready for data transfer.

  --`MEDIA_LNK_FL_IMMUTABLE`: This flag means the link enabled state can't be modified at runtime.

  --`MEDIA_LNK_FL_DYNAMIC`: This flag means the state of the link can be modified during streaming. However, this flag is set by drivers but is read-only for applications.

- `reverse`: Pointer to the link (the backlink, actually) for the reverse direction of a pad-to-pad link.

- `is_backlink`: Tells whether this link is a backlink.

Each entity has a list that points to all links originating at or targeting any of its pads. A given link is thus stored twice, once in the source entity and once in the target entity. When you want to link `A` to `B`, two links are actually created:

- One that corresponds to what was expected; the link is stored in the source entity, and the source entity's `num_links` field is incremented.

- Another one is stored in the sink entity. The sink and source remain the same, with the difference being that the `is_backlink` member is set to `true`. This corresponds to the reverse of the link you created. The sink entity's `num_backlinks` and `num_links` fields will be incremented. This backlink is then assigned to the original link's `reverse` member.

At the end, the `mdev->topology_version` member is incremented twice. This principle of link and backlink allows the media controller to numerate entities, along with the possible and current links between entities, such as in the following diagram:

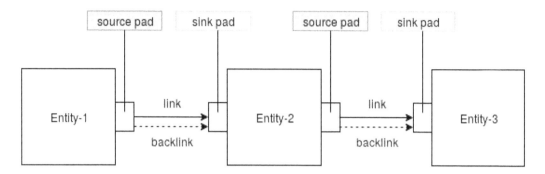

Figure 8.2 – Media controller entity description

In the preceding diagram, if we consider **Entity-1** and **Entity-2**, then **link** and **backlink** are essentially the same, except that **link** belongs to **Entity-1** and **backlink** belongs to **Entity-2**. You should then consider the backlink as a backup link. We can see that an entity can be either a sink, a source, or both.

The data structures we have introduced so far may make the media controller framework sound a bit scary. However, most of those data structures will be managed under the hood by the framework by means of the APIs it offers. That being said, the complete framework's documentation can be found in `Documentation/media-framework.txt` in the kernel sources.

Integrating media controller support in the driver

When the support of the media controller is needed, the V4L2 driver must first initialize `struct media_device` within `struct v4l2_device` using the `media_device_init()` function. Each entity driver must initialize its entities (actually `video_device->entity` or `v4l2_subdev->entity`) and its pad arrays using the `media_entity_pads_init()` function and, if needed, create pad-to-pad links using `media_create_pad_link()`. After that, entities can be registered. However, the V4L2 framework will handle this registration for you through either the `v4l2_device_register_subdev()` or the `video_register_device()` methods. In both cases, the underlying registration function that is invoked is `media_device_register_entity()`.

As a final step, the media device has to be registered using `media_device_register()`. It's worth mentioning that the media device registration should be postponed to later in the future when we are sure that every sub-device (or should I say entities) is registered and ready to be used. It definitely makes sense registering the media device in the root notifier's `.complete` callback.

Initializing and registering pads and entities

The same function is used to initialize both the entity and its pad array:

```
int media_entity_pads_init(struct media_entity *entity,
                        u16 num_pads, struct media_pad *pads);
```

In the preceding prototype, `*entity` is the entity to which the pads to be registered belong, `*pads` is the array of pads to be registered, and `num_pads` is the number of entities in the array that should be registered. The driver must have set the type of every pad in the pads array before calling:

```
struct mydrv_state_struct {
    struct v4l2_subdev sd;
    struct media_pad pad;
[...]
};
static int my_probe(struct i2c_client *client,
                    const struct i2c_device_id *id)
{
    struct v4l2_subdev *sd;
    struct mydrv_state_struct *my_struct;
[...]
    sd = &my_struct->sd;
    my_struct->pad.flags = MEDIA_PAD_FL_SINK |
                            MEDIA_PAD_FL_MUST_CONNECT;
    ret = media_entity_pads_init(&sd->entity, 1,
                            &my_struct->pad);
[...]
    return 0;
}
```

Drivers that need to unregister entities must call the following function on the entity to be unregistered:

```
media_device_unregister_entity(struct media_entity *entity);
```

Then, in order for a driver to free resources associated with an entity, it should call the following:

```
media_entity_cleanup(struct media_entity *entity);
```

When a media device is unregistered, all of its entities are unregistered automatically. No unregistration of manual entities is then required.

Media entity operations

An entity may be provided link-related callbacks, so that these can be invoked by the media framework upon link creation and validation:

```
struct media_entity_operations {
    int (*get_fwnode_pad)(struct fwnode_endpoint *endpoint);
    int (*link_setup)(struct media_entity *entity,
                      const struct media_pad *local,
                      const struct media_pad *remote,
                      u32 flags);
    int (*link_validate)(struct media_link *link);
};
```

Providing the preceding structure is optional. However, there may be situations where additional stuff needs to be done or checked either at link setup or link validation. In this case, note the following descriptions:

- get_fwnode_pad: Returns the pad number based on a fwnode endpoint or a negative value on error. This operation can be used to map a fwnode to a media pad number (optional).

- link_setup: Notifies the entity of link changes. This operation can return an error, in which case the link setup will be canceled (optional).

- `link_validate`: Returns whether a link is valid from the entity point of view. The `media_pipeline_start()` function validates all the links this entity is involved in by calling this operation. This member is optional. However, if it has not been set, then `v4l2_subdev_link_validate_default` will be used as the default callback function, which ensures that the source pad and sink pad width, height, and media bus pixels code are consistent; otherwise, it will return an error.

The concept of a media bus

The main purpose of the media framework is to configure and control the pipeline and its entities. Video sub-devices, such as cameras and decoders, connect to video bridges or other sub-devices over specialized buses. Data is being transferred over these buses in various formats. That being said, in order for two entities to actually exchange data, their pad configs need to be the same.

Applications are responsible for configuring coherent parameters on the whole pipeline and ensuring that connected pads have compatible formats. The pipeline is checked for formats that are mismatching at `VIDIOC_STREAMON` time.

The driver is responsible for applying the configuration of every block in the video pipeline according to the requested (from the user) format at the pipeline input and/or output.

Take the following simple data flow, `sensor ---> CPHY ---> csi ---> isp ---> stream`.

In order for the media framework to be able to configure the bus prior to streaming data, the driver needs to provide some pad-level setter and getter for the media bus properties, which are present in the `struct v4l2_subdev_pad_ops` structure. This structure implements pad-level operations that have to be defined if the sub-device driver intends to process the video and integrate with the media framework. The following is its definition:

```
struct v4l2_subdev_pad_ops {
[...]
    int (*enum_mbus_code)(struct v4l2_subdev *sd,
                    struct v4l2_subdev_pad_config *cfg,
                    struct v4l2_subdev_mbus_code_enum *code);
    int (*enum_frame_size)(struct v4l2_subdev *sd,
                    struct v4l2_subdev_pad_config *cfg,
                    struct v4l2_subdev_frame_size_enum *fse);
    int (*enum_frame_interval)(struct v4l2_subdev *sd,
                struct v4l2_subdev_pad_config *cfg,
```

```
                    struct v4l2_subdev_frame_interval_enum *fie);
    int (*get_fmt)(struct v4l2_subdev *sd,
                    struct v4l2_subdev_pad_config *cfg,
                    struct v4l2_subdev_format *format);
    int (*set_fmt)(struct v4l2_subdev *sd,
                    struct v4l2_subdev_pad_config *cfg,
                    struct v4l2_subdev_format *format);
#ifdef CONFIG_MEDIA_CONTROLLER
    int (*link_validate)(struct v4l2_subdev *sd,
                        struct media_link *link,
                        struct v4l2_subdev_format *source_fmt,
                        struct v4l2_subdev_format *sink_fmt);
#endif /* CONFIG_MEDIA_CONTROLLER */
[...]
};
```

The following are the meanings of the members in this structure:

- `init_cfg`: Initializes the pad config to default values. This is the right place to initialize `cfg->try_fmt`, which can be grabbed through `v4l2_subdev_get_try_format()`.

- `enum_mbus_code`: Callback for the `VIDIOC_SUBDEV_ENUM_MBUS_CODE` ioctl handler code. Enumerates the currently supported data format. This callback handles pixel format enumeration.

- `enum_frame_size`: Callback for the `VIDIOC_SUBDEV_ENUM_FRAME_SIZE` ioctl handler code. Enumerates the frame (image) size supported by the sub-device. Enumerates the currently supported resolution.

- `enum_frame_interval`: Callback for the `VIDIOC_SUBDEV_ENUM_FRAME_INTERVAL` ioctl handler code.

- `get_fmt`: Callback for the `VIDIOC_SUBDEV_G_FMT` ioctl handler code.

- `set_fmt`: Callback for the `VIDIOC_SUBDEV_S_FMT` ioctl handler code. Sets the output data format and resolution.

- `get_selection`: Callback for the `VIDIOC_SUBDEV_G_SELECTION` ioctl handler code.

- `set_selection`: Callback for the `VIDIOC_SUBDEV_S_SELECTION` ioctl handler code.

- `link_validate`: Used by the media controller code to check whether the links that belong to a pipeline can be used for the stream.

The argument that all of these callbacks have in common is `cfg`, which is of the `struct v4l2_subdev_pad_config` type and is used for storing sub-device pad information. This structure is defined in `include/uapi/linux/v4l2-mediabus.h` as follows:

```
struct v4l2_subdev_pad_config {
    struct v4l2_mbus_framefmt try_fmt;
    struct v4l2_rect try_crop;
[...]
};
```

In the preceding code block, the main field we are interested in is `try_fmt`, which is of the `struct v4l2_mbus_framefmt` type. This data structure is used to describe the pad-level media bus format and is defined as follows:

```
struct v4l2_subdev_format {
    __u32 which;
    __u32 pad;
    struct v4l2_mbus_framefmt format;
[...]
};
```

In the preceding structure, `which` is the format type (try or active) and `pad` is the pad number as reported by the media API. This field is set by user space. `format` represents the frame format on the bus. The `format` term here means a combination of the media bus data format, frame width, and frame height. It is of the `struct v4l2_mbus_framefmt` type and its turn is defined as follows:

```
struct v4l2_mbus_framefmt {
    __u32  width;
    __u32  height;
    __u32  code;
    __u32  field;
    __u32  colorspace;
[...]
};
```

In the preceding bus frame format data structure, only the fields that are relevant to us have been listed. `width` and `height`, respectively, represent the image width and height. `code` is from enum `v4l2_mbus_pixelcode` and represents the data format code. `field` indicates the used interlacing type, which should be from enum `v4l2_field`, and `colorspace` represents the color space of the data from enum `v4l2_colorspace`.

Now, let's pay more attention to the `get_fmt` and `set_fmt` callbacks. They get and set, respectively, the data format on a sub-device pad. These ioctl handlers are used to negotiate the frame format at specific sub-device pads in the image pipeline. To set the current format applications, set the `.pad` field of `struct v4l2_subdev_format` to the desired pad number as reported by the media API and the `which` field (which is from enum `v4l2_subdev_format_whence`) to either `V4L2_SUBDEV_FORMAT_TRY` or `V4L2_SUBDEV_FORMAT_ACTIVE`, and issue a `VIDIOC_SUBDEV_S_FMT` ioctl with a pointer to this structure. This ioctl ends up calling the `v4l2_subdev_pad_ops->set_fmt` callback. If `which` is set to `V4L2_SUBDEV_FORMAT_TRY`, then the driver should set the `.try_fmt` field of the requested pad config with the values of the `try` format given in the argument. However, if `which` is set to `V4L2_SUBDEV_FORMAT_ACTIVE`, the driver must then apply the config to the device. It is common in this case to store the requested "active" format in a driver-state structure and apply it to the underlying device when the pipeline starts the stream. This way, the right place to actually apply the format config to the device is from within a callback invoked at the start of the streaming, such as `v4l2_subdev_video_ops.s_stream`, for example. The following is an example from the RCAR CSI driver:

```
static int rcsi2_set_pad_format(struct v4l2_subdev *sd,
                                struct v4l2_subdev_pad_config *cfg,
                                struct v4l2_subdev_format *format)
{
    struct v4l2_mbus_framefmt *framefmt;

    /* retrieve the private data structure */
    struct rcar_csi2 *priv = sd_to_csi2(sd);
    [...]

    /* Store the requested format so that it can be applied to
     * the device when the pipeline starts
     */
    if (format->which == V4L2_SUBDEV_FORMAT_ACTIVE) {
        priv->mf = format->format;
```

```
    } else { /* V4L2_SUBDEV_FORMAT_TRY */

        /* set the .try_fmt of this pad config with the
         * value of the requested "try" format
         */
        framefmt = v4l2_subdev_get_try_format(sd, cfg, 0);
        *framefmt = format->format;

        /* driver is free to update any format->* field */
        [...]
    }
    return 0;
}
```

Also, note that the driver is free to change the values in the requested format to the one it actually supports. It is then up to the application to check for it and adapt its logic according to the format granted by the driver. Modifying those try formats leaves the device state untouched.

On the other hand, when it comes to retrieving the current format, applications should do the same as the preceding and issue a VIDIOC_SUBDEV_G_FMT ioctl. This ioctl will end up calling the v4l2_subdev_pad_ops->get_fmt callback. The driver fills the members of the format field either with the currently active format values or with the last try format stored (most of the time in the driver-state structure):

```
static int rcsi2_get_pad_format(struct v4l2_subdev *sd,
                                struct v4l2_subdev_pad_config *cfg,
                                struct v4l2_subdev_format *format)
{
    struct rcar_csi2 *priv = sd_to_csi2(sd);

    if (format->which == V4L2_SUBDEV_FORMAT_ACTIVE)
        format->format = priv->mf;
    else
        format->format = *v4l2_subdev_get_try_format(sd, cfg, 0);

    return 0;
}
```

It is obvious that the .try_fmt field of the pad config should have been initialized before it can be passed to the get callback for the first time, and the v4l2_subdev_pad_ops. init_cfg callback is the right place for this initialization, as in the following example:

```
/*
 * Initializes the TRY format to the ACTIVE format on all pads
 * of a subdev. Can be used as the .init_cfg pad operation.
 */
int imx_media_init_cfg(struct v4l2_subdev *sd,
                       struct v4l2_subdev_pad_config *cfg)
{
    struct v4l2_mbus_framefmt *mf_try;
    struct v4l2_subdev_format format;
    unsigned int pad;
    int ret;

    for (pad = 0; pad < sd->entity.num_pads; pad++) {
        memset(&format, 0, sizeof(format));

        format.pad = pad;
        format.which = V4L2_SUBDEV_FORMAT_ACTIVE;
        ret = v4l2_subdev_call(sd, pad, get_fmt, NULL, &format);
        if (ret)
            continue;

        mf_try = v4l2_subdev_get_try_format(sd, cfg, pad);
        *mf_try = format.format;
    }

    return 0;
}
```

> **Important note**
>
> The list of supported formats can be found in `include/uapi/linux/videodev2.h` from the kernel source, and part of their documentation is available at this link: `https://linuxtv.org/downloads/v4l-dvb-apis/userspace-api/v4l/subdev-formats.html`.

Now that we are familiar with the concept of media, we can learn how to finally make the media device part of the system by using the appropriate API to register it.

Registering the media device

Drivers register media device instances by calling `__media_device_register()` via the `media_device_register()` macro and unregister them by calling `media_device_unregister()`. Upon successful registration, a character device named `media[0-9]` + will be created. The device major and minor numbers are dynamic. `media_device_register()` accepts a pointer to the media device to be registered and returns 0 on success or a negative error code on error.

As we said earlier, you're better off registering the media device from within the root notifier's `.complete` callback in order to make sure that the actual media device is registered only after all its entities have been probed. The following is an excerpt from the TI OMAP3 ISP media driver (the whole code can be found in `drivers/media/platform/omap3isp/isp.c` in the kernel sources):

```c
static int isp_subdev_notifier_complete(
                        struct v4l2_async_notifier *async)
{
    struct isp_device *isp =
            container_of(async, struct isp_device, notifier);
[...]
    return media_device_register(&isp->media_dev);
}

static const
struct v4l2_async_notifier_operations isp_subdev_notifier_ops =
{
    .complete = isp_subdev_notifier_complete,
};
```

The preceding code shows how you can take benefit of the root notifier's `.complete` callback to register the final media device, by means of the `media_device_register()` method.

Now that the media device is part of the system, the time has come to leverage it, particularly from user space. Let's now see how, from the command line, we can take control of and interact with the media device.

Media controller from user space

Though it remains the streaming interface, `/dev/video0` is not the default pipeline centerpiece anymore since it is wrapped by `/dev/mediaX`. The pipeline can be configured through the media node (`/dev/media*`), and the control operations, such as stream on/off, can be performed through the video node (`/dev/video*`).

Using media-ctl (the v4l-utils package)

The `media-ctl` application from the `v4l-utils` package is a user space application that uses the Linux media controller API to configure pipelines. The following are the flags to use with it:

- `--device <dev>` specifies the media device (`/dev/media0` by default).
- `--entity <name>` prints the device name associated with the given entity.
- `--set-v4l2 <v4l2>` provides a comma-separated list of formats to set up.
- `--get-v4l2 <pad>` prints an active format on a given pad.
- `--set-dv <pad>` configures DV timings on a given pad.
- `--interactive` modifies links interactively.
- `--links <linux>` provides a comma-separated list of link descriptors to set up.
- `--known-mbus-fmts` lists known formats and their numeric values.
- `--print-topology` prints the device topology, or the short version, `-p`.
- `--reset` resets all links to inactive.

That being said, the basic configuration steps for a hardware media pipeline are as follows:

1. Reset all links with `media-ctl --reset`.
2. Configure links with `media-ctl --links`.

3. Configure pad formats with `media-ctl --set-v4l2`.

4. Configure sub-device properties with `v4l2-ctl` capture frames on the `/dev/video*` device.

Using `media-ctl --links` to link an entity source pad to an entity sink pad should follow the following pattern:

```
media-ctl --links\
 "<entitya>:<srcpadn> -> <entityb>:<sinkpadn>[<flags>]
```

In the preceding line, `flags` can be either `0` (inactive) or `1` (active). Additionally, to see the current settings of the media bus, use the following:

```
$ media-ctl --print-topology
```

On some systems, media device `0` may not be the default one, in which case you should use the following:

```
$ media-ctl --device /dev/mediaN --print-topology
```

The previous command would print the media topology associated with the specified media device.

Do note that `--print-topology` just dumps the media topology on the console in an ASCII format. However, this topology can be better represented by generating its `dot` representation, changing this representation into a graphic image that is more human-friendly. The following are the commands to use:

```
$ media-ctl --print-dot > graph.dot
$ dot -Tpng graph.dot > graph.png
```

For example, in order to set up a media pipe, the following commands have been run on an UDOO QUAD board. The board has been shipped with an i.MX6 quad core and an OV5640 camera plugged into the MIPI CSI-2 connector:

```
# media-ctl -l "'ov5640 2-003c':0 -> 'imx6-mipi-csi2':0[1]"
# media-ctl -l "'imx6-mipi-csi2':2 -> 'ipu1_csi1':0[1]"
# media-ctl -l "'ipu1_csi1':1 -> 'ipu1_ic_prp':0[1]"
# media-ctl -l "'ipu1_ic_prp':1 -> 'ipu1_ic_prpenc':0[1]"
# media-ctl -l "'ipu1_ic_prpenc':1 -> 'ipu1_ic_prpenc
capture':0[1]"
```

The following is a diagram representing the preceding setup:

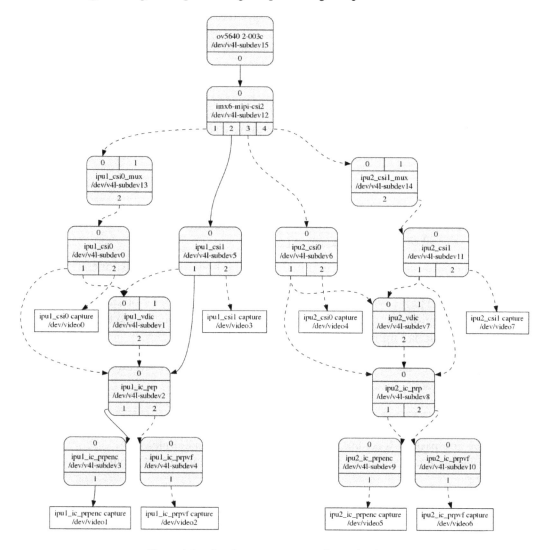

Figure 8.3 – Graph representation of a media device

As you can see, it helps to visualize what the hardware components are. The following are descriptions of these generated images:

- Dashed lines show possible connections. You can use these to determine the possibilities.
- Solid lines show active connections.

- Green boxes show media entities.

- Yellow boxes show **Video4Linux** (**V4L**) endpoints.

After that, you can see that solid lines correspond exactly to the setup that was done earlier. We have five solid lines, which correspond to the number of commands used to configure the media device. The following are the meanings of these commands:

- `media-ctl -l "'ov5640 2-003c':0 -> 'imx6-mipi-csi2':0[1]"` means linking output pad number 0 of the camera sensor (`'ov5640 2-003c':0`) to MIPI CSI-2 input pad number 0 (`'imx6-mipi-csi2':0`) and setting this link active (`[1]`).

- `media-ctl -l "'imx6-mipi-csi2':2 -> 'ipu1_csi1':0[1]"` means linking output pad number 2 of the MIPI CSI-2 entity (`'imx6-mipi-csi2':2`) to the input pad number 0 of the IPU capture sensor interface #1 (`'ipu1_csi1':0`) and setting this link active (`[1]`).

- The same decoding rules apply to other command lines, until the last one, `media-ctl -l "'ipu1_ic_prpenc':1 -> 'ipu1_ic_prpenc capture':0[1]"`, which means linking output pad number 1 of ipu1's image converter preprocessing encode entity (`'ipu1_ic_prpenc':1`) to the capture interface input pad number 0 and setting this link to active.

Do not hesitate to go back to the image and read those descriptions several times in order to understand the concepts of entity, link, and pad.

Important note

If the `dot` package is not installed on your target, you can download the `.dot` file on your host (assuming it has the package installed) and convert it into an image.

WaRP7 with an OV2680 example

The WaRP7 is an i.MX7-based board, which, unlike the i.MX5/6 family, does not contain an IPU. Because of this, there are fewer capabilities to perform operations or manipulation of the capture frames. The i.MX7 image capture chain is made up of three units: the camera censor interface, the video multiplexer, and the MIPI CSI-2 receiver, which represent the media entities, described as follows:

- `imx7-mipi-csi2`: This is the MIPI CSI-2 receiver entity. It has one sink pad to receive the pixel data from the MIPI CSI-2 camera sensor. It has one source pad, corresponding to virtual channel 0.

- csi_mux: This is the video multiplexer. It has two sink pads to select from either camera sensors with a parallel interface or MIPI CSI-2 virtual channel 0. It has a single source pad that routes to the CSI.

- csi: The CSI allows the chip to connect directly to the external CMOS image sensor. The CSI can interface directly with parallel and MIPI CSI-2 buses. It has 256 x 64 FIFO to store received image pixel data and embedded DMA controllers to transfer data from the FIFO through the AHB bus. This entity has one sink pad that receives from the csi_mux entity and a single source pad that routes video frames directly to memory buffers. This pad is routed to a capture device node:

```
                                   |\
MIPI Camera Input --> MIPI CSI-2 -- > |  \
                                   |   \
                                   | M |
                                   | U |  --> CSI --> Capture
                                   | X |
                                   |  /
Parallel Camera Input --------------> |  /
                                   |/
```

On this platform, an OV2680 MIPI CSI-2 module is connected to the internal MIPI CSI-2 receiver. The following example configures a video capture pipeline with an output of 800 x 600 in BGGR 10-bit Bayer format:

```
# Setup links
media-ctl --reset
media-ctl -l "'ov2680 1-0036':0 -> 'imx7-mipi-csis.0':0[1]"
media-ctl -l "'imx7-mipi-csis.0':1 -> 'csi_mux':1[1]"
media-ctl -l "'csi_mux':2 -> 'csi':0[1]"
media-ctl -l "'csi':1 -> 'csi capture':0[1]"
```

The preceding lines could be merged into one single command, as follows:

```
media-ctl -r -l '"ov2680 1-0036":0->"imx7-mipi-csis.0":0[1], \
                "imx7-mipi-csis.0":1 ->"csi_mux":1[1], \
                "csi_mux":2->"csi":0[1], \
                "csi":1->"csi capture":0[1]'
```

In the preceding commands, note the following:

- `-r` means reset all links to inactive.

- `-l` sets up links in a comma-separated list of the links' descriptors.

- `"ov2680 1-0036":0->"imx7-mipi-csis.0":0[1]` links output pad number 0 of the camera sensor to MIPI CSI-2 input pad number 0 and sets this link to active.

- `"csi_mux":2->"csi":0[1]` links output pad number 2 of `csi_mux` to `csi` input pad number 0 and sets this link to active.

- `"csi":1->"csi capture":0[1]` links output pad number 1 of `csi` to capture the interface's input pad number 0 and sets this link to active.

In order to configure the format on each pad, we can use the following commands:

```
# Configure pads for pipeline
media-ctl -V "'ov2680 1-0036':0 [fmt:SBGGR10_1X10/800x600
field:none]"
media-ctl -V "'csi_mux':1 [fmt:SBGGR10_1X10/800x600
field:none]"
media-ctl -V "'csi_mux':2 [fmt:SBGGR10_1X10/800x600
field:none]"
media-ctl \
      -V "'imx7-mipi-csis.0':0 [fmt:SBGGR10_1X10/800x600
field:none]"
media-ctl -V "'csi':0 [fmt:SBGGR10_1X10/800x600 field:none]"
```

Once again, the preceding command lines could be merged into a single command, as follows:

```
media-ctl \
    -f '"ov2680 1-0036":0 [SGRBG10 800x600 (32,20)/800x600], \
        "csi_mux":1 [SGRBG10 800x600], \
        "csi_mux":2 [SGRBG10 800x600], \
        "mx7-mipi-csis.0":2 [SGRBG10 800x600], \
        "imx7-mipi-csi.0":0 [SGRBG10 800x600], \
        "csi":0 [UYVY 800x600]'
```

The preceding command lines could be translated as follows:

- -f: Sets up pad formats into a comma-separated list of format descriptors.

- "ov2680 1-0036":0 [SGRBG10 800x600 (32,20)/800x600]: Sets up the camera sensor pad number 0 format to a RAW Bayer 10-bit image with a resolution (capture size) of 800 x 600. Sets the maximum allowed sensor window width by specifying the crop rectangle.

- "csi_mux":1 [SGRBG10 800x600]: Sets up the csi_mux pad number 1 format to a RAW Bayer 10-bit image with a resolution of 800 x 600.

- "csi_mux":2 [SGRBG10 800x600]: Sets up the csi_mux pad number 2 format to a RAW Bayer 10-bit image with a resolution of 800 x 600.

- "csi":0 [UYVY 800x600]: Sets up the csi pad number 0 format to a YUV4:2:2 image with a resolution of 800 x 600.

video_mux, csi, and mipi-csi-2 are all part of the SoC, so they are declared in the vendor dtsi file (that is, arch/arm/boot/dts/imx7s.dtsi in the kernel sources). video_mux is declared as follows:

```
gpr: iomuxc-gpr@30340000 {
[...]
    video_mux: csi-mux {
        compatible = "video-mux";
        mux-controls = <&mux 0>;
        #address-cells = <1>;
        #size-cells = <0>;
        status = "disabled";

        port@0 {
            reg = <0>;
        };
        port@1 {
            reg = <1>;
            csi_mux_from_mipi_vc0: endpoint {
                remote-endpoint = <&mipi_vc0_to_csi_mux>;
            };
        };
        port@2 {
```

```
            reg = <2>;
          csi_mux_to_csi: endpoint {
              remote-endpoint = <&csi_from_csi_mux>;
          };
        };
      };
    };
```

In the preceding code block, we have three ports, where ports 1 are 2 are connected to remote endpoints. `csi` and `mipi-csi-2` are declared as follows:

```
mipi_csi: mipi-csi@30750000 {
    compatible = "fsl,imx7-mipi-csi2";
[...]
    status = "disabled";

    port@0 {
        reg = <0>;
    };

    port@1 {
        reg = <1>;

        mipi_vc0_to_csi_mux: endpoint {
            remote-endpoint = <&csi_mux_from_mipi_vc0>;
        };
    };
};

[...]
csi: csi@30710000 {
    compatible = "fsl,imx7-csi"; [...]
    status = "disabled";

    port {
        csi_from_csi_mux: endpoint {
            remote-endpoint = <&csi_mux_to_csi>;
```

```
        };
    };
};
```

From the `csi` and `mipi-csi-2` nodes, we can see how they are linked to their remote ports in the `video_mux` node.

> **Important note**
>
> More information on `video_mux` binding can be found in `Documentation/devicetree/bindings/media/video-mux.txt` in the kernel sources.

However, most of the vendor-declared nodes are disabled by default, and need to be enabled from within the board file (the `dts` file, actually). This is what is done in the following code block. Moreover, the camera sensor is part of the board, not the SoC. So, it needs to be declared in the board `dts` file, which is `arch/arm/boot/dts/imx7s-warp.dts` in kernel sources. The following is an excerpt:

```
&video_mux {
    status = "okay";
};


&mipi_csi {
    clock-frequency = <166000000>;
    fsl,csis-hs-settle = <3>;
    status = "okay";

    port@0 {
        reg = <0>;

        mipi_from_sensor: endpoint {
            remote-endpoint = <&ov2680_to_mipi>;
            data-lanes = <1>;
        };

    };
};
```

```
&i2c2 {
    [...]
    status = "okay";

    ov2680: camera@36 {
        compatible = "ovti,ov2680";
        [...]

    port {
        ov2680_to_mipi: endpoint {
            remote-endpoint = <&mipi_from_sensor>;
            clock-lanes = <0>;
            data-lanes = <1>;
        };
    };
};
```

> **Important note**
>
> More on i.MX7 entity binding can be found in both `Documentation/devicetree/bindings/media/imx7-csi.txt` and `Documentation/devicetree/bindings/media/imx7-mipi-csi2.txt` in the kernel sources.

After this, the streaming can start. The `v4l2-ctl` tool can be used to select any of the resolutions supported by the sensor:

```
root@imx7s-warp:~# media-ctl -p
Media controller API version 4.17.0

Media device information
------------------------
driver          imx7-csi
model           imx-media
serial
bus info
hw revision     0x0
```

```
driver version   4.17.0

Device topology
- entity 1: csi (2 pads, 2 links)
          type V4L2 subdev subtype Unknown flags 0
          device node name /dev/v4l-subdev0
      pad0: Sink
            [fmt:SBGGR10_1X10/800x600 field:none]
            <- "csi-mux":2 [ENABLED]
      pad1: Source
            [fmt:SBGGR10_1X10/800x600 field:none]
            -> "csi capture":0 [ENABLED]

- entity 4: csi capture (1 pad, 1 link)
          type Node subtype V4L flags 0
          device node name /dev/video0
      pad0: Sink
            <- "csi":1 [ENABLED]

- entity 10: csi-mux (3 pads, 2 links)
          type V4L2 subdev subtype Unknown flags 0
          device node name /dev/v4l-subdev1
      pad0: Sink
            [fmt:unknown/0x0]
      pad1: Sink
            [fmt:unknown/800x600 field:none]
            <- "imx7-mipi-csis.0":1 [ENABLED]
      pad2: Source
            [fmt:unknown/800x600 field:none]
            -> "csi":0 [ENABLED]

- entity 14: imx7-mipi-csis.0 (2 pads, 2 links)
          type V4L2 subdev subtype Unknown flags 0
          device node name /dev/v4l-subdev2
      pad0: Sink
            [fmt:SBGGR10_1X10/800x600 field:none]
```

```
                <- "ov2680 1-0036":0 [ENABLED]
        pad1: Source
                [fmt:SBGGR10_1X10/800x600 field:none]
                -> "csi-mux":1 [ENABLED]

- entity 17: ov2680 1-0036 (1 pad, 1 link)
             type V4L2 subdev subtype Sensor flags 0
             device node name /dev/v4l-subdev3
        pad0: Source
                [fmt:SBGGR10_1X10/800x600 field:none]
                -> "imx7-mipi-csis.0":0 [ENABLED]
```

As data streams from left to right, we can interpret the preceding console logs as follows:

- `-> "imx7-mipi-csis.0":0 [ENABLED]`: This source pad feeds data to the entity on its right, which is `"imx7-mipi-csis.0":0`.

- `<- "ov2680 1-0036":0 [ENABLED]`: This sink pad is fed by (that is, it queries data from) the entity to its left, which is `"ov2680 1-0036":0`.

We are now done with all the aspects of the media controller framework. We started with its architecture, then described the data structure it is made of, and then learned about its API in detail. We ended with its use from user space in order to leverage the mode media pipe.

Summary

In this chapter, we went through the V4L2 asynchronous interface, which eases video bridge and sub-device driver probing. This is useful for intrinsically asynchronous and unordered device registration systems, such as flattened device tree driver probing. Moreover, we dealt with the media controller framework, which allows leveraging V4L2 video pipelines. What we have seen so far lies in the kernel space.

In the next chapter, we will see how to deal with V4L2 devices from user space, thus leveraging features exposed by their device drivers.

9
Leveraging the V4L2 API from the User Space

The main purpose of device drivers is controlling and leveraging the underlying hardware while exposing functionalities to users. These users may be applications running in user space or other kernel drivers. While the two previous chapters dealt with V4L2 device drivers, in this chapter, we will learn how to take advantage of V4L2 device functionalities exposed by the kernel. We will start by describing and enumerating user space V4L2 APIs, and then we will learn how to leverage those APIs to grab video data from the sensor, including mangling the sensor properties.

This chapter will cover the following topics:

- V4L2 user space APIs
- Video device property management from user space
- Buffer management from user space
- V4L2 user space tools

Technical requirements

In order to make the most out of this chapter, you will need the following:

- Advanced computer architecture knowledge and C programming skills

- Linux kernel v4.19.X sources, available at `https://git.kernel.org/pub/scm/linux/kernel/git/stable/linux.git/refs/tags`

Introduction to V4L2 from user space

The main purpose of writing device drivers is to ease the control and usage of the underlying device by the application. There are two ways for user space to deal with V4L2 devices: either by using all-in-one utilities such as `GStreamer` and its `gst-*` tools or by writing a dedicated application using user space V4L2 APIs. In this chapter, we only deal with the code, thus we will cover how to write applications that use the V4L2 API.

The V4L2 user space API

The V4L2 user space API has a reduced number of functions and a lot of data structures, all defined in `include/uapi/linux/videodev2.h`. In this section, we will try to describe the most important of them—or, better said, the most commonly used. Your code should include the following header:

```
#include <linux/videodev2.h>
```

This API relies on the following functions:

- `open()`: To open a video device

- `close()`: To close a video device

- `ioctl()`: To send ioctl commands to the display driver

- `mmap()`: To memory map a driver-allocated buffer to user space

- `read()` or `write()`, depending on the streaming method

This reduced set of APIs is extended by a very large number of ioctl commands, the most important of which are as follows:

- `VIDIOC_QUERYCAP`: This is used to query the capabilities of the driver. People used to say it is used to query the device's capabilities, but this is not true as the device may be capable of things that are not implemented in the driver. User space passes a `struct v4l2_capability` structure, which will be filled by the video driver with the relevant information.

- VIDIOC_ENUM_FMT: This is used to enumerate the image formats that are supported by the driver. The driver user space passes a `struct v4l2_fmtdesc` structure, which will be filled by the driver with the relevant information.

- VIDIOC_G_FMT: For a capture device, this is used to get the current image format. However, for a display device, you use this to get the current display window. In either case, user space passes a `struct v4l2_format` structure, which will be filled by the driver with the relevant information.

- VIDIOC_TRY_FMT should be used when you are unsure about the format to be submitted to the device. This is used to validate a new image format for a capture device or a new display window depending on an output (display) device. User space passes a `struct v4l2_format` structure with the properties it would like to apply, and the driver may change the given values if they are not supported. The application should then check what is granted.

- VIDIOC_S_FMT is used to set a new image format for a capture device or a new display window for a display (output device). The driver may change the values passed by user space if they are not supported. The application should check what is granted if VIDIOC_TRY_FMT is not used first.

- VIDIOC_CROPCAP is used to get the default cropping rectangle based on the current image size and the current display panel size. The driver fills a `struct v4l2_cropcap` structure.

- VIDIOC_G_CROP is used to get the current cropping rectangle. The driver fills a `struct v4l2_crop` structure.

- VIDIOC_S_CROP is used to set a new cropping rectangle. The driver fills a `struct v4l2_crop` structure. The application should check what is granted.

- VIDIOC_REQBUFS: This ioctl is used to request a number of buffers that can later be memory mapped. The driver fills a `struct v4l2_requestbuffers` structure. As the driver may allocate more or less than the actual number of buffers requested, the application should check how many buffers are really granted. No buffer is queued yet after this.

- The VIDIOC_QUERYBUF ioctl is used to get a buffer's information, which can be used by the `mmap()` system call in order to map that buffer to user space. The driver fills a `struct v4l2_buffer` structure.

- VIDIOC_QBUF is used to queue a buffer by passing a struct v4l2_buffer structure associated with that buffer. On the execution path of this ioctl, the driver will add this buffer to its list of buffers so that it is filled when there are no more pending queued buffers before it. Once the buffer is filled, it is passed to the V4L2 core, which maintains its own list (that is, a ready buffer list) and it is moved off the driver's list of DMA buffers.

- VIDIOC_DQBUF is used to dequeue a filled buffer (from the V4L2's list of ready buffers for the input device) or a displayed (output device) buffer by passing a struct v4l2_buffer structure associated with that buffer. This will block if no buffer is ready unless O_NONBLOCK was used with open(), in which case VIDIOC_DQBUF will immediately return with an EAGAIN error code. You should call VIDIOC_DQBUF only after STREAMON has been called. In the meantime, calling this ioctl after STREAMOFF would return -EINVAL.

- VIDIOC_STREAMON is used to turn on streaming. After that, any VIDIOC_QBUF results in an image are rendered.

- VIDIOC_STREAMOFF is used to turn off streaming. This ioctl removes all buffers. It actually flushes the buffer queue.

There are many more ioctl commands than those we have just enumerated. There are actually at least as many ioctls as there are ops in the kernel's v4l2_ioctl_ops data structure. However, the preceding ioctls are enough to go deeper into the V4L2 user space API. In this section, we will not go into detail about each data structure. You should then keep open the include/uapi/linux/videodev2.h file, also available at https://elixir.bootlin.com/linux/v4.19/source/include/uapi/linux/videodev2.h, as it contains all the V4L2 APIs and data structures. That being said, the following pseudo-code shows a typical ioctl sequence to grab video from user space using V4L2 APIs:

```
open()
int ioctl(int fd, VIDIOC_QUERYCAP,
        struct v4l2_capability *argp)
int ioctl(int fd, VIDIOC_S_FMT, struct v4l2_format *argp)
int ioctl(int fd, VIDIOC_S_FMT, struct v4l2_format *argp)
/* requesting N buffers */
int ioctl(int fd, VIDIOC_REQBUFS,
        struct v4l2_requestbuffers *argp)
/* queueing N buffers */
int ioctl(int fd, VIDIOC_QBUF, struct v4l2_buffer *argp)
/* start streaming */
```

```
int ioctl(int fd, VIDIOC_STREAMON, const int *argp)
read_loop: (for i=0; I < N; i++)
    /* Dequeue buffer i */
    int ioctl(int fd, VIDIOC_DQBUF, struct v4l2_buffer *argp)
    process_buffer(i)
    /* Requeue buffer i */
    int ioctl(int fd, VIDIOC_QBUF, struct v4l2_buffer *argp)
end_loop
    releases_memories()
    close()
```

The preceding sequence will serve as a guideline to deal with the V4L2 API in user space.

Be aware that it is possible for the ioctl system call to return a -1 value while errno = EINTR. In this case, it would not mean an error but simply that the system call was interrupted, in which case it should be tried again. To address this (rare but possible) issue, we can consider writing our own wrapper for ioctl, such as the following:

```
static int xioctl(int fh, int request, void *arg)
{
        int r;

        do {
                r = ioctl(fh, request, arg);
        } while (-1 == r && EINTR == errno);

        return r;
}
```

Now that we are done with the video grabbing sequence overview, we can figure out what steps are required to proceed to video streaming from device opening to closing, through format negotiation. We can now jump to the code, starting with the device opening, from which everything begins.

Video device opening and property management

Drivers expose node entries in the /dev/ directory corresponding to the video interfaces they are responsible for. These file nodes correspond to the /dev/videoX special files for capture devices (in our case). The application must open the appropriate file node prior to any interaction with the video device. It uses the open() system call for that, which will return a file descriptor that will be the entry point for any command sent to the device, as in the following example:

```
static const char *dev_name = "/dev/video0";
fd = open (dev_name, O_RDWR);
if (fd == -1) {
    perror("Failed to open capture device\n");
    return -1;
}
```

The preceding snippet is an opening in blocking mode. Passing O_NONBLOCK to open() would prevent the application from being blocked if there is no ready buffer while trying to dequeue. Once you're done with the video device, it should be closed using the close() system call:

```
close (fd);
```

After we are able to open the video device, we can start our interaction with it. Generally, the first action that takes place once the video device is opened is to query its capabilities, through which we can make it operate optimally.

Querying the device capabilities

It is common to query the capabilities of the device in order to make sure it supports the mode we need to work with. You do this using the VIDIOC_QUERYCAP ioctl command. To achieve this, the application passes a struct v4l2_capability structure (defined in include/uapi/linux/videodev2.h), which will be filled by the driver. This structure has a .capabilities field that has to be checked. That field contains the capabilities of the whole device. The following excerpt from the kernel source shows the possible values:

```
/* Values for 'capabilities' field */
#define V4L2_CAP_VIDEO_CAPTURE 0x00000001 /*video capture
device*/
```

```
#define V4L2_CAP_VIDEO_OUTPUT 0x00000002   /*video output
device*/
#define V4L2_CAP_VIDEO_OVERLAY 0x00000004 /*Can do video
overlay*/
[...] /* VBI device skipped */


/* video capture device that supports multiplanar formats */
#define V4L2_CAP_VIDEO_CAPTURE_MPLANE    0x00001000
/* video output device that supports multiplanar formats */
#define V4L2_CAP_VIDEO_OUTPUT_MPLANE     0x00002000
/* mem-to-mem device that supports multiplanar formats */
#define V4L2_CAP_VIDEO_M2M_MPLANE 0x00004000
/* Is a video mem-to-mem device */
#define V4L2_CAP_VIDEO_M2M   0x00008000
[...] /* radio, tunner and sdr devices skipped */


#define V4L2_CAP_READWRITE   0x01000000 /*read/write
systemcalls */
#define V4L2_CAP_ASYNCIO     0x02000000  /* async I/O */
#define V4L2_CAP_STREAMING   0x04000000  /* streaming I/O
ioctls */
#define V4L2_CAP_TOUCH 0x10000000   /* Is a touch device */
```

The following code block shows a common use case that shows how to query the device capabilities from the code using the VIDIOC_QUERYCAP ioctl:

```
#include <linux/videodev2.h>
[...]
struct v4l2_capability cap;
memset(&cap, 0, sizeof(cap));

if (-1 == xioctl(fd, VIDIOC_QUERYCAP, &cap)) {
    if (EINVAL == errno) {
        fprintf(stderr, "%s is no V4L2 device\n", dev_name);
        exit(EXIT_FAILURE);
    } else {
        errno_exit("VIDIOC_QUERYCAP"

    }
}
```

In the preceding code, `struct v4l2_capability` is first zeroed thanks to `memset()` prior to being given to the `ioctl` command. At this step, if no error occurs, then our `cap` variable now contains the device capabilities. You can use the following to check for the device type and the I/O methods:

```
if (!(cap.capabilities & V4L2_CAP_VIDEO_CAPTURE)) {
    fprintf(stderr, "%s is not a video capture device\n",
            dev_name);
    exit(EXIT_FAILURE);
}
if (!(cap.capabilities & V4L2_CAP_READWRITE))
    fprintf(stderr, "%s does not support read i/o\n",
            dev_name);


/* Check whether USERPTR and/or MMAP method are supported */
if (!(cap.capabilities & V4L2_CAP_STREAMING))
    fprintf(stderr, "%s does not support streaming i/o\n",
            dev_name);


/* Check whether driver support read/write i/o */
if (!(cap.capabilities & V4L2_CAP_READWRITE))
    fprintf (stderr, "%s does not support read i/o\n",
             dev_name);
```

You may have noticed that we first zeroed our `cap` variable prior to using it. It is good practice to always clear parameters that will be given to V4L2 APIs in order to avoid stale content. Let's then define a macro—say, CLEAR—that will zero any variable given as a parameter, and use it in the rest of the chapter:

```
#define CLEAR(x) memset(&(x), 0, sizeof(x))
```

Now, we are done with querying the video device capabilities. This allows us to configure the device and tweak the image format according to what we need to achieve. By negotiating the appropriate image format, we can leverage the video device, as we will see in the next section.

Buffer management

You should consider that in V4L2, two buffer queues are maintained: one for the driver (referred to as the **INPUT queue**) and one for the user (referred to as the **OUTPUT queue**). Buffers are queued into the driver's queue by the user space application in order to be filled with data (the application uses the VIDIOC_QBUF ioctl for this). Buffers are filled by the driver in the order they have been enqueued. Once filled, each buffer is moved off the INPUT queue and put into the OUTPUT queue, which is the user queue.

Whenever the user application calls VIDIOC_DQBUF in order to dequeue a buffer, this buffer is looked for in the OUTPUT queue. If it's in there, the buffer will be dequeued and *pushed* to the user application; otherwise, the application will wait until a filled buffer is there. After the user finishes using the buffer, it must call VIDIOC_QBUF on this buffer in order to enqueue it back in the INPUT queue so that it can be filled again.

After driver initialization, the application calls the VIDIOC_REQBUFS ioctl to set the number of buffers it needs to work with. Once this is granted, the application queues all the buffers using VIDIOC_QBUF, and then calls the VIDIOC_STREAMON ioctl. Then, the driver goes ahead on its own and fills all the queued buffers. If there are no more queued buffers, then the driver will be waiting for a buffer to be queued in by the application. If such a case arises, then it means that some frames are lost in the capture itself.

Image (buffer) format

After making sure that the device is of the correct type and supports the modes it can work with, the application must negotiate the video format it needs. The application has to make sure that the video device is configured to send video frames in a format that the application can deal with. It has to do this before starting to grab and gather data (or video frames). The V4L2 API uses struct v4l2_format to represent the buffer format, whatever the type of the device is. This structure is defined as follows:

```
struct v4l2_format {
 u32 type;
 union {
  struct v4l2_pix_format pix; /* V4L2_BUF_TYPE_VIDEO_CAPTURE */
  struct v4l2_pix_format_mplane pix_mp; /* _CAPTURE_MPLANE */
  struct v4l2_window win;     /* V4L2_BUF_TYPE_VIDEO_OVERLAY */
  struct v4l2_vbi_format vbi; /* V4L2_BUF_TYPE_VBI_CAPTURE */
  struct v4l2_sliced_vbi_format sliced;/*_SLICED_VBI_CAPTURE */
  struct v4l2_sdr_format sdr;   /* V4L2_BUF_TYPE_SDR_CAPTURE */
```

```
    struct v4l2_meta_format meta;/* V4L2_BUF_TYPE_META_CAPTURE */
        [...]
    } fmt;
};
```

In the preceding structure, the `type` field represents the type of the data stream and should be set by the application. Depending on its value, the `fmt` field will be of the appropriate type. In our case, `type` must be `V4L2_BUF_TYPE_VIDEO_CAPTURE` as we are dealing with video capture devices. `fmt` will then be of the `struct v4l2_pix_format` type.

> **Important note**
> Almost all (if not all) ioctls playing directly or indirectly with the buffer (such as cropping, buffer requesting/queue/dequeue/querying) need to specify the buffer type, which makes sense. We will use `V4L2_BUF_TYPE_VIDEO_CAPTURE` as it is the only choice we have for our device type. The whole list of buffer types is of the `enum v4l2_buf_type` type defined in `include/uapi/linux/videodev2.h`. You should have a look.

It is common for applications to query the current format of the video device and then only change the properties of interest in it, and send back the new, mangled buffer format to the video device. However, this is not mandatory. We have only done it here to demonstrate how you can either get or set the current format. The application queries the current buffer format using the `VIDIOC_G_FMT` ioctl command. It has to pass a fresh (by fresh, I mean zeroed) `struct v4l2_format` structure with the `type` field set. The driver will fill the rest in the return path of the ioctl. The following is an example:

```
struct v4l2_format fmt;
CLEAR(fmt);

/* Get the current format */
fmt.type = V4L2_BUF_TYPE_VIDEO_CAPTURE;
if (ioctl(fd, VIDIOC_G_FMT, &fmt)) {
    printf("Getting format failed\n");
    exit(2);
}
```

Once we have the current format, we can change the relevant properties and send back the new format to the device. These properties may be pixel format, memory organization for each color component, and interlaced capture memory organization for each field. We can also describe the size and pitch of the buffer. Common (but not the only) pixel formats supported by devices are as follows:

- `V4L2_PIX_FMT_YUYV`: YUV422 (interleaved)

- `V4L2_PIX_FMT_NV12`: YUV420 (semi-planar)

- `V4L2_PIX_FMT_NV16`: YUV422 (semi-planar)

- `V4L2_PIX_FMT_RGB24`: RGB888 (packed)

Now, let's write the piece of code that changes the properties we need. However, sending the new format to the video device requires using a new ioctl command—that is, `VIDIOC_S_FMT`:

```
#define WIDTH    1920
#define HEIGHT   1080
#define PIXFMT   V4L2_PIX_FMT_YUV420


/* Changing required properties and set the format */
fmt.fmt.pix.width = WIDTH;

fmt.fmt.pix.height = HEIGHT;

fmt.fmt.pix.bytesperline = fmt.fmt.pix.width * 2u;

fmt.fmt.pix.sizeimage = fmt.fmt.pix.bytesperline *
fmt.fmt.pix.height;

fmt.fmt.pix.colorspace = V4L2_COLORSPACE_REC709;

fmt.fmt.pix.field = V4L2_FIELD_ANY;

fmt.fmt.pix.pixelformat = PIXFMT;

fmt.type = V4L2_BUF_TYPE_VIDEO_CAPTURE;


if (xioctl(fd, VIDIOC_S_FMT, &fmt)) {
    printf("Setting format failed\n");
    exit(2);
}
```

> **Important note**
> We could have used the preceding code without needing the current format.

The ioctl may succeed. However, this does not mean your parameters have been applied as is. By default, a device may not support every combination of image width and height, or even the required pixel format. In this case, the driver will apply the closest values it supports according to the ones you requested. You then have to check whether your parameters have been accepted or whether the ones that are granted are good enough for you to proceed:

```
if (fmt.fmt.pix.pixelformat != PIXFMT)
    printf("Driver didn't accept our format. Can't proceed.\n");

/* because VIDIOC_S_FMT may change width and height */
if ((fmt.fmt.pix.width != WIDTH) ||
    (fmt.fmt.pix.height != HEIGHT))
    fprintf(stderr, "Warning: driver is sending image at %dx%d\n",
            fmt.fmt.pix.width, fmt.fmt.pix.height);
```

We can even go further by changing the streaming parameters, such as the number of frames per second. We can achieve this by doing the following:

- Using the `VIDIOC_G_PARM` ioctl to query the video device's streaming parameters. This ioctl accepts as a parameter a fresh `struct v4l2_streamparm` structure with its `type` member set. This type should be one of the enum `v4l2_buf_type` values.

- Checking `v4l2_streamparm.parm.capture.capability` and making sure the `V4L2_CAP_TIMEPERFRAME` flag is set. This means that the driver allows changing the capture frame rate.

  If so, we can (optionally) use the `VIDIOC_ENUM_FRAMEINTERVALS` ioctl in order to get the list of possible frame intervals (the API uses the frame interval, which is the inverse of the frame rate).

- Using the `VIDIOC_S_PARM` ioctl and filling in the `v4l2_streamparm.parm.capture.timeperframe` members with the appropriate values. That should allow setting the capture-side frame rate. It's your task to make sure you're reading fast enough to not get frame drops.

The following is an example:

```
#define FRAMERATE 30

struct v4l2_streamparm parm;
int error;

CLEAR(parm);
parm.type = V4L2_BUF_TYPE_VIDEO_CAPTURE;

/* first query streaming parameters */
error = xioctl(fd, VIDIOC_G_PARM, &parm);
if (!error) {
    /* Now determine if the FPS selection is supported */
    if (parm.parm.capture.capability & V4L2_CAP_TIMEPERFRAME) {
        /* yes we can */
        CLEAR(parm);
        parm.type = V4L2_BUF_TYPE_VIDEO_CAPTURE;
        parm.parm.capture.capturemode = 0;
        parm.parm.capture.timeperframe.numerator = 1;
        parm.parm.capture.timeperframe.denominator = FRAMERATE;
        error = xioctl(fd, VIDIOC_S_PARM, &parm);

        if (error)
            printf("Unable to set the FPS\n");
        else
            /* once again, driver may have changed our requested
             * framerate */
            if (FRAMERATE !=
                    parm.parm.capture.timeperframe.denominator)
                printf ("fps coerced ......: from %d to %d\n",
                        FRAMERATE,
                    parm.parm.capture.timeperframe.denominator);
```

Now, we can negotiate image formats and set the streaming parameter. The next logical continuation would be requesting buffers and proceeding to further processing.

Requesting buffers

Once done with format preparation, it is time to instruct the driver to allocate memory that is to be used to store video frames. The `VIDIOC_REQBUFS` ioctl is there to achieve this. This ioctl takes a fresh `struct v4l2_requestbuffers` structure as an argument. Prior to being given to the ioctl, `v4l2_requestbuffers` must have some of its fields set:

- `v4l2_requestbuffers.count`: This member should be set with the number of memory buffers to be allocated. This member should be set with a value ensuring that frames won't be dropped because of a lack of queued buffers in the INPUT queue. Most of the time, 3 or 4 are correct values. Therefore, the driver may not be comfortable with the requested number of buffers. In this case, the driver will set `v4l2_requestbuffers.count` with the granted number of buffers on the return path of the ioctl. The application should then check this value in order to make sure this granted value fits its needs.

- `v4l2_requestbuffers.type`: This must be set with the video buffer type, of the `enum 4l2_buf_type` type. Here, again, we use `V4L2_BUF_TYPE_VIDEO_CAPTURE`. This would be `V4L2_BUF_TYPE_VIDEO_OUTPUT` for an output device, for example.

- `v4l2_requestbuffers.memory`: This must be one of the possible `enum v4l2_memory` values. Possible values of interest are `V4L2_MEMORY_MMAP`, `V4L2_MEMORY_USERPTR`, and `V4L2_MEMORY_DMABUF`. These are all streaming methods. However, depending on the value of this member, the application may have additional tasks to perform.

Unfortunately, the `VIDIOC_REQBUFS` command is the only way for an application to discover which types of streaming I/O buffer are supported by a given driver. The application can then try `VIDIOC_REQBUFS` with each of these values and adapt its logic according to which one failed or succeeded.

Requesting user pointer buffers – VIDIOC_REQBUFS and malloc

This step involves the driver supporting streaming mode, especially user pointer I/O mode. Here, the application informs the driver that it is about to allocate a given number of buffers:

```
#define BUF_COUNT 4

struct v4l2_requestbuffers req;
CLEAR (req);
req.count  = BUF_COUNT;
req.type   = V4L2_BUF_TYPE_VIDEO_CAPTURE;
req.memory = V4L2_MEMORY_USERPTR;

if (-1 == xioctl (fd, VIDIOC_REQBUFS, &req)) {
    if (EINVAL == errno)
        fprintf(stderr,
                "%s does not support user pointer i/o\n",
                dev_name);
    else
        fprintf("VIDIOC_REQBUFS failed \n");
}
```

Then, the application allocates the buffer memory from user space:

```
struct buffer_addr {
    void  *start;
    size_t length;
};

struct buffer_addr *buf_addr;
int i;

buf_addr = calloc(BUF_COUNT, sizeof (*buffer_addr));
if (!buf_addr) {
    fprintf(stderr, "Out of memory\n");
    exit (EXIT_FAILURE);
}
```

```
for (i = 0; i < BUF_COUNT; ++i) {
    buf_addr[i].length = buffer_size;
    buf_addr[i].start = malloc(buffer_size);

    if (!buf_addr[i].start) {
        fprintf(stderr, "Out of memory\n");
        exit(EXIT_FAILURE);
    }
}
```

This is the first type of streaming, where buffers are malloced in user space and given to the kernel in order to be filled with video data: the so-called user pointer I/O mode. There is another fancy streaming mode, where almost everything is done from the kernel. Without delay, let's introduce it.

Requesting the memory mappable buffer – VIDIOC_REQBUFS, VIDIOC_QUERYBUF, and mmap

In driver buffer mode, this ioctl also returns the actual number of buffers allocated in the count member of the v4l2_requestbuffer structure. This streaming method also requires a new data structure, struct v4l2_buffer. After buffers are allocated by the driver in the kernel, this structure is used along with the VIDIOC_QUERYBUFS ioctl in order to query the physical address of each allocated buffer, which can be used with the mmap() system call. The physical address returned from the driver will be stored in buffer.m.offset.

The following code excerpt instructs the driver to allocate memory buffers and check the number of buffers granted:

```
#define BUF_COUNT_MIN 3

struct v4l2_requestbuffers req; CLEAR (req);

req.count  = BUF_COUNT;
req.type   = V4L2_BUF_TYPE_VIDEO_CAPTURE;
req.memory = V4L2_MEMORY_MMAP;

if (-1 == xioctl (fd, VIDIOC_REQBUFS, &req)) {
    if (EINVAL == errno)
```

```
            fprintf(stderr, "%s does not support memory mapping\n",
                    dev_name);
        else
            fprintf("VIDIOC_REQBUFS failed \n");
    }


    /* driver may have granted less than the number of buffers we
     * requested let's then make sure it is not less than the
     * minimum we can deal with
     */
    if (req.count < BUF_COUNT_MIN) {
        fprintf(stderr, "Insufficient buffer memory on %s\n",
                dev_name);
        exit (EXIT_FAILURE);
    }
```

After this, the application should call the `VIDIOC_QUERYBUF` ioctl on each allocated buffer in order to get their corresponding physical addresses, as the following example shows:

```
struct buffer_addr {
    void *start;
    size_t length;
};


struct buffer_addr *buf_addr;
buf_addr = calloc(BUF_COUNT, sizeof (*buffer_addr));


if (!buf_addr) {
    fprintf (stderr, "Out of memory\n");
    exit (EXIT_FAILURE);
}


for (i = 0; i < req.count; ++i) {
    struct v4l2_buffer buf;
    CLEAR (buf);
```

```
    buf.type = V4L2_BUF_TYPE_VIDEO_CAPTURE;
    buf.memory = V4L2_MEMORY_MMAP; buf.index   = i;


    if (-1 == xioctl (fd, VIDIOC_QUERYBUF, &buf))
        errno_exit("VIDIOC_QUERYBUF");


    buf_addr[i].length = buf.length;
    buf_addr[i].start =
        mmap (NULL /* start anywhere */, buf.length,
              PROT_READ | PROT_WRITE /* required */,
              MAP_SHARED /* recommended */, fd, buf.m.offset);


    if (MAP_FAILED == buf_addr[i].start)
        errno_exit("mmap");
}
```

In order for the application to internally track the memory mapping (obtained with mmap()) of each buffer, we defined a custom data structure, struct buffer_addr, allocated for each granted buffer, which will hold the mapping corresponding to this buffer.

Requesting DMABUF buffers – VIDIOC_REQBUFS, VIDIOC_EXPBUF, and mmap

DMABUF is mostly used on mem2mem devices and introduces the concept of the **exporter** and **importer**. Say driver **A** wants to use buffers created by driver **B**; then we call **B** as the exporter and **A** as the buffer user/importer.

The export method instructs the driver to export its DMA buffers to user space by means of the file descriptor. The application achieves this using the VIDIOC_EXPBUF ioctl and requires a new data structure, struct v4l2_exportbuffer. On the return path of this ioctl, the driver will set the v4l2_requestbuffers.md member with the file descriptor corresponding to the given buffer. This is a DMABUF file descriptor:

```
/* V4L2 DMABuf export */
struct v4l2_requestbuffers req;
CLEAR (req);
req.count = BUF_COUNT;
req.type = V4L2_BUF_TYPE_VIDEO_CAPTURE;
```

```
req.memory = V4L2_MEMORY_DMABUF;


if (-1 == xioctl(fd, VIDIOC_REQBUFS, &req))
    errno_exit ("VIDIOC_QUERYBUFS");
```

It is possible for the application to export those buffers as DMABUF file descriptors so that they can be memory mapped to access the captured video content. The application should use the VIDIOC_EXPBUF ioctl for this. This ioctl extends the memory mapping I/O method, so it is only available for V4L2_MEMORY_MMAP buffers. However, it is actually useless at exporting capture buffers using VIDIOC_EXPBUF and then mapping them. You should use V4L2_MEMORY_MMAP instead.

VIDIOC_EXPBUF becomes very interesting when it comes to V4L2 output devices. This way, the application allocates buffers on both capture and output devices using the VIDIOC_REQBUFS ioctl, and then the application exports the output device's buffers as DMABUF file descriptors and uses these file descriptors to set the v4l2_buffer.m.fd field before the enqueueing ioctl on the capture device. The queued buffer will then have its counterpart (the output device buffer corresponding to v4l2_buffer.m.fd) filled.

In the following example, we export output device buffers as DMABUF file descriptors. This assumes buffers for this output device have been allocated using the VIDIOC_REQBUFS ioctl with req.type set to V4L2_BUF_TYPE_VIDEO_OUTPUT and req.memory set to V4L2_MEMORY_DMABUF:

```
int outdev_dmabuf_fd[BUF_COUNT] = {-1};
int i;
for (i = 0; i < req.count; i++) {
    struct v4l2_exportbuffer expbuf;
    CLEAR (expbuf);
    expbuf.type = V4L2_BUF_TYPE_VIDEO_OUTPUT;
    expbuf.index = i;

    if (-1 == xioctl(fd, VIDIOC_EXPBUF, &expbuf)
        errno_exit ("VIDIOC_EXPBUF");

    outdev_dmabuf_fd[i] = expbuf.fd;
}
```

Now, we have learned about DMABUF-based streaming and introduced the concepts it comes with. The next and last streaming method is much simpler and requires less code. Let's jump to it.

Requesting read/write I/O memory

This is the simpler streaming mode from a coding point of view. In the case of **read/write I/O**, there is nothing to do except to allocate the memory location where the application will store the read data, as in the following example:

```
struct buffer_addr {
    void *start;
    size_t length;
};
struct buffer_addr *buf_addr;
buf_addr = calloc(1, sizeof(*buf_addr));
if (!buf_addr) {
    fprintf(stderr, "Out of memory\n");
    exit(EXIT_FAILURE);
}
buf_addr[0].length = buffer_size;
buf_addr[0].start = malloc(buffer_size);

if (!buf_addr[0].start) {
    fprintf(stderr, "Out of memory\n");
    exit(EXIT_FAILURE);
}
```

In the previous code snippet, we have defined the same custom data structure, `struct buffer_addr`. However, there is no real buffer request here (`VIDIOC_REQBUFS` is not used) because nothing goes to the kernel yet. The buffer memory is simply allocated and that is all.

Now, we are done with buffer requests. The next step is to enqueue the requested buffers so that they can be filled with video data by the kernel. Let's now see how to do this.

Enqueueing the buffer and enabling streaming

Prior to a buffer being accessed and its data being read, this buffer must be enqueued. This consists of using the `VIDIOC_QBUF` ioctl on the buffer when using the streaming I/O method (everything except read/write I/O). Enqueueing a buffer will lock the memory pages of that buffer in physical memory. This way, those pages cannot be swapped out to disk. Do note that those buffers remain locked until they are dequeued, until the `VIDIOC_STREAMOFF` or `VIDIOC_REQBUFS` ioctls are called, or until the device is closed.

In the V4L2 context, locking a buffer means passing this buffer to the driver for hardware access (usually DMA). If an application accesses (reads/writes) a locked buffer, then the result is undefined.

To enqueue a buffer, the application must prepare `struct v4l2_buffer`, and `v4l2_buffer.type`, `v4l2_buffer.memory`, and `v4l2_buffer.index` should be set according to the buffer type, the streaming mode, and the index of the buffer when it has been allocated. Other fields depend on the streaming mode.

> **Important note**
> The *read/write I/O* method does not require enqueueing.

The concept of prime buffers

For capturing applications, it is customary to enqueue a number (most of the time, the number of allocated buffers) of empty buffers before you start capturing and enter the read loop. This helps improve the smoothness of the application and prevent it from being blocked because of the lack of a filled buffer. This should be done right after the buffers are allocated.

Enqueuing user pointer buffers

To enqueue a user pointer buffer, the application must set the `v4l2_buffer.memory` member to `V4L2_MEMORY_USERPTR`. The particularity here is the `v4l2_buffer.m.userptr` field, which must be set with the address of the buffer previously allocated and `v4l2_buffer.length` set to its size. When the multi-planar API is used, the `m.userptr` and `length` members of the passed array of `struct v4l2_plane` have to be used instead:

```
/* Prime buffers */
for (i = 0; i < BUF_COUNT; ++i) {
    struct v4l2_buffer buf;
```

```
    CLEAR(buf);

    buf.type = V4L2_BUF_TYPE_VIDEO_CAPTURE;
    buf.memory = V4L2_MEMORY_USERPTR; buf.index = i;
    buf.m.userptr = (unsigned long)buf_addr[i].start;
    buf.length = buf_addr[i].length;

    if (-1 == xioctl(fd, VIDIOC_QBUF, &buf))
        errno_exit("VIDIOC_QBUF");
}
```

Enqueuing memory mappable buffers

To enqueue a memory mappable buffer, the application must fill struct v4l2_buffer by setting the type, memory (which must be V4L2_MEMORY_MMAP), and index members, just as in the following excerpt:

```
/* Prime buffers */
for (i = 0; i < BUF_COUNT; ++i) {
    struct v4l2_buffer buf; CLEAR (buf);
    buf.type = V4L2_BUF_TYPE_VIDEO_CAPTURE;
    buf.memory = V4L2_MEMORY_MMAP;
    buf.index = i;

    if (-1 == xioctl (fd, VIDIOC_QBUF, &buf))
        errno_exit ("VIDIOC_QBUF");
}
```

Enqueuing DMABUF buffers

To enqueue an output device's DMABUF buffer into the capture device's one, applications should fill struct v4l2_buffer, setting the memory field to V4L2_MEMORY_DMABUF, the type field to V4L2_BUF_TYPE_VIDEO_CAPTURE, and the m.fd field to a file descriptor associated with a DMABUF buffer of the output device, as in the following excerpt:

```
/* Prime buffers */
for (i = 0; i < BUF_COUNT; ++i) {
    struct v4l2_buffer buf; CLEAR (buf);
```

```
    buf.type = V4L2_BUF_TYPE_VIDEO_CAPTURE;
    buf.memory = V4L2_MEMORY_DMABUF; buf.index = i;
    buf.m.fd = outdev_dmabuf_fd[i];
    /* enqueue the dmabuf to capture device */
    if (-1 == xioctl (fd, VIDIOC_QBUF, &buf))
        errno_exit ("VIDIOC_QBUF");
}
```

The preceding code excerpt shows how a V4L2 DMABUF import works. The `fd` argument in the ioctl is the file descriptor associated with the capture device, obtained at the `open()` syscall. `outdev_dmabuf_fd` is the array that contains the output device's DMABUF file descriptors. You may wonder how this can work on output devices that are not V4L2 but are DRM-compatible, for example. The following is a brief explanation.

First, the DRM subsystem provides APIs in a driver-dependent way, which you can use to allocate a (dumb) buffer on the GPU, which will return a GEM handle. The DRM also provides the `DRM_IOCTL_PRIME_HANDLE_TO_FD` ioctl, which allows exporting a buffer into the DMABUF file descriptor through `PRIME`, then the `drmModeAddFB2()` API to create a `framebuffer` object (which is something that will be read and displayed onscreen, or should I say, the CRT controller, to be exact) corresponding to this buffer so that it can finally be rendered using the `drmModeSetPlane()` or `drmModeSetPlane()` APIs. The application can then set the `v4l2_requestbuffers.m.fd` field with the file descriptor returned by the `DRM_IOCTL_PRIME_HANDLE_TO_FD` ioctl. Then, in the read loop, after each `VIDIOC_DQBUF` ioctl, the application can change the plane's frame buffer and position using the `drmModeSetPlane()` API.

> **Important note**
> **PRIME** is the name of the `drm dma-buf` interface layer integrated with GEM, which is one of the memory managers supported by the DRM subsystem

Enabling streaming

Enabling streaming is kind of like informing V4L2 that the *OUTPUT* queue will be accessed as of now. The application should use `VIDIOC_STREAMON` to achieve this. The following is an example:

```
/* Start streaming */
int ret;
int a = V4L2_BUF_TYPE_VIDEO_CAPTURE;
```

```
ret = xioctl(capt.fd, VIDIOC_STREAMON, &a);
if (ret < 0) {
    perror("VIDIOC_STREAMON\n");
    return -1;
}
```

The preceding excerpt is short but is mandatory to enable streaming, without which buffers can't be dequeued later.

Dequeuing buffers

This is actually part of the application's read loop. The application dequeues buffers using the VIDIOC_DQBUF ioctl. This is only possible if the streaming has been enabled before. When the application calls the VIDIOC_DQBUF ioctl, it instructs the driver to check whether there are any filled buffers in the **OUTPUT queue**, and if there are, it outputs one filled buffer and the ioctl immediately returns. However, if there is no buffer in the **OUTPUT queue**, then the application will block (unless the O_NONBLOCK flag has been set during the open() system call) until a buffer is queued and filled.

> **Important note**
>
> Trying to dequeue a buffer without queuing it first is an error, and the VIDIOC_DQBUF ioctl should return -EINVAL. When the O_NONBLOCK flag is given to the open() function, VIDIOC_DQBUF returns immediately with an EAGAIN error code when no buffer is available.

After dequeuing a buffer and processing its data, the application must immediately queue back this buffer again so that it can be refilled for the next reading, and so on.

Dequeuing memory-mapped buffers

The following is an example of dequeuing a buffer that has been memory mapped:

```
struct v4l2_buffer buf;

CLEAR (buf);
buf.type = V4L2_BUF_TYPE_VIDEO_CAPTURE;
buf.memory = V4L2_MEMORY_MMAP;

if (-1 == xioctl (fd, VIDIOC_DQBUF, &buf)) {
    switch (errno) {
```

```
    case EAGAIN:
        return 0;
    case EIO:
    default:
        errno_exit ("VIDIOC_DQBUF");
    }
}
/* make sure the returned index is coherent with the number
 * of buffers allocated
 */
assert (buf.index < BUF_COUNT);


/* We use buf.index to point to the correct entry in our
 * buf_addr
 */
process_image(buf_addr[buf.index].start);


/* Queue back this buffer again, after processing is done */
if (-1 == xioctl (fd, VIDIOC_QBUF, &buf))
    errno_exit ("VIDIOC_QBUF");
```

This could have been done in a loop. For example, let's say you need 200 images. The read loop could look as follows:

```
#define MAXLOOPCOUNT 200

/* Start the loop of capture */
for (i = 0; i < MAXLOOPCOUNT; i++) {
    struct v4l2_buffer buf;
    CLEAR (buf);
    buf.type = V4L2_BUF_TYPE_VIDEO_CAPTURE;
    buf.memory = V4L2_MEMORY_MMAP;

    if (-1 == xioctl (fd, VIDIOC_DQBUF, &buf)) {
        [...]
    }

    /* Queue back this buffer again, after processing is done */
```

```
    [...]
}
```

This preceding snippet is just a reimplementation of the buffer dequeuing using a loop, where the counter represents the number of images needed to grab.

Dequeuing user pointer buffers

The following is an example of dequeuing a buffer using the **user pointer**:

```
struct v4l2_buffer buf; int i;


CLEAR (buf);
buf.type = V4L2_BUF_TYPE_VIDEO_CAPTURE;
buf.memory = V4L2_MEMORY_USERPTR;


/* Dequeue a captured buffer */
if (-1 == xioctl (fd, VIDIOC_DQBUF, &buf)) {
    switch (errno) {
    case EAGAIN:
        return 0;
    case EIO:
        [...]
    default:
        errno_exit ("VIDIOC_DQBUF");
    }
}


/*
 * We may need the index to which corresponds this buffer
 * in our buf_addr array. This is done by matching address
 * returned by the dequeue ioctl with the one stored in our
 * array
 */
for (i = 0; i < BUF_COUNT; ++i)
    if (buf.m.userptr == (unsigned long)buf_addr[i].start &&
                    buf.length == buf_addr[i].length)
        break;
```

```
/* the corresponding index is used for sanity checks only */
assert (i < BUF_COUNT);
process_image ((void *)buf.m.userptr);

/* requeue the buffer */
if (-1 == xioctl (fd, VIDIOC_QBUF, &buf))
    errno_exit ("VIDIOC_QBUF");
```

The preceding code shows how to dequeue a user pointer buffer, and is well commented enough to not require any further explanation. However, this could be implemented in a loop if many buffers were needed.

Read/write I/O

This is the last example, showing how to dequeue a buffer using the read() system call:

```
if (-1 == read (fd, buffers[0].start, buffers[0].length)) {
    switch (errno) {
    case EAGAIN:
        return 0;
    case EIO:
        [...]
    default:
        errno_exit ("read");
    }
}
process_image (buffers[0].start);
```

None of the previous examples have been discussed in detail because each of them uses a concept that was already introduced in the *The V4L2 user space API* section. Now that we are familiar with writing V4L2 user space code, let's see how not to write any code by using dedicated tools that can be used for quickly prototyping your camera system.

V4L2 user space tools

So far, we have learned how to write user space code to interact with the driver in the kernel. For rapid prototyping and testing, we could leverage some community-provided V4L2 user space tools. By using those tools, we can focus on the system design and validate the camera system. The most well-known tool is `v4l2-ctl`, which we will focus on; it is shipped with the `v4l-utils` package.

Though it is not discussed in this chapter, there is also the **yavta** tool (which stands for **Yet Another V4L2 Test Application**), which can be used to test, debug, and control the camera subsystem.

Using v4l2-ctl

`v4l2-utils` is a user space application that can be used to query or configure V4L2 devices (including subdevices). This tool can help in setting up and designing fine-grained V4L2-based systems as it helps tweak and leverage the device's features.

> **Important note**
>
> `qv4l2` is the Qt GUI equivalent of `v4l2-ctl`. `v4l2-ctl` is ideal for embedded systems, while `qv4l2` is ideal for interactive testing.

Listing the video devices and their capabilities

First of all, we would need to list all the available video devices using the `--list-devices` option:

```
# v4l2-ctl --list-devices
Integrated Camera: Integrated C (usb-0000:00:14.0-8):
    /dev/video0
    /dev/video1
```

If several devices are available, we can use the `-d` option after any `v4l2-ctl` commands in order to target a specific device. Do note that if the `-d` option is not specified, `/dev/video0` is targeted by default.

In order to have information on a specific device, you must use the `-D` option, as follows:

```
# v4l2-ctl -d /dev/video0 -D
Driver Info (not using libv4l2):
    Driver name    : uvcvideo
    Card type      : Integrated Camera: Integrated C
```

```
    Bus info       : usb-0000:00:14.0-8
    Driver version: 5.4.60
    Capabilities  : 0x84A00001
      Video Capture
      Metadata Capture
      Streaming
      Extended Pix Format
      Device Capabilities
    Device Caps   : 0x04200001
      Video Capture
      Streaming
      Extended Pix Format
```

The preceding command shows the device information (such as the driver and its version) as well as its capabilities. That being said, the --all command gives better verbosity. You should give it a try.

Changing the device properties (controlling the device)

Before we look at changing the device properties, we first need to know what controls the device supports, what their value types (integer, Boolean, string, and so on) are, what their default values are, and what values are accepted.

In order to get the list of controls supported by the device, we can use v4l2-ctl with the -L option, as follows:

```
# v4l2-ctl -L
                   brightness 0x00980900 (int)    : min=0 max=255
step=1 default=128 value=128
                     contrast 0x00980901 (int)    : min=0 max=255
step=1 default=32 value=32
                   saturation 0x00980902 (int)    : min=0 max=100
step=1 default=64 value=64
                          hue 0x00980903 (int)    : min=-180 max=180
step=1 default=0 value=0
  white_balance_temperature_auto 0x0098090c (bool)   : default=1
value=1
                        gamma 0x00980910 (int)    : min=90 max=150
step=1 default=120 value=120
        power_line_frequency 0x00980918 (menu)    : min=0 max=2
```

```
default=1 value=1
                    0: Disabled
                    1: 50 Hz
                    2: 60 Hz
        white_balance_temperature 0x0098091a (int)   : min=2800
max=6500 step=1 default=4600 value=4600 flags=inactive
                    sharpness 0x0098091b (int)       : min=0 max=7
step=1 default=3 value=3
        backlight_compensation 0x0098091c (int)      : min=0 max=2
step=1 default=1 value=1
                 exposure_auto 0x009a0901 (menu)     : min=0 max=3
default=3 value=3
                    1: Manual Mode
                    3: Aperture Priority Mode
        exposure_absolute 0x009a0902 (int)      : min=5 max=1250
step=1 default=157 value=157 flags=inactive
        exposure_auto_priority 0x009a0903 (bool)   : default=0
value=1
jma@labcsmart:~$
```

In the preceding output, the `"value="` field returns the current value of the control, and the other fields are self-explanatory.

Now that we are aware of the list of controls supported by the device, a control value can be changed thanks to the `--set-ctrl` option, as in the following example:

```
# v4l2-ctl --set-ctrl brightness=192
```

After that, we can check the current value with the following:

```
# v4l2-ctl -L
                    brightness 0x00980900 (int)      : min=0 max=255
step=1 default=128 value=192
                         [...]
```

Or, we could have used the `--get-ctrl` command, as follows:

```
# v4l2-ctl --get-ctrl brightness
brightness: 192
```

Now it may be time to tweak the device. Before that, let's check the video characteristics of the device.

Setting the pixel format, resolution, and frame rate

Before selecting a specific format or resolution, we need to enumerate what is available for the device. In order to get the supported pixel format, as well as the resolution and frame rate, the `--list-formats-ext` option needs to be given to `v4l2-ctl`, as follows:

```
# v4l2-ctl --list-formats-ext
ioctl: VIDIOC_ENUM_FMT
    Index        : 0
    Type         : Video Capture
    Pixel Format: 'MJPG' (compressed)
    Name         : Motion-JPEG
      Size: Discrete 1280x720
            Interval: Discrete 0.033s (30.000 fps)
      Size: Discrete 960x540
            Interval: Discrete 0.033s (30.000 fps)
      Size: Discrete 848x480
            Interval: Discrete 0.033s (30.000 fps)
      Size: Discrete 640x480
            Interval: Discrete 0.033s (30.000 fps)
      Size: Discrete 640x360
            Interval: Discrete 0.033s (30.000 fps)
      Size: Discrete 424x240
            Interval: Discrete 0.033s (30.000 fps)
      Size: Discrete 352x288
            Interval: Discrete 0.033s (30.000 fps)
      Size: Discrete 320x240
            Interval: Discrete 0.033s (30.000 fps)
      Size: Discrete 320x180
            Interval: Discrete 0.033s (30.000 fps)

    Index        : 1
    Type         : Video Capture
    Pixel Format: 'YUYV'
    Name         : YUYV 4:2:2
```

```
Size: Discrete 1280x720
        Interval: Discrete 0.100s (10.000 fps)
Size: Discrete 960x540
        Interval: Discrete 0.067s (15.000 fps)
Size: Discrete 848x480
        Interval: Discrete 0.050s (20.000 fps)
Size: Discrete 640x480
        Interval: Discrete 0.033s (30.000 fps)
Size: Discrete 640x360
        Interval: Discrete 0.033s (30.000 fps)
Size: Discrete 424x240
        Interval: Discrete 0.033s (30.000 fps)
Size: Discrete 352x288
        Interval: Discrete 0.033s (30.000 fps)
Size: Discrete 320x240
        Interval: Discrete 0.033s (30.000 fps)
Size: Discrete 320x180
        Interval: Discrete 0.033s (30.000 fps)
```

From the previous output, we can see what is supported by the target device, which is the **MJPG** (mjpeg) compressed format and the YUYV raw format.

Now, in order to change the camera configuration, first select the frame rate using the --set-parm option, as follows:

```
# v4l2-ctl --set-parm=30
Frame rate set to 30.000 fps
#
```

Then, you can select the required resolution and/or pixel format using the --set-fmt-video option, as follows:

```
# v4l2-ctl --set-fmt-video=width=640,height=480,
  pixelformat=MJPG
```

When it comes to the frame rate, you would want to use v4l2-ctl with the --set-parm option, giving the frame rate numerator only—the denominator is fixed to 1 (only integer frame rate values are allowed)—as follows:

```
# v4l2-ctl --set-parm=<framerate numerator>
```

Capturing frames and streaming

`v4l2-ctl` supports many more options than you can imagine. In order to see the possible options, you can print the help message of the appropriate section. Common help commands related to streaming and video capture are the following:

- `--help-streaming`: Prints the help message for all options that deal with streaming
- `--help-subdev`: Prints the help message for all options that deal with `v4l-subdevX` devices
- `--help-vidcap`: Prints the help message for all options that get/set/list video capture formats

From those help commands, I've built the following command in order to capture a QVGA MJPG compressed frame on disk:

```
# v4l2-ctl --set-fmt-video=width=320,height=240,
  pixelformat=MJPG \
    --stream-mmap --stream-count=1 --stream-to=grab-320x240.mjpg
```

I've also managed to capture a raw YUV image with the same resolution with the following command:

```
# v4l2-ctl --set-fmt-video=width=320,height=240,
  pixelformat=YUYV \
    --stream-mmap --stream-count=1 --stream-to=grab-320x240-yuyv.
 raw
```

The raw YUV image cannot be displayed unless you use a decent raw image viewer. In order to do so, the raw image must be converted using the `ffmpeg` tool, for example, as follows:

```
# ffmpeg -f rawvideo -s 320x240 -pix_fmt yuyv422 \
         -i grab-320x240-yuyv.raw grab-320x240.png
```

You can notice a big difference in terms of the size between the raw and the compressed image, as in the following snippet:

```
# ls -hl grab-320x240.mjpg
-rw-r--r-- 1 root root 8,0K oct.  21 20:26 grab-320x240.mjpg
# ls -hl grab-320x240-yuyv.raw
-rw-r--r-- 1 root root 150K oct.  21 20:26 grab-320x240-yuyv.
raw
```

Do note that it is good practice to include the image format in the filename of a raw capture (such as yuyv in grab-320x240-yuyv.raw) so that you can easily convert from the right format. This rule is not necessary with compressed image formats because these formats are image container formats with a header that describes the pixel data that follows, and can be easily read with the gst-typefind-1.0 tool. JPEG is such a format and the following is how its header can be read:

```
# gst-typefind-1.0 grab-320x240.mjpg
grab-320x240.mjpg - image/jpeg, width=(int)320,
height=(int)240, sof-marker=(int)0
# gst-typefind-1.0 grab-320x240-yuyv.raw
grab-320x240-yuyv.raw - FAILED: Could not determine type of
stream.
```

Now that we are done with tool usages, let's see how to go deeper and learn about V4L2 debugging and from user space.

Debugging V4L2 in user space

Since our video system setup may not be free of bugs, V4L2 provides a simple but large backdoor for debugging from user space, in order to track and shoot down trouble coming from either the VL4L2 framework core or the user space API.

Framework debugging can be enabled as follows:

```
# echo 0x3 > /sys/module/videobuf2_v4l2/parameters/debug
# echo 0x3 > /sys/module/videobuf2_common/parameters/debug
```

The preceding commands will instruct V4L2 to add core traces to the kernel log message. This way, it will easily track where the trouble is coming from, assuming it's coming from the core. Run the following command:

```
# dmesg
[831707.512821] videobuf2_common:  __setup_offsets: buffer 0,
plane 0 offset 0x00000000
[831707.512915] videobuf2_common:  __setup_offsets: buffer 1,
plane 0 offset 0x00097000
[831707.513003] videobuf2_common:  __setup_offsets: buffer 2,
plane 0 offset 0x0012e000
[831707.513118] videobuf2_common:  __setup_offsets: buffer 3,
plane 0 offset 0x001c5000
[831707.513119] videobuf2_common:  __vb2_queue_alloc: allocated
4 buffers, 1 plane(s) each
[831707.513169] videobuf2_common: vb2_mmap: buffer 0, plane 0
successfully mapped
[831707.513176] videobuf2_common: vb2_core_qbuf: qbuf of buffer
0 succeeded
[831707.513205] videobuf2_common: vb2_mmap: buffer 1, plane 0
successfully mapped
[831707.513208] videobuf2_common: vb2_core_qbuf: qbuf of buffer
1 succeeded
[...]
```

In the previous kernel log messages, we can see the kernel-related V4L2 core functions call, along with some other details. If for any reason the V4L2 core tracing is not necessary or not enough for you, you can also enable V4L2 userland API tracing with the following command:

```
$ echo 0x3 > /sys/class/video4linux/video0/dev_debug
```

After running the command, allowing you to capture a raw image, we can see the following in the kernel log messages:

```
$ dmesg
[833211.742260] video0: VIDIOC_QUERYCAP: driver=uvcvideo,
card=Integrated Camera: Integrated C, bus=usb-0000:00:14.0-8,
version=0x0005043c, capabilities=0x84a00001,
device_caps=0x04200001
[833211.742275] video0: VIDIOC_QUERY_EXT_CTRL: id=0x980900,
```

```
type=1, name=Brightness, min/max=0/255, step=1, default=128,
flags=0x00000000, elem_size=4, elems=1, nr_of_dims=0,
dims=0,0,0,0
[...]
[833211.742318] video0: VIDIOC_QUERY_EXT_CTRL: id=0x98090c,
type=2, name=White Balance Temperature, Auto, min/max=0/1,
step=1, default=1, flags=0x00000000, elem_size=4, elems=1,
nr_of_dims=0, dims=0,0,0,0
[833211.742365] video0: VIDIOC_QUERY_EXT_CTRL: id=0x98091c,
type=1, name=Backlight Compensation, min/max=0/2, step=1,
default=1, flags=0x00000000, elem_size=4, elems=1,
nr_of_dims=0, dims=0,0,0,0
[833211.742376] video0: VIDIOC_QUERY_EXT_CTRL: id=0x9a0901,
type=3, name=Exposure, Auto, min/max=0/3, step=1, default=3,
flags=0x00000000, elem_size=4, elems=1, nr_of_dims=0,
dims=0,0,0,0
[...]
[833211.756641] videobuf2_common: vb2_mmap: buffer 1, plane 0
successfully mapped
[833211.756646] videobuf2_common: vb2_core_qbuf: qbuf of buffer
1 succeeded
[833211.756649] video0: VIDIOC_QUERYBUF: 00:00:00.00000000
index=2, type=vid-cap, request_fd=0, flags=0x00012000,
field=any, sequence=0, memory=mmap, bytesused=0, offset/
userptr=0x12e000, length=614989
[833211.756657] timecode=00:00:00 type=0, flags=0x00000000,
frames=0, userbits=0x00000000
[833211.756698] videobuf2_common: vb2_mmap: buffer 2, plane 0
successfully mapped
[833211.756704] videobuf2_common: vb2_core_qbuf: qbuf of buffer
2 succeeded
[833211.756706] video0: VIDIOC_QUERYBUF: 00:00:00.00000000
index=3, type=vid-cap, request_fd=0, flags=0x00012000,
field=any, sequence=0, memory=mmap, bytesused=0, offset/
userptr=0x1c5000, length=614989
[833211.756714] timecode=00:00:00 type=0, flags=0x00000000,
frames=0, userbits=0x00000000
[833211.756751] videobuf2_common: vb2_mmap: buffer 3, plane 0
successfully mapped
[833211.756755] videobuf2_common: vb2_core_qbuf: qbuf of buffer
3 succeeded
```

```
[833212.967229] videobuf2_common: vb2_core_streamon: successful
[833212.967234] video0: VIDIOC_STREAMON: type=vid-cap
```

In the preceding output, we can trace the different V4L2 userland API calls, which correspond to the different `ioctl` commands and their parameters.

V4L2 compliance driver testing

In order for a driver to be V4L2-compliant, it must meet some criteria, which includes passing the `v4l2-compliance` tool test, which is used to test V4L devices of all kinds. `v4l2-compliance` attempts to test almost all aspects of a V4L2 device and it covers almost all V4L2 ioctls.

As with other V4L2 tools, a video device can be targeted using the `-d` or `--device=` commands. If a device is not specified, `/dev/video0` is targeted. The following is an output excerpt:

```
# v4l2-compliance
v4l2-compliance SHA   : not available

Driver Info:
    Driver name    : uvcvideo
    Card type      : Integrated Camera: Integrated C
    Bus info       : usb-0000:00:14.0-8
    Driver version : 5.4.60
    Capabilities   : 0x84A00001
      Video Capture
      Metadata Capture
      Streaming
      Extended Pix Format
      Device Capabilities
    Device Caps    : 0x04200001
      Video Capture
      Streaming
      Extended Pix Format

Compliance test for device /dev/video0 (not using libv4l2):

Required ioctls:
```

```
    test VIDIOC_QUERYCAP: OK

Allow for multiple opens:
    test second video open: OK
    test VIDIOC_QUERYCAP: OK
    test VIDIOC_G/S_PRIORITY: OK
    test for unlimited opens: OK

Debug ioctls:
    test VIDIOC_DBG_G/S_REGISTER: OK (Not Supported)
    test VIDIOC_LOG_STATUS: OK (Not Supported)
[]

Output ioctls:
    test VIDIOC_G/S_MODULATOR: OK (Not Supported)
    test VIDIOC_G/S_FREQUENCY: OK (Not Supported)
[...]

Test input 0:

    Control ioctls:
       fail: v4l2-test-controls.cpp(214): missing control class
for class 00980000
       fail: v4l2-test-controls.cpp(251): missing control class
for class 009a0000
       test VIDIOC_QUERY_EXT_CTRL/QUERYMENU: FAIL
       test VIDIOC_QUERYCTRL: OK
       fail: v4l2-test-controls.cpp(437): s_ctrl returned an
error (84)
       test VIDIOC_G/S_CTRL: FAIL
       fail: v4l2-test-controls.cpp(675): s_ext_ctrls returned an
error (
```

In the preceding logs, we can see that /dev/video0 has been targeted. Additionally, we notice that Debug ioctls and Output ioctls are not supported by our driver (these are not failures). Though the output is verbose enough, it is better to use the --verbose command as well, which makes the output more user friendly and much more detailed. It then goes without saying that if you want to submit a new V4L2 driver, that driver must pass the V4L2 compliance tests.

Summary

In this chapter, we walked through the user space implementation of V4L2. We started with V4L2 buffer management, from video streaming. We also learned how to deal with video device property management, all from user space. However, V4L2 is a heavy framework, not just in terms of code but also in terms of power consumption. So, in the next chapter, we will address Linux kernel power management in order to keep the system at the lowest consumption level possible without degrading the system properties.

10
Linux Kernel Power Management

Mobile devices are becoming increasingly complex with more and more features in order to follow commercial trends and satisfy consumers. While a few parts of such devices run proprietary or bare metal software, most of them run Linux-based operating systems (embedded Linux distributions, Android, to name but a few), and all of them are battery powered. In addition to full functionality and performance, consumers require the longest possible autonomy and long-lasting batteries. It goes without saying that full performance and autonomy (power saving) are two totally incompatible concepts, and that a compromise must be found at all times when using the device. This compromise comes with Power Management, which allows us to deal with the lower consumption possible and device performance without ignoring the time needed for the device to wake up (or to be fully operational) after it has been put in a low-power state.

The Linux kernel comes with several power management capabilities, ranging from allowing you to save power during brief idle periods (or execution of tasks with lower power demands) to putting the whole system into a sleep state when it is not actively in use.

Additionally, as and when devices are added to the system, they can participate in this power management effort thanks to the generic Power Management APIs that the Linux kernel offers in order to allow device driver developers to benefit from power management mechanisms implemented in devices, whatever they are. This allows either a per-device or system-wide power parameters to be adjusted in order to extend not only the autonomy of the device, but also the lifetime of the battery.

In this chapter, we will walk through the Linux kernel power management subsystem, leveraging its APIs and managing its options from user space. Hence, the following topics will be covered:

- The concept of power management on Linux-based systems
- Adding power management capabilities to device drivers
- Being a source of system wakeup

Technical requirements

For a better understanding of this chapter, you'll require the following:

- Basic electrical knowledge
- Basic C programming skills
- Good knowledge of computer architecture
- Linux Kernel 4.19 sources available at `https://github.com/torvalds/linux`

The concept of power management on Linux-based systems

Power management (**PM**) entails consuming as little power as possible at any time. There are two types of power management that the operating system must handle: **Device Power Management** and **System Power Management**.

- **Device Power Management**: This is device specific. It allows a device to be put in a low-power state while the system is running. This may allow, among other things, part of the device not currently in use to be turned off in order to conserve power, such as the keyboard backlight when you are not typing. Individual device power management may be invoked explicitly on devices regardless of the power management activity, or may happen automatically after a device has been idle for a set amount of time. Device power management is an alias for the so-called *Runtime Power Management*.

- **System Power Management**, also known as *Sleep States*: This enables platforms to enter a system-wide low-power state. In other words, entering a sleep state is the process by which the entire system is placed in a low-power state. There are several low-power states (or sleep states) that a system may enter, depending on the platform, its capabilities, and the target wake up latency. This happens, for example, when the lid is closed on a laptop computer, when turning off the screen of the phone, or when some critical state has been reached (such as battery level). Many of these states are similar across platforms (such as freezing, which is purely software, and hence not device or system dependent), and will be discussed in detail later. The general concept is that the state of the running system is saved before the system is powered down (or put into a sleep state, which is different from a shutdown), and restored once the system has regained power. This prevents the system from performing an entire shutdown and startup sequence.

Although system PM and runtime PM deals with different scenarios for idle management, deploying both is important to prevent wasting power for a platform. You should think of them as complementary, as we will see in forthcoming sections.

Runtime power management

This is the part of Linux PM that manages power for individual devices without taking the whole system into a low-power state. In this mode, actions take effect while the system is running, hence its name, Runtime Power Management. In order to adapt device power consumption, its properties are changed on the fly with the system still in operation, hence its other name, **dynamic power management**.

A tour of some dynamic power management interfaces

Aside from per-device power management capabilities that driver developers can implement in device drivers, the Linux kernel provides user space interfaces to add/remove/modify power policies. The most well-known of these are listed here:

- **CPU Idle**: This assists in managing CPU power consumption when it has no task to execute.

- **CPUFreq**: This allows CPU power properties (that is, voltage and frequency, which are related) to be changed depending on the system load.

- **Thermal**: This allows power properties to be adjusted according to temperatures sensed in predefined zones of the system, most of the time areas close to the CPU.

You may have noticed that the preceding policies deal with the CPU. This is because the CPU is one of the principal sources of power dissipation on mobile devices (or embedded systems). While only three interfaces are introduced in the next sections, other interfaces exist, too, such as QoS and DevFreq. Readers are free to explore these to satisfy their curiosity.

CPU Idle

Whenever a logical CPU in the system has no task to execute, it may need to be put in a particular state in order to save power. In this situation, most operating systems simply schedule a so-called *idle thread*. While executing this thread, the CPU is said to be idle, or in an idle state. **CPU Idle** is a framework that manages idle threads. There are several levels (or modes or states) of idling. It depends on the built-in power-saving hardware embedded in the CPU. CPU idle modes are sometimes referred to as C-modes or even C-states, which is an **Advanced Configuration and Power Interface** (**ACPI**) term. Those states usually start from C0, which is the normal CPU operating mode; in other words, the CPU is 100% turned on. As the C number increases, the CPU sleep mode becomes deeper; in other words, more circuits and signals are turned off and the longer the amount of time the CPU will require to return to C0 mode, that is, to wake up. C1 is the first C-state, C2 is the second one, and so on. When a logical processor is idle (any C-state except C0), its frequency is typically 0.

The next event (in time) determines how long the CPU can sleep for. Each idle state is described by three characteristics:

- An exit latency, in µS: This is the latency to get out of this state.

- A power consumption, in mW: This is not always reliable.

- A target residency, in µS: This is the idle duration from which it becomes interesting to use this state.

CPU Idle drivers are platform specific, and the Linux kernel expects CPU drivers to support at most 10 states (see CPUIDLE_STATE_MAX in the kernel source code). However, the real number of states depends on the underlying CPU hardware (which embeds built-in power-saving logic), and the majority of ARM platforms only provide one or two idle states. The choice of the state to enter is based on policies managed by governors.

A governor in this context is a simple module implementing an algorithm enabling the best C-state choice to be made, depending on some properties. In other words, the governor is the one that decides the target C-state of the system. Though multiple governors can exist on the system, only one will be in control of a given CPU at any time. It is designed in a way that, if the scheduler run queue is empty (which means the CPU has nothing else to do) and it needs to idle the CPU, it will request CPU idling to the CPU idle framework. The framework will then rely on the currently selected governor to select the appropriate *C-state*. There are two CPU Idle governors: ladder (for periodic timer tick-based systems) and menu (for tick-less systems). While the ladder governor is always available, if CONFIG_CPU_IDLE is selected, the menu governor additionally requires CONFIG_NO_HZ_IDLE (or CONFIG_NO_HZ on older kernels) to be set. The governor is selected while configuring the kernel. Roughly speaking, which of them to use depends on the configuration of the kernel and, in particular, on whether or not the scheduler tick can be stopped by the idle loop, hence CONFIG_NO_HZ_IDLE. You can refer to Documentation/timers/NO_HZ.txt for further reading on this.

The governor may decide whether to continue in the current state or transition to a different state, in which case it will instruct the current driver to transition to the selected state. The current idle driver can be identified by reading the content of the /sys/devices/system/cpu/cpuidle/current_driver file, and the current governor from /sys/devices/system/cpu/cpuidle/current_governor_ro:

```
$ cat /sys/devices/system/cpu/cpuidle/current_governor_ro menu
```

On a given system, each directory in /sys/devices/system/cpu/cpuX/cpuidle/ corresponds to a C-state, and the contents of each C-state directory attribute files describing this C-state:

```
$ ls /sys/devices/system/cpu/cpu0/cpuidle/
state0 state1 state2 state3 state4 state5 state6 state7 state8
$ ls /sys/devices/system/cpu/cpu0/cpuidle/state0/
above below desc disable latency name power residency time
usage
```

On ARM platforms, idle states can be described in the device tree. You can consult the `Documentation/devicetree/bindings/arm/idle-states.txt` file in kernel sources for more reading on this.

> **Important note**
>
> Unlike other power management frameworks, CPU Idle requires no user intervention for it to work.

There is a framework slightly similar to this one, that is, `CPU Hotplug`, which allows the dynamic enabling and disabling of CPUs at runtime without having to reboot the system. For example, to hotplug CPU #2 out of the system, you can use the following command:

```
# echo 0 > /sys/devices/system/cpu/cpu2/online
```

We can make sure that CPU #2 is actually disabled by reading `/proc/cpuinfo`:

```
# grep processor /proc/cpuinfo
processor : 0
processor : 1
processor : 3
processor : 4
processor : 5
processor : 6
processor : 7
```

The preceding confirms that CPU2 is now offline. In order to hotplug that CPU back into the system, we can execute the following command:

```
# echo 1 > /sys/devices/system/cpu/cpu2/online
```

What CPU hotplugging does under the hood will depend on your particular hardware and drivers. It may simply result in the CPU being put into idle on some systems, whereas other systems may physically remove power from the specified core.

CPUfreq or dynamic voltage and frequency scaling (DVFS)

This framework allows dynamic voltage selection and frequency scaling for the CPU, based on constraints and requirements, user preferences, or other factors. Because this framework deals with frequency, it unconditionally involves a clock framework. This framework uses the concept of **Operating Performance Points (OPPs)**, which consists of representing the performance state of a system with `{Frequency,voltage}` tuples.

OPPs can be described in the device tree, and its binding documentation in the kernel sources can be a good starting point for more information on it: `Documentation/devicetree/bindings/opp/opp.txt`.

> **Important note**
> You'll occasionally come across the term **P-state**. This is also an ACPI term (as is C-state) to designate CPU built-in hardware OPPs. This is the case with some Intel CPUs, and the operating system uses the policy objects to deal with these. You can check the result of `ls /sys/devices/system/cpu/cpufreq/` on an Intel-based machine. Thus, C-states are idle power-saving states, in contrast to P-states, which are execution power-saving states.

CPUfreq also uses the concept of governors (which implement scaling algorithms), and the governors in this framework are as follows:

- `ondemand`: This governor samples the load of the CPU and scales it up aggressively in order to provide the proper amount of processing power, but resets the frequency to the maximum when necessary.

- `conservative`: This is similar to `ondemand`, but uses a less aggressive method of increasing the OPP. For example, it will never skip from the lowest OPP to the highest one even if the system suddenly requires high performance. It will do it progressively.

- `performance`: This governor always selects the OPP with the highest frequency possible. This governor prioritizes performance.

- `powersave`: In contrast to performance, this governor always selects the OPP with the lowest frequency possible. This governor prioritizes power saving.

- `userspace`: This governor allows the user to set the desired OPP using any value found within `/sys/devices/system/cpu/cpuX/cpufreq/scaling_available_frequencies` by echoing it into `/sys/devices/system/cpu/cpuX/cpufreq/scaling_setspeed`.

- `schedutil`: This governor is part of the scheduler, so it can access the scheduler data structure internally, allowing it to grab more reliable and accurate stats about the system load, for the purpose of better selecting the appropriate OPP.

The `userspace` governor is the only one that allows users to select the OPP. For other governors, OPP change happens automatically based on the system load of their algorithm. That said, from `userspace`, the governors available are listed here:

```
$ cat /sys/devices/system/cpu/cpu0/cpufreq/scaling_available_
governors
performance powersave
```

To view the current governor, implement the following command:

```
$ cat /sys/devices/system/cpu/cpu0/cpufreq/scaling_governor
powersave
```

To set a governor, the following command can be used:

```
$ echo userspace > /sys/devices/system/cpu/cpu0/cpufreq/
scaling_governor
```

To view the current OPP (frequency in kHz), implement the following command:

```
$ cat /sys/devices/system/cpu/cpu0/cpufreq/scaling_cur_freq
800031
```

To view supported OPPs (frequency in kHz), implement the following command:

```
$ cat /sys/devices/system/cpu/cpu0/cpufreq/scaling_available_
frequencies
275000 500000 600000 800031
```

To change the OPP, you can use the following command:

```
$ echo 275000 > /sys/devices/system/cpu/cpu0/cpufreq/scaling_
setspeed
```

> **Important note**
>
> There is also the `devfreq` framework, which is a generic **Dynamic Voltage and Frequency Scaling (DVFS)** framework for non-CPU devices, and with governors such as `Ondemand`, `performance`, `powersave`, and `passive`.

Note that the preceding command only works when the `ondemand` governor is selected, as it is the only one that allows the OPP to be changed. However, in all the preceding commands, `cpu0` has been used only for didactic purposes. Think of it like *cpuX*, where *X* is the index of the CPU as seen by the system.

Thermal

This framework is dedicated to monitoring the system temperature. It has dedicated profiles according to temperature thresholds. Thermal sensors sense the hot points and report. This framework works in conjunction with cooling devices, which aid in power dissipation to control/limit overheating.

The thermal framework uses the following concepts:

- **Thermal zones**: You can think of a thermal zone as hardware whose temperature needs to be monitored.

- **Thermal sensors**: These are components used to take temperature measurements. Thermal sensors provide temperature sensing capabilities in thermal zones.

- **Cooling devices**: These devices provide control in terms of power dissipation. Typically, there are two cooling methods: passive cooling, which consists of regulating device performance, in which case DVFS is used; and active cooling, which consists of activating special cooling devices, such as fans (GPIO-fan, PWM-fan).

- **Trip points**: These describe key temperatures (actually thresholds) at which a cooling action is recommended. Those sets of points are chosen based on hardware limits.

- **Governors**: These comprise algorithms to choose the best cooling according to some criteria.

- **Cooling maps**: These are used to describe links between trip points and cooling devices.

The thermal framework can be divided into four parts, these being `thermal zone`, `thermal governor`, `thermal cooling`, and `thermal core`, which is the glue between the three previous parts. It can be managed in user space from within the `/sys/class/thermal/` directory:

```
$ ls /sys/class/thermal/
cooling_device0   cooling_device4 cooling_device8   thermal_zone3
thermal_zone7
cooling_device1   cooling_device5 thermal_zone0     thermal_zone4
```

```
cooling_device2   cooling_device6 thermal_zone1    thermal_zone5
cooling_device3   cooling_device7 thermal_zone2    thermal_zone6
```

In the preceding, each `thermal_zoneX` file represents a thermal zone driver, or a thermal driver. A thermal zone driver is the driver of the thermal sensor associated with a thermal zone. This driver exposes trip points at which cooling is necessary, but also provides a list of cooling devices associated with the sensor. The thermal workflow is designed to obtain the temperature through the thermal zone driver, and then make decisions through the thermal governor, and finally perform temperature control by means of thermal cooling. Further reading on this is available in the thermal sysfs documentation in the kernel sources, `Documentation/thermal/sysfs- api.txt`. Moreover, thermal zone description, trip point definitions, and cooling device binding can be performed in the device tree, and its associated documentation in the sources is `Documentation/devicetree/bindings/thermal/thermal.txt`.

System power management sleep states

System power management targets the entire system. Its aim is to put it into a low-power state. In this low-power state, the system is consuming a small, but minimal amount of power, yet maintaining a relatively low response latency to the user. The exact amount of power and response latency depends on how deep is the sleep state that the system is in. This is also referred to as Static Power Management because it is activated when the system is inactive for an extended period.

The states a system can enter are dependent on the underlying platform, and differ across architectures and even generations or families of the same architecture. There are, however, four sleep states that are commonly found on most platforms. These are suspend to idle (also known as freeze), power-on standby (standby), suspend to ram (mem), and suspend to disk (hibernation). These are also occasionally referred to by their ACPI states: `S0`, `S1`, `S3`, and `S4`, respectively:

```
# cat /sys/power/state
freeze mem disk standby
```

`CONFIG_SUSPEND` is the kernel configuration option that must be set in order for the system to support the system's power management sleep state. That said, except for *freeze*, each sleep state is platform specific. Thus, for a platform to support any of the three remaining states, it must explicitly register for each state with the core system suspend subsystem. However, the support for hibernation depends on other kernel configuration options, as we will see later.

> **Important note**
>
> Because only the user knows when the system is not going to be used (or
> even user code, such as GUI), system power management actions are always
> initiated from the user space. The kernel has no idea of that. This is why most
> of the content in this section deals with sysfs and the command line.

Suspend to idle (freeze)

This is the most basic and lightweight. This state is purely software driven and involves
keeping the CPUs in their deepest idle state as much as possible. To achieve this, the user
space is frozen (all user space tasks are) and all I/O devices are put into low-power states
(possibly lower power than available at runtime) so that the processors can spend more
time in their idle states. The following is the command that idles the system:

```
$ echo freeze > /sys/power/state
```

The preceding command puts the system in an idle state. Because it is purely software, this
state is always supported (assuming the CONFIG_SUSPEND kernel configuration option
is set). This state can be used for platforms without power-on-suspend or suspend-to-ram
support. However, as we will see later, it can be used in addition to suspend-to-ram to
provide reduced resume latency.

> **Important note**
>
> Suspend to idle equals frozen processes + suspended devices + idle processors

Power-on standby (standby or power-on suspend)

In addition to freezing the user space and putting all I/O devices into a low-power state,
another action performed by this state is to power off all non-boot CPUs. The following is
the command that puts the system in standby, assuming it is supported by the platform:

```
$ echo standby > /sys/power/state
```

As this state goes further than the freeze state, it also allows more energy to be saved
relative to *Suspend-to-Idle*, but the resume latency will generally be greater than for the
freeze state, although it is quite low.

Suspend-to-ram (suspend, or mem)

In addition to putting everything in the system into a low-power state, this state goes further by powering off all CPUs and putting the memory into self-refresh so that its contents are not lost, although additional operations may take place depending on the platform's capabilities. Response latency is higher than standby, yet still quite low. In this state, the system and device state is saved and kept in memory. This is the reason why only the RAM is fully operational, hence the state name:

```
# echo mem > /sys/power/state
```

The preceding command is supposed to put the system into a Suspend-to-RAM state. However, the real actions performed while writing the mem string are controlled by the /sys/power/mem_sleep file. This file contains a list of strings where each string represents a mode the system can enter after mem has been written to /sys/power/state. Although not all are always available (it depends on the platform), possible modes include the following:

- s2idle: This is the equivalent of Suspend-to-Idle. For this reason, it is always available.

- shallow: This is equivalent to Power-On Suspend, or standby. Its availability depends on the platform's support of the standby mode.

- deep: This is the real Suspend-To-RAM state and its availability depends on the platform.

An example of querying the content can be seen here:

```
$ cat /sys/power/mem_sleep
[s2idle] deep
```

The selected mode is enclosed in square brackets, [ ]. If a mode is not supported by the platform, the string that corresponds to it will still not be present in /sys/power/mem_sleep. Writing one of the other strings present in /sys/power/mem_sleep to it causes the suspend mode to be used subsequently to change to the one represented by that string.

When the system is booted, the default suspend mode (in other words, the one to be used without writing anything into /sys/power/mem_sleep) is either deep (if Suspend-To-RAM is supported) or s2idle, but it can be overridden by the value of the mem_sleep_default parameter in the kernel command line.

One method for testing this is to use an RTC available on the system, assuming it supports the `wakeup alarm` feature. You can identify available RTCs on your system using `ls /sys/class/rtc/`. There will be a directory for each RTC (in other words, `rtc0` and `rtc1`). For an `rtc` that supports the `alarm` feature, there will be a `wakealarm` file in that `rtc` directory, which can be used as follows to configure an alarm and then suspend the system to RAM:

```
/* No value returned means no alarms are set */
$ cat /sys/class/rtc/rtc0/wakealarm
/* Set the wakeup alarm for 20s */
# echo +20 > /sys/class/rtc/rtc0/wakealarm
/* Now Suspend system to RAM */
# echo mem > /sys/power/state
```

You should see no further activity on the console until wakeup.

Suspend to disk (hibernation)

This state gives the greatest power savings as a result of powering off as much of the system as possible, including the memory. Memory contents (a snapshot) are written to persistent media, usually a disk. After this, memory is powered down, along with the entire system. Upon resumption, the snapshot is read back into memory and the system boots from this hibernation image. However, this state is also the longest to resume but still quicker than performing a full (re)boot sequence:

```
$ echo disk > /sys/power/state
```

Once the memory state is written to disk, several actions can take place. The action to be performed is controlled by the `/sys/power/disk` file and its contents. This file contains a list of strings where each string represents an action that can be performed once the system state is saved on persistent storage media (after the hibernation image has actually been saved). Possible actions include the following:

- `platform`: Custom- and platform-specific, which may require firmware (BIOS) intervention.
- `shutdown`: Power off the system.
- `reboot`: Reboots the system (useful for diagnostics mostly).

- `suspend`: Puts the system into the suspend sleep state selected through the `mem_sleep` file described earlier. If the system is successfully woken up from that state, then the hibernation image is simply discarded and everything continues. Otherwise, the image is used to restore the previous state of the system.

- `test_resume`: This is for system resumption diagnostic purposes. Loads the image as if the system had just woken up from hibernation and the currently running kernel instance was a restore kernel and follows up with full system resumption.

However, supported actions on a given platform depend on the content of the `/sys/power/disk` file:

```
$ cat /sys/power/disk
[platform] shutdown reboot suspend test_resume
```

The selected action is enclosed in square brackets, `[ ]`. Writing one of the listed strings to this file causes the option represented by it to be selected. Hibernation is such a complex operation that it has its own configuration option, `CONFIG_HIBERNATION`. This option has to be set in order to enable the hibernation feature. That said, this option can only be set if support for the given CPU architecture includes the low-level code for system resumption (refer to the `ARCH_HIBERNATION_POSSIBLE` kernel configuration option).

For suspend-to-disk to work, and depending on where the hibernation image should be stored, a dedicated partition may be required on the disk. This partition is also known as a swap partition. This partition is used to write memory contents to free swap space. In order to check whether hibernation works as expected, it is common to try to hibernate in `reboot` mode as follows:

```
$ echo reboot > /sys/power/disk
# echo disk > /sys/power/state
```

The first command informs the power management core of what action should be performed when the hibernation image has been created. In this case, it is a reboot. Upon reboot, the system is restored from the hibernation image and you should get back to the command prompt where you started the transition. The success of this test may show that hibernation is most likely to work correctly. That said, it should be done several times in order to reinforce the test.

Now that we are done with sleep state management from a running system, we can see how to implement its support in driver code.

Adding power management capabilities to device drivers

Device drivers on their own can implement a distinct power management capability, which is known as runtime power management. Not all devices support runtime power management. However, those that do must export some callbacks for controlling their power state depending on the user or system's policy decisions. As we have seen earlier, this is device-specific. In this section, we will learn how to extend device driver capabilities with power management support.

Though device drivers provide runtime power management callbacks, they also facilitate and participate in the system sleep state by providing another set of callbacks, where each set participates in a particular system sleep state. Whenever the system needs to enter or resume from a given set, the kernel will walk through each driver that provided callbacks for this state and then invoke them in a precise order. Simply speaking, device power management consists of a description of the state a device is in, and a mechanism for controlling those states. This is facilitated by the kernel providing the `struct dev_pm_ops` that each device driver/class/bus interested in power management must fill. This allows the kernel to communicate with every device in the system, regardless of the bus the device resides on or the class it belongs to. Let's take a step back and remember what a `struct device` looks like:

```
struct device {
    [...]
    struct device *parent;
    struct bus_type *bus;
    struct device_driver *driver;
    struct dev_pm_info power;
    struct dev_pm_domain *pm_domain;
}
```

In the preceding `struct device` data structure, we can see that a device can be either a child (its `.parent` field points to another device) or a device parent (when the `.parent` field of another device points to it), can sit behind a given bus, or can belong to a given class, or can belong indirectly to a given subsystem. Moreover, we can see that a device can be part of a given power domain. The `.power` field is of the `struct dev_pm_info` type. It mainly saves PM-related states, such as the current power state, whether it can be awakened, whether it has been prepared, and whether it has been suspended. Since there is so much content involved, we will explain these in detail when we use them.

In order for devices to participate in power management, either at the subsystem level or at the device driver level, their drivers need to implement a set of device power management operations by defining and populating objects of the `struct dev_pm_ops` type defined in `include/linux/pm.h` as follows:

```
struct dev_pm_ops {
    int (*prepare)(struct device *dev);
    void (*complete)(struct device *dev);
    int (*suspend)(struct device *dev);
    int (*resume)(struct device *dev);
    int (*freeze)(struct device *dev);
    int (*thaw)(struct device *dev);
    int (*poweroff)(struct device *dev);
    int (*restore)(struct device *dev);
    [...]
    int (*suspend_noirq)(struct device *dev);
    int (*resume_noirq)(struct device *dev);
    int (*freeze_noirq)(struct device *dev);
    int (*thaw_noirq)(struct device *dev);
    int (*poweroff_noirq)(struct device *dev);
    int (*restore_noirq)(struct device *dev);
    int (*runtime_suspend)(struct device *dev);
    int (*runtime_resume)(struct device *dev);
    int (*runtime_idle)(struct device *dev);
};
```

In the preceding data structure, `*_early()` and `*_late()` callbacks have been removed for the sake of readability. I suggest you have a look at the full definition. That said, given the huge number of callbacks in there, we will describe them in due course in the sections of the chapter where their use will be necessary.

> **Important note**
>
> Device power states are sometimes referred to as *D* states, inspired by the PCI device and ACPI specifications. Those states range from state D0 to D3, inclusive. Although not all device types define power states in this way, this representation can map to all known device types.

Implementing runtime PM capability

Runtime power management is a per-device power management feature allowing a particular device to have its states controlled when the system is running, regardless of the global system. For a driver to implement runtime power management, it should provide only a subset of the whole list of callbacks in `struct dev_pm_ops`, shown as follows:

```
struct dev_pm_ops {
    [...]
    int (*runtime_suspend)(struct device *dev);
    int (*runtime_resume)(struct device *dev);
    int (*runtime_idle)(struct device *dev);
};
```

The kernel also provides `SET_RUNTIME_PM_OPS()`, which accepts the three callbacks to be populated in the structure. This macro is defined as follows:

```
#define SET_RUNTIME_PM_OPS(suspend_fn, resume_fn, idle_fn) \
        .runtime_suspend = suspend_fn, \
        .runtime_resume = resume_fn, \
        .runtime_idle = idle_fn,
```

The preceding callbacks are the only ones involved in runtime power management, and here are descriptions of what they must do:

- `.runtime_suspend()` must record the device's current state if necessary and put the device in a quiescent state. This method is invoked by the PM when the device is not used. In its simple form, this method must put the device in a state in which it won't be able to communicate with the CPU(s) and RAM.

- `.runtime_resume()` is invoked when the device must be put in a fully functional state. This may be the case if the system needs to access this device. This method must restore power and reload any required device state.

- .runtime_idle() is invoked when the device is no longer used based on the device usage counter (actually when it reaches 0), as well as the number of active children. However, the action performed by this callback is driver-specific. In most cases, the driver invokes runtime_suspend() on the device if some conditions are met, or invokes pm_schedule_suspend() (given a delay in order to set up a timer to submit a suspend request in future), or pm_runtime_autosuspend() (to schedule a suspend request in the future based on a delay that has already been set using pm_runtime_set_autosuspend_delay()). If the .runtime_idle callback doesn't exist or if it returns 0, the PM core will immediately invoke the .runtime_suspend() callback. For the PM core to do nothing, .runtime_idle() must return a non-zero value. It is common for drivers to return -EBUSY, or 1 in this case.

After callbacks have been implemented, they can be fed in struct dev_pm_ops, as in the following example:

```
static const struct dev_pm_ops bh1780_dev_pm_ops = {
    SET_SYSTEM_SLEEP_PM_OPS(pm_runtime_force_suspend,
                            pm_runtime_force_resume)
    SET_RUNTIME_PM_OPS(bh1780_runtime_suspend,
                       bh1780_runtime_resume, NULL)
};
[...]
static struct i2c_driver bh1780_driver = {
    .probe = bh1780_probe,
    .remove = bh1780_remove,
    .id_table = bh1780_id,
    .driver = {
        .name = "bh1780",
        .pm = &bh1780_dev_pm_ops,
        .of_match_table = of_match_ptr(of_bh1780_match),
    },
};
module_i2c_driver(bh1780_driver);
```

The preceding is an excerpt from `drivers/iio/light/bh1780.c`, an IIO ambient light sensor driver. In this excerpt, we can see how `struct dev_pm_ops` is populated, using convenient macros. `SET_SYSTEM_SLEEP_PM_OPS` is used here to populate system sleep-related macros, as we will see in the next sections. `pm_runtime_force_suspend` and `pm_runtime_force_resume` are special helpers that the PM core exposes to force device suspension and resumption, respectively.

Runtime PM anywhere in the driver

In fact, the PM core keeps track of the activity of each device using two counters. The first counter is `power.usage_count`, which counts active references to the device. These may be external references, such as open file handles, or other devices that are making use of this one, or they may be internal references used to keep the device active for the duration of an operation. The other counter is `power.child_count`, which counts the number of children that are active.

These counters define the active/idle conditions of a given device from the PM point of view. The active/idle condition of a device is the only reliable means for the PM core to determine whether a device is accessible. An idle condition is when the device usage count is decremented until 0, and an active condition (also known as a resume condition) occurs whenever the device usage count is incremented.

In the event of an idle condition, the PM core sends/performs an idle notification (that is, setting the device's `power.idle_notification` field to `true`, invoking the bus type/class/device `->runtime_idle()` callback, and setting the `.idle_notification` field back to `false` again) in order to check whether the device can be suspended. If the `->runtime_idle()` callback doesn't exist or if it returns 0, the PM core will immediately invoke the `->runtime_suspend()` callback to suspend the device, after which the device's `power.runtime_status` field is set to `RPM_SUSPENDED`, which means the device is suspended. Upon a resume condition (the device usage count is incremented), the PM core will carry out a resumption (under certain conditions only) of this device, either synchronously or asynchronously. Have a look at the `rpm_resume()` function and its description in `drivers/base/power/runtime.c`.

Initially, the runtime PM is disabled for all devices. This means invoking most PM-related helpers on the device will fail until `pm_runtime_enable()` is called for the device, which enables a runtime PM of this device. Though the initial runtime PM status of all devices is suspended, it need not reflect the actual physical state of the device. Thus, if the device is initially active (in other words, it is able to process I/O), its runtime PM status must be changed to active with the help of `pm_runtime_set_active()` (which will set `power.runtime_status` to `RPM_ACTIVE`), and if possible, its usage count must be increased using `pm_runtime_get_noresume()` before `pm_runtime_enable()` is called for the device. Once the device is fully initialized, you can call `pm_runtime_put()` on it.

The reason for invoking `pm_runtime_get_noresume()` here is that, if there is a call to `pm_runtime_put()`, the device usage count will come back to zero, which corresponds to an idle condition, and then an idle notification will be carried out. At this time, you'll be able to check whether necessary conditions have been met and suspend the device. However, if the initial device state is *disabled*, there is no need to do so.

There are also `pm_runtime_get()`, `pm_runtime_get_sync()`, `pm_runtime_put_noidle()`, and `pm_runtime_put_sync()` helpers. The difference between `pm_runtime_get_sync()`, `pm_runtime_get()`, and `pm_runtime_get_noresume()` is that the former will synchronously (immediately) carry out a resumption of the device if the active/resume condition is matched after the device usage count has been incremented, while the second helper will do it asynchronously (submitting a request for it). The third and final one will return immediately after having decremented the device usage count (without even checking the resume condition). The same mechanism applies to `pm_runtime_put_sync()`, `pm_runtime_put()`, and `pm_runtime_put_noidle()`.

The number of active children of a given device affects the usage count of this device. Normally, the parent is needed to access the child, so powering down the parent while children are active would be counterproductive. Sometimes, however, it might be necessary to ignore active children of a device when determining whether this device is idle. One good example is the I2C bus, where the bus can be reported as idle while devices sitting on this bus (children) are active. For such cases, `pm_suspend_ignore_children()` can be invoked to allow a device to report as idle even when it has active children(s).

Runtime PM synchronous and asynchronous operations

In the previous section, we introduced the fact that the PM core could carry out synchronous or asynchronous PM operations. While things are straightforward for synchronous operations (method calls are serialized), we need to pay some attention to what steps are performed while invoking things asynchronously in a PM context.

You should keep in mind that, in asynchronous mode, a request for the action is submitted instead or invoking this action's handler immediately. It works as follows:

1. The PM core sets the device's `power.request` field (which is of the `enum rpm_request` type) with the type of request to be submitted (in other words, `RPM_REQ_IDLE` for an idle notification request, `RPM_REQ_SUSPEND` for a suspend request, or `RPM_REQ_AUTOSUSPEND` for an autosuspend request), which corresponds to the action to be performed.

2. The PM core sets the device's `power.request_pending` field to `true`.

3. The PM core queues (schedules for a later execution) the device's RPM-related work (`power.work`, whose work function is `pm_runtime_work()`; see `pm_runtime_init()` where it is initialized) in the Global PM-related work queue.

4. When this work has the chance to run, the work function (that is, `pm_runtime_work()`) will first check whether there is still a request pending on the device (`if (dev->power.request_pending)`) and perform a `switch ... case` on the device's `power.request_pending` field in order to invoke the underlying request handler.

Do note that a work queue manages its own thread(s), which can run scheduled works. Because, in asynchronous mode, the handler is scheduled in a work queue, asynchronous PM-related helpers are totally safe to be invoked in an atomic context. If invoked within an IRQ handler, for example, it would be equivalent to deferring the PM request handling.

Autosuspend

Autosuspend is a mechanism used by drivers that do not want their device to suspend as soon as it becomes idle at runtime, but they rather want the device to remain inactive for a certain minimum period of time first.

In the context of RPM, the term *autosuspend* does not mean the device automatically suspends itself. It is instead based on a timer that, upon expiration, will queue a suspend request. This timer is actually the device's `power.suspend_timer` field (see `pm_runtime_init()` where it is set up). Calling `pm_runtime_put_autosuspend()` will start the timer, while `pm_runtime_set_autosuspend_delay()` will set the timeout (though that can be set via `sysfs` in the `/sys/devices/.../power/autosuspend_delay_ms` attribute) represented by the device's `power.autosuspend_delay` field.

This timer can be used by the `pm_schedule_suspend()` helper as well, with a delay in argument (which in this case will take precedence on the one set in the `power.autosuspend_delay` field), after which a suspend request will be submitted. You can regard this timer as something that can be used to add a delay between the counters reaching zero and the device being considered to be idle. This is useful for devices with a high cost associated with turning on or off.

In order to use `autosuspend`, subsystems or drivers must call `pm_runtime_use_autosuspend()` (preferably before registering the device). This helper will set the device's `power.use_autosuspend` field to `true`. After soliciting a device on which autosuspend is enabled, you should invoke `pm_runtime_mark_last_busy()` on this device, which lets it set the `power.last_busy` field to the current time (in `jiffies`), because this field is used in calculating inactivity periods for autosuspend (for example, `new_expire_time = last_busy + msecs_to_jiffies(autosuspend_delay)`).

Given all the runtime PM concepts introduced, let's put it all together now and see how things are done in a real driver.

Putting it all together

The preceding theoretical studies of the runtime PM core would be less significant without a genuine case study. Now is the time to see how the previous concept is applied. For this case study, we will pick the `bh1780` Linux driver, which is a **digital 16-bit I2C** ambient light sensor. The driver of this device is `drivers/iio/light/bh1780.c` in the Linux kernel sources.

To start with, let's see an excerpt of the `probe` method:

```
static int bh1780_probe(struct i2c_client *client,
                    const struct i2c_device_id *id)
{
    [...]
    /* Power up the device */ [...]
    pm_runtime_get_noresume(&client->dev);
    pm_runtime_set_active(&client->dev);
    pm_runtime_enable(&client->dev);

    ret = bh1780_read(bh1780, BH1780_REG_PARTID);
    dev_info(&client->dev, "Ambient Light Sensor, Rev : %lu\n",
                (ret & BH1780_REVMASK));

    /*
     * As the device takes 250 ms to even come up with a fresh
     * measurement after power-on, do not shut it down
     * unnecessarily.
     * Set autosuspend to five seconds.
     */
    pm_runtime_set_autosuspend_delay(&client->dev, 5000);
    pm_runtime_use_autosuspend(&client->dev);
    pm_runtime_put(&client->dev);
    [...]
    ret = iio_device_register(indio_dev);
    if (ret)
        goto out_disable_pm; return 0;

out_disable_pm:
    pm_runtime_put_noidle(&client->dev);
    pm_runtime_disable(&client->dev); return ret;
}
```

In the preceding snippet, only the power management-related calls are left, for the sake of readability. First, pm_runtime_get_noresume() will increment the device usage count without carrying an idle notification of the device (the _noidle suffix). You may use the pm_runtime_get_noresume() interface to turn off the runtime suspend function or to make the usage count positive even while the device is suspended, so as to avoid issues that do not wake up normally due to the runtime suspension. Then, the next line in the driver is pm_runtime_set_active(). This helper marks the device as active (power.runtime_status = RPM_ACTIVE) and clears the device's power. runtime_error field. Additionally, the device parent's counter of unsuspended (active) children is modified to reflect the new status (it is incremented actually). Invoking pm_ runtime_set_active() on a device will prevent this device's parent from suspending at runtime (assuming the parent's runtime PM is enabled), unless the parent's power. ignore_children flag is set. For this reason, once pm_runtime_set_active() has been called for the device, pm_runtime_enable() should be called for it too, as soon as is reasonably possible. Invoking this function is not mandatory; it has to be coherent with the PM core and the status of the device, assuming the initial status is RPM_ SUSPENDED.

> **Important note**
>
> The opposite of pm_runtime_set_active() is pm_runtime_set_ suspended(), which changes the device status to RPM_SUSPENDED, and decrements the parent's counter of active children. An idle notification request for the parent is submitted.

pm_runtime_enable() is the mandatory runtime PM helper, which enables the runtime PM of a device, that is, de-increments the device's power.disable_depth value in case its value is greater than 0. For information, the device's power.disable_ depth value is checked on each runtime PM helper call, and its value must be 0 for the helper to progress. Its initial value is 1, and this value is decremented upon a call to pm_ runtime_enable(). On the error path, pm_runtime_put_noidle() is invoked in order to make the PM runtime counter balance, and pm_runtime_disable() completely disables the runtime PM on the device.

As you may have guessed, this driver also deals with the IIO framework, which means it exposes entries in sysfs, which correspond to its physical conversion channels. Reading the sysfs file corresponding to a channel will report the digital value of the conversion resulting from this channel. However, for the bh1780, the channel read entry point in its driver is bh1780_read_raw(). An excerpt of this method can be seen here:

```
static int bh1780_read_raw(struct iio_dev *indio_dev,
                            struct iio_chan_spec const *chan,
                            int *val, int *val2, long mask)
{
    struct bh1780_data *bh1780 = iio_priv(indio_dev);
    int value;

    switch (mask) {
    case IIO_CHAN_INFO_RAW:
        switch (chan->type) {
        case IIO_LIGHT:
            pm_runtime_get_sync(&bh1780->client->dev);
            value = bh1780_read_word(bh1780, BH1780_REG_DLOW);
            if (value < 0)
                return value;
            pm_runtime_mark_last_busy(&bh1780->client->dev);
            pm_runtime_put_autosuspend(&bh1780->client->dev);
            *val = value;
            return IIO_VAL_INT;
        default:
            return -EINVAL;
    case IIO_CHAN_INFO_INT_TIME:
        *val = 0;
        *val2 = BH1780_INTERVAL * 1000;
        return IIO_VAL_INT_PLUS_MICRO;

    default:
        return -EINVAL;
    }
}
```

Here again, only runtime PM-related function calls are worthy of our attention. In the event of a channel read, the preceding function is invoked. The device driver has to instruct the device to sample the channel, to perform the conversion whose result will be read by the device driver and reported to the reader. The thing is, the device may be in a suspended state. Thus, because the driver needs immediate access to the device, the driver calls pm_runtime_get_sync() on it. If you recall, this method increments the device usage count and carries out a synchronous (_sync suffix) resumption of the device. After the device resumes, the driver can talk with the device and read the conversion value. Because the driver supports autosuspend, pm_runtime_mark_last_busy() is called in order to mark the last time the device was active. This will update the timeout value of the timer used for autosuspend. Finally, the driver invokes pm_runtime_put_ autosuspend(), which will carry out a runtime suspend of the device following the autosuspend timer expiration, unless this timer is restarted again by pm_runtime_ mark_last_busy() being invoked somewhere or when entering the read function again (the reading of the channel, in sysfs, for example) prior to expiration.

To summarize, before accessing the hardware, the driver can resume the device with pm_ runtime_get_sync(), and when it's finished with the hardware, the driver can notify the device as being idle with either pm_runtime_put_sync(), pm_runtime_put(), or pm_runtime_put_autosuspend() (assuming autosuspend is enabled, in which case pm_runtime_mark_last_busy() must be invoked beforehand in order to update the autosuspend timer's timeout).

Finally, let's focus on the method invoked when the module is being unloaded. The following is an excerpt in which only PM-related calls are of interest:

```
static int bh1780_remove(struct i2c_client *client)
{
    int ret;
    struct iio_dev *indio_dev = i2c_get_clientdata(client);
    struct bh1780_data *bh1780 = iio_priv(indio_dev);

    iio_device_unregister(indio_dev);
    pm_runtime_get_sync(&client->dev);
    pm_runtime_put_noidle(&client->dev);
    pm_runtime_disable(&client->dev);

    ret = bh1780_write(bh1780, BH1780_REG_CONTROL,
                       BH1780_POFF);
    if (ret < 0) {
```

```
        dev_err(&client->dev, "failed to power off\n");
        return ret;
    }
    return 0;
}
```

The first runtime PM method invoked here is `pm_runtime_get_sync()`. This call gets us guessing that the device is going to be used, that is, the driver needs to access the hardware. Thus, this helper immediately resumes the device (it actually increments the device usage counter and carries out a synchronous resumption of the device). After this, `pm_runtime_put_noidle()` is called in order to de-increment the device usage count without carrying an idle notification. Next, `pm_runtime_disable()` is called in order to disable runtime PM on the device. This will increment `power.disable_depth` for the device and if it was zero previously, cancel all pending runtime PM requests for the device and wait for all operations in progress to complete, so that with regard to the PM core, the device no longer exists (remember, `power.disable_depth` will not match what the PM core expects, meaning that any further runtime PM helper invoked on this device will fail). Finally, the device is powered off thanks to an i2c command, after which its hardware status will reflect its runtime PM status.

The following are general rules that apply to runtime PM callback and execution:

- `->runtime_idle()` and `->runtime_suspend()` can only be executed for active devices (those whose status is active).

- `->runtime_idle()` and `->runtime_suspend()` can only be executed for a device with the usage counter equal to zero and either with the counter of active children equal to zero, or with the `power.ignore_children` flag set.

- `->runtime_resume()` can only be executed for suspended devices (those whose status is *suspended*).

Additionally, the helper functions provided by the PM core obey the following rules:

- If `->runtime_suspend()` is about to be executed or there's a pending request to execute it, `->runtime_idle()` will not be executed for the same device.

- A request to execute or to schedule the execution of `->runtime_suspend()` will cancel any pending requests to execute `->runtime_idle()` for the same device.

- If `->runtime_resume()` is about to be executed or there's a pending request to execute it, the other callbacks will not be executed for the same device.

- A request to execute `->runtime_resume()` will cancel any pending or scheduled requests to execute the other callbacks for the same device, except for scheduled autosuspends.

The preceding rules are good indicators of reasons why any invocation of these callbacks may fail. From these, we can also observe that a resumption, or a request to resume, outperforms any other callback or request.

The concept of power domain

Technically, a power domain is a set of devices sharing power resources (for example, clocks or power planes). From the kernel's perspective, a power domain is a set of devices whose power management uses the same set of callbacks with common PM data at the subsystem level. From the hardware perspective, a power domain is a hardware concept for managing devices whose power voltages are correlated; for example, the video core IP sharing a power rail with the display IP.

Because of SoC designs being more complex, an abstraction method needed to be found so that drivers remain as generic as possible; then, `genpd` came out. This stands for Generic Power Domain. It is a Linux Kernel abstraction that extends per-device runtime power management to a group of devices sharing power rails. Moreover, power domains are defined as part of a device tree in which relationships between devices and power controllers are described. This allows power domains to be redesigned on the fly and drivers to adapt without having to reboot the whole system or rebuild a new kernel.

It is designed so that if a power domain object exists for a device, its PM callbacks take precedence over the bus type (or device class or type) callback. Generic documentation on this is available in `Documentation/devicetree/bindings/power/power_domain.txt` in kernel sources, and documentation related to your SoC can be found in the same directory.

System suspend and resume sequences

The introduction of the `struct dev_pm_ops` data structure has somehow facilitated the understanding of the steps and actions performed by the PM core during a suspension or resumption phase, which can be summarized as follows:

```
"prepare -> Suspend -> suspend_late -> suspend_noirq"
          |---------- Wakeup ----------|
 "resume_noirq -> resume_early -> resume -> complete"
```

The preceding is the full system PM chain, as enumerated in enum `suspend_stat_step`, defined in `include/linux/suspend.h`. This flow should remind you of the `struct dev_pm_ops` data structure.

In the Linux kernel code, `enter_state()` is the function invoked by the system power management core to enter a system sleep state. Let's now spend some time on what really goes on during system suspension and resumption.

Suspend stages

The following are the steps that `enter_state()` goes through when suspended:

1. It first invokes `sync()` on the filesystem (see `ksys_sync()`) if the `CONFIG_SUSPEND_SKIP_SYNC` kernel configuration option is not set.

2. It invokes suspend notifiers (while the user space is still there). Refer to `register_pm_notifier()`, which is the helper used for their registration.

3. It freezes tasks (see `suspend_freeze_processes()`), which freezes the user space as well as kernel threads. This step is skipped if `CONFIG_SUSPEND_FREEZER` is not set in a kernel configuration.

4. Devices are suspended by invoking every `.suspend()` callbacks registered by drivers. This is the first phase of suspending (see `suspend_devices_and_enter()`).

5. It disables device interrupts (see `suspend_device_irqs()`). This prevents device drivers from receiving interrupts.

6. Then, the second phase of suspending devices happens (`.suspend_noirq` callbacks are invoked). This step is known as the *noirq* stage.

7. It disables non-boot CPUs (using a CPU hotplug). The CPU scheduler is told not to schedule anything on those CPUs before they go offline (see `disable_nonboot_cpus()`).

8. It turns interrupts off.

9. It executes system core callbacks (see `syscore_suspend()`).

10. It puts the system to sleep.

This is a rough description of the actions performed before the system goes to sleep. The behavior of certain actions may vary slightly according to the sleep state the system is going to enter.

Resume stages

Once a system is suspended (however deep it is), once a wakeup event occurs, the system needs to resume. The following are the steps and actions the PM core performs in order to wake up the system:

1. (Wakeup signal.)

2. Run the CPU's wakeup code.

3. Execute system core callbacks.

4. Turn the interrupts on.

5. Enable non-boot CPUs (using the CPU hotplug).

6. The first phase of resuming devices (`.resume_noirq()` callbacks).

7. Enable device interrupts.

8. The second phase of suspending devices (`.resume()` callbacks).

9. Thaw tasks.

10. Call notifiers (when the user space is back).

I will let you discover in the PM code which functions are invoked at each step of the resumption process. From within the driver, however, these steps are all transparent. The only thing the driver needs to do is to fill `struct dev_pm_ops` with the appropriate callbacks according to the steps it wishes to be involved in, as we will see in the next section.

Implementing system sleep capability

System sleep and runtime PM are different things, though they are related to one another. There are cases where, by doing it in different ways, they bring the system to the same physical state. Thus, it is generally not a good idea to replace one with the other.

We have seen how device drivers participate in the system sleep by populating some callbacks in the `struct dev_pm_ops` data structure according to the sleep state they need to participate in. Commonly provided callbacks, irrespective of the sleep state, are `.suspend`, `.resume`, `.freeze`, `.thaw`, `.poweroff`, and `.restore`. They are quite generic callbacks and are defined as follows:

- `.suspend`: This is executed before the system is put into a sleep state in which the contents of the main memory are preserved.

- .resume: This callback is invoked after waking the system up from a sleep state in which the contents of the main memory were preserved, and the state of the device at the time this callback is run depends on the platform and subsystem the device belongs to.

- .freeze: Hibernation-specific, this callback is executed before creating a hibernation image. It's analogous to .suspend, but it should not enable the device to signal wakeup events or change its power state. Most device drivers implementing this callback only have to save the device setting in memory so that it can be used back during subsequent .resume from hibernation.

- .thaw: This callback is hibernation-specific, and it is executed after creating a hibernation image OR if the creation of an image has failed. It is also executed after a failed attempt to restore the contents of main memory from such an image. It must undo the changes made by the preceding .freeze in order to make the device operate in the same way as immediately prior to the call to .freeze.

- .poweroff: Also hibernation-specific, this is executed after saving a hibernation image. It's analogous to .suspend, but it need not save the device's settings in memory.

- .restore: This is the last hibernation-specific callback, which is executed after restoring the contents of the main memory from a hibernation image. It's analogous to .resume.

Most of the preceding callbacks are quite similar or perform roughly similar operations. While the .resume, .thaw, and .restore trio may perform similar tasks, the same is true for the other trio – ->suspend, ->freeze, and ->poweroff. Thus, in order to improve code readability or facilitate callback population, the PM core provides the SET_SYSTEM_SLEEP_PM_OPS macro, which takes suspend and resume functions and populates system-related PM callbacks as follows:

```
#define SET_SYSTEM_SLEEP_PM_OPS(suspend_fn, resume_fn) \
        .suspend = suspend_fn, \
        .resume = resume_fn, \
        .freeze = suspend_fn, \
        .thaw = resume_fn, \
        .poweroff = suspend_fn, \
        .restore = resume_fn,
```

The same is true for `_noirq()`-related callbacks. In case the driver only needs to participate in the `noirq` phase of the system suspend, the `SET_NOIRQ_SYSTEM_SLEEP_PM_OPS` macro can be used in order to automatically populate `_noirq()`-related callbacks in the `struct dev_pm_ops` data structure. The following is a definition of the macro:

```
#define SET_NOIRQ_SYSTEM_SLEEP_PM_OPS(suspend_fn, resume_fn) \
        .suspend_noirq = suspend_fn, \
        .resume_noirq = resume_fn, \
        .freeze_noirq = suspend_fn, \
        .thaw_noirq = resume_fn, \
        .poweroff_noirq = suspend_fn, \
        .restore_noirq = resume_fn,
```

The preceding macro takes only two parameters, which represent, as in the former macro, the `suspend` and `resume` callbacks, but for the `noirq` phase this time. You should remember that such callbacks are invoked with IRQs disabled on the system.

Finally, there is the `SET_LATE_SYSTEM_SLEEP_PM_OPS` macro, which will point `->suspend_late`, `-> freeze_late`, and `-> poweroff_late` to the same function, and vice versa for `->resume_early`, `->thaw_early`, and `->restore_early`:

```
#define SET_LATE_SYSTEM_SLEEP_PM_OPS(suspend_fn, resume_fn) \
        .suspend_late = suspend_fn, \
        .resume_early = resume_fn, \
        .freeze_late = suspend_fn, \
        .thaw_early = resume_fn, \
        .poweroff_late = suspend_fn, \
        .restore_early = resume_fn,
```

In addition to reducing the coding effort, all the preceding macros are conditioned with the `#ifdef CONFIG_PM_SLEEP` kernel configuration option so that they are not built if the PM is not needed. Finally, if you want to use the same suspend and resume callbacks for suspension to RAM and hibernation, you can use the following command:

```
#define SIMPLE_DEV_PM_OPS(name, suspend_fn, resume_fn) \
const struct dev_pm_ops name = { \
    SET_SYSTEM_SLEEP_PM_OPS(suspend_fn, resume_fn) \
}
```

In the preceding snippet, `name` represents the name with which the device PM ops structure will be instantiated. `suspend_fn` and `resume_fn` are the callbacks to be invoked when the system is entering a suspend state or when it resumes from a sleep state.

Now that we are able to implement system sleep capabilities in our driver code, let's see how to behave a system wakeup source, which allows the sleep state to be exited.

Being a source of system wakeup

The PM core allows the system to be awoken following a system suspend. A device capable of system wakeup is known as a **wakeup source** in PM language. For a wakeup source to operate normally, it needs a so-called **wakeup event**, which, most of time, is assimilated to an IRQ line. In other words, a wakeup source generates wakeup events. When a wakeup source generates a wakeup event, the wakeup source is set to the activated state through the interface provided by the wakeup event's framework. When the event processing ends, it is set to the deactivated state. The interval between activate and deactivate indicates that the event is being processed. In this section, we will see how to make your device be a source of system wakeup in driver code.

Wakeup sources work so that when there is any wakeup event being processed in the system, suspension is not allowed. If suspension is in progress, it is terminated. The kernel abstracts wakeup sources by means of `struct wakeup_source`, which is also used for collecting statistics related to them. The following is a definition of this data structure in `include/linux/pm_wakeup.h`:

```
struct wakeup_source {
    const char *name;
    struct list_head entry;
    spinlock_t lock;
    struct wake_irq *wakeirq;
    struct timer_list timer;
    unsigned long timer_expires;
    ktime_t total_time;
    ktime_t max_time;
    ktime_t last_time;
    ktime_t start_prevent_time;
    ktime_t prevent_sleep_time;
    unsigned long event_count;
    unsigned long active_count;
    unsigned long relax_count;
```

```
    unsigned long expire_count;
    unsigned long wakeup_count;
    bool active:1;
    bool autosleep_enabled:1;
};
```

This structure is absolutely useless for you in terms of code, but studying it will help you understand what wakeup source `sysfs` attributes mean:

- `entry` is used to track all wakeup sources in a linked list.

- `timer` goes hand in hand with `timer_expires`. When a wakeup source generates a wakeup event and that event is being processed, the wakeup source is said to be *active*, and this prevents system suspension. After the wakeup event is processed (the system is no longer required to be active for this purpose), it returns to being inactive. Both activate and deactivate operations can be performed by the driver, or the driver can decide otherwise by specifying a timeout during activation. This timeout will be used by the PM wakeup core to configure a timer that will automatically set the event to the inactive state after it expires. `timer` and `timer_expires` are used for this purpose.

- `total_time` is the total time this wakeup source has been active. It sums up the total amount of time the wakeup source spent in the active state. It is a good indicator of the busy level and power consumption level of the device corresponding to the wakeup source.

- `max_time` is the longest amount of time that the wakeup source remained (or was continuously) in the active state. The longer it is, the more abnormal it is.

- `last_time` indicates the start time of the last time this wakeup source was active.

- `start_prevent_time` is the point in time when the wakeup source started to prevent the system from autosleeping.

- `prevent_sleep_time` is the total time this wakeup source prevented the system from autosleeping.

- `event_count` represents the number of events reported by the wakeup source. In other words, it indicates the number of signaled wakeup events.

- `active_count` represents the number of times the wakeup source was activated. This value may not be relevant or coherent in certain situations. For example, when a wakeup event occurs, the wakeup source needs to be switched to the active state. However, this is not always the case because the event may occur while the wakeup source is already activated. Therefore `active_count` may be less than `event_count`, in which case, it would mean it is likely that another wakeup event was generated before the previous wakeup event was processed until the end. This reflects the business of the equipment represented by the wakeup source to some extent.

- `relax_count` represents the number of times the wakeup source was deactivated.

- `expire_count` represents the number of times the wakeup source timeout has expired.

- `wakeup_count` is the number of times the wakeup source has terminated the suspend process. If the wakeup source generates a wakeup event during the suspend process, the suspend process will be aborted. This variable records the number of times the wakeup source has terminated the suspend process. This may be a good indicator for checking whether you established that the system always fails to suspend.

- `active` represents the activated state of the wakeup source.

- `autosleep_enabled`, for me, records the state of the system's autosleep status, whether it is enabled or not.

In order for a device to be a wakeup source, its driver must call `device_init_wakeup()`. This function sets the device's `power.can_wakeup` flag (so that the `device_can_wakeup()` helper returns the current device's capability of being a wakeup source) and adds its wakeup-related attributes to sysfs. Additionally, it creates a wakeup source object, registers it, and attaches it to the device (`dev->power.wakeup`). However, `device_init_wakeup()` only turns the device into a wakeup-capable device without assigning a wakeup event to it.

> **Important note**
> Note that only devices with wake-up capability will have a power directory in sysfs to provide all wakeup information.

In order to assign a wakeup event, the driver must call `enable_irq_wake()`, giving as a parameter the IRQ line that will be used as a wakeup event. What `enable_irq_wake()` does may be platform-specific (Among other things it invokes the `irq_chip.irq_set_wake` callback exposed by the underlying irqchip driver). In addition to turning on the platform logic for handling the given IRQ as a system wakeup interrupt line, it instructs `suspend_device_irqs()` (which is invoked on the system suspend path: refer to the *Suspend stages* section, *step 5*) to treat the given IRQ differently. As a result, the IRQ will remain enabled for the next interrupt, after which it will be disabled, marked as pending, and suspended so that it will be re-enabled by `resume_device_irqs()` during the subsequent system resumption. This makes the driver's `->suspend` method the right place to invoke `enable_irq_wake()`, so that the wakeup event is always rearmed at the right moment. On the other hand, the driver's `->resume` callback is the right place for invoking `disable_irq_wake()`, which would turn off that platform configuration for the system wakeup capability of the IRQ.

While the device's capability of being a wakeup source is a matter of hardware, whether or not a wakeup-capable device should issue wakeup events is a policy decision and is managed by the user space through a `sysfs` attribute, `/sys/devices/.../power/wakeup`. This file allows the user space to check or decide whether the device (through its wakeup event) is enabled to wake up the system from sleep states. This file can be read and written to. When read, either `enabled` or `disabled` can be returned. If `enabled` is returned, this would mean the device is able to issue the events; if instead `disabled` is returned, this would mean the device is not able to do so. Writing `enabled` or `disabled` strings to it will indicate whether or not, respectively, the device is supposed to signal system wakeup (the kernel `device_may_wakeup()` helper will return `true` or `false`, respectively). Do note that this file is not present for the devices that are not capable of generating system wakeup events.

Let's see in an example how drivers make use of the wakeup capability of the device. The following is an excerpt of the *i.MX6 SNVS* powerkey driver, in `drivers/input/keyboard/snvs_pwrkey.c`:

```
static int imx_snvs_pwrkey_probe(struct platform_device *pdev)
{
    [...]
    error = devm_request_irq(&pdev->dev, pdata->irq,
    imx_snvs_pwrkey_interrupt, 0, pdev->name, pdev);
    pdata->wakeup = of_property_read_bool(np, "wakeup-source");
    [...]
    device_init_wakeup(&pdev->dev, pdata->wakeup);
    return 0;
```

```
}

static int
    maybe_unused imx_snvs_pwrkey_suspend(struct device *dev)
{
    [...]
    if (device_may_wakeup(&pdev->dev))
        enable_irq_wake(pdata->irq);
    return 0;
}

static int maybe_unused imx_snvs_pwrkey_resume(struct
                                               device *dev)
{
    [...]
    if (device_may_wakeup(&pdev->dev))
        disable_irq_wake(pdata->irq);
    return 0;
}
```

In the preceding code excerpt, from top to bottom, we have the driver probe method, which first enables the device wakeup capability using the `device_init_wakeup()` function. Then, in the PM resume callback, it checks whether the device is allowed to issue a wakeup signal thanks to the `device_may_wakeup()` helper, prior to enabling the wakeup event by calling `enable_irq_wake()`, with the associated IRQ number as a parameter. The reason for using `device_may_wakeup()` for conditioning wakeup event enabling/disabling is because the user space may have changed the wakeup policy for this device (thanks to the `/sys/devices/.../power/wakeup` sysfs file), in which case this helper will return the current enabled/disabled status. This helper enables coherence with the user space decision. The same is true for the resume method, which does the same checks prior to disabling the wakeup event's IRQ line.

Next, at the bottom of the driver code, we can see the following:

```
static SIMPLE_DEV_PM_OPS(imx_snvs_pwrkey_pm_ops,
                         imx_snvs_pwrkey_suspend,
                         imx_snvs_pwrkey_resume);
static struct platform_driver imx_snvs_pwrkey_driver = {
    .driver = {
```

```
        .name = "snvs_pwrkey",
        .pm    = &imx_snvs_pwrkey_pm_ops,
        .of_match_table = imx_snvs_pwrkey_ids,
    },
    .probe = imx_snvs_pwrkey_probe,
};
```

The preceding shows the usage of the famous `SIMPLE_DEV_PM_OPS` macro, which means the same suspend callback (that is, `imx_snvs_pwrkey_suspend`) will be used for Suspend-to-RAM or hibernation sleep states, and the same resume callback (`imx_snvs_pwrkey_resume` actually) will be used to resume from these states. The device PM structure is named `imx_snvs_pwrkey_pm_ops` as we can see in the macro, and fed to the driver later. Populating PM ops is as simple as that.

Before ending this section, let's pay attention to the IRQ handler in this device driver:

```
static irqreturn_t imx_snvs_pwrkey_interrupt(int irq,
                                             void *dev_id)
{
    struct platform_device *pdev = dev_id;
    struct pwrkey_drv_data *pdata = platform_get_drvdata(pdev);

    pm_wakeup_event(pdata->input->dev.parent, 0);
    [...]
    return IRQ_HANDLED;
}
```

The key function here is `pm_wakeup_event()`. Roughly speaking, it reports a wakeup event. Additionally, this will halt the current system state transition. For example, on the suspend path, it will abort the suspend operation and prevent the system from going to sleep. The following is the prototype of this function:

```
void pm_wakeup_event(struct device *dev, unsigned int msec)
```

The first parameter is the device to which the wakeup source belongs, and `msec`, the second parameter, is the number of milliseconds to wait before the wakeup source is automatically switched to an inactive state by the PM wakeup core. If `msec` equals 0, then the wakeup source is immediately disabled after the event has been reported. If `msec` is different from 0, then the wakeup source deactivation is scheduled `msec` milliseconds later in the future.

This is where the wakeup source's `timer` and `timer_expires` field are used. Roughly speaking, wakeup event reporting consists of the following steps:

- It increments the wakeup source's `event_count` counter and increments the wakeup source's `wakeup_count`, which is the number of times the wakeup source might abort the suspend operation.

- If the wakeup source is not yet active (the following are the steps performed on the activation path):

  – It marks the wakeup source as active and increments the wakeup source's `active_count` element.

  – It updates the wakeup source's `last_time` field to the current time.

  – It updates the wakeup source's `start_prevent_time` field if the other field, `autosleep_enabled`, is `true`.

Then, wakeup source deactivation consists of the following steps:

- It sets the wakeup source's `active` field to `false`.

- It updates the wakeup source's `total_time` field by adding the time spent in the active state to its old value.

- It updates the wakeup source's `max_time` field with the duration spent in the active state if this duration is greater than the value of the old `max_time` field.

- It updates the wakeup source's `last_time` field with the current time, deletes the wakeup source's timer, and clears `timer_expires`.

- It updates the wakeup source's `prevent_sleep_time` field if the other field, `prevent_sleep_time`, is `true`.

Deactivation may occur either immediately if `msec == 0`, or scheduled `msec` milliseconds later in the future if different to zero. All the this should remind you of `struct wakeup_source`, which we introduced earlier, most of whose elements are updated by this function call. The IRQ handler is a good place for invoking it because the interrupt triggering also marks the wakeup event. You should also note that each property of any wakeup source can be inspected from the sysfs interface, as we will see in the next section.

Wakeup source and sysfs (or debugfs)

There is something else that needs to be mentioned here, at least for debugging purposes. The whole list of wakeup sources in the system can be listed by printing the content of `/sys/kernel/debug/wakeup_sources` (assuming `debugfs` is mounted on the system):

```
# cat /sys/kernel/debug/wakeup_sources
```

This file also reports the statistics for each wakeup source, which may be gathered individually thanks to the device's power-related sysfs attributes. Some of these sysfs file attributes are as follows:

```
#ls /sys/devices/.../power/wake*
wakeup wakeup_active_count  wakeup_last_time_ms autosuspend_
delay_ms
wakeup_abort_count  wakeup_count  wakeup_max_time_ms wakeup_
active
wakeup_expire_count   wakeup_total_time_ms
```

I used the `wake*` pattern in order to filter out runtime PM-related attributes, which are also in this same directory. Instead of describing what each attribute is, it would be more worthwhile indicating in which fields in the `struct wakeup_source` structure the preceding attributes are mapped:

- `wakeup` is an RW attribute and has already been described earlier. Its content determines the return value of the `device_may_wakeup()` helper. Only this attribute is both readable and writable. The others here are all read-only.

- `wakeup_abort_count` and `wakeup_count` are read-only attributes that point to the same field, that is, `wakeup->wakeup_count`.

- The `wakeup_expire_count` attribute is mapped to the `wakeup->expire_count` field.

- `wakeup_active` is read-only and mapped to the `wakeup->active` element.

- `wakeup_total_time_ms` is a read-only attribute that returns the `wakeup->total_time` value, and its unit is ms.

- `wakeup_max_time_ms` returns the `power.wakeup->max_time` value in ms.

- `wakeup_last_time_ms`, a read-only attribute, corresponds to the `wakeup->last_time` value; the unit is ms.

- `wakeup_prevent_sleep_time_ms` is also read-only and is mapped onto the `wakeup->prevent_sleep_time` value, whose unit is ms.

Not all devices are wakeup capable, but those that are can roughly follow this guideline.

Now that we are done and familiar with wakeup source management from sysfs, we can introduce the special IRQF_NO_SUSPEND flag, which assists in helps in preventing an IRQ from being disabled in the system suspend path.

The IRQF_NO_SUSPEND flag

There are interrupts that need to be able to trigger even during the entire system suspend-resume cycle, including the noirq phases of suspending and resuming devices, as well as during the time when non-boot CPUs are taken offline and brought back online. This is the case for timer interrupts, for example. This flag has to be set on such interrupts. Although this flag helps to keep the interrupt enabled during the suspend phase, it does not guarantee that the IRQ will wake the system from a suspended state – for such cases, it is necessary to use enable_irq_wake(), which once again, is platform-specific. Thus, you should not confuse or mix the usage of the IRQF_NO_SUSPEND flag and enable_irq_wake().

If an IRQ with this flag is shared by several users, every user will be affected, not just the one that has set the flag. In other words, every handler registered with the interrupt will be invoked as usual, even after suspend_device_irqs(). This is probably not what you need. For this reason, you should avoid mixing IRQF_NO_SUSPEND and IRQF_SHARED flags.

Summary

In this chapter, we have learned to manage the power consumption of the system, both from within the code in the driver as from the user space with the command line), either at runtime by acting on individual devices, or by acting on the whole system by playing with sleep states. We have also learned how other frameworks can help to reduce the power consumption of the system (such as CPUFreq, Thermal, and CPUIdle).

In the next chapter, we will move onto PCI device drivers, which deal with the devices sitting on this famous bus that needs no introduction.

Section 3: Staying Up to Date with Other Linux Kernel Subsystems

This section delves into some useful Linux kernel subsystems that are not discussed enough or for which the available documentation is not up to date. The section takes a step-by-step approach to PCI device driver development, leveraging NVMEM and the watchdog frameworks, and improving efficiency with some tips and best practices.

This section contains the following chapters:

- *Chapter 11, Writing PCI Device Drivers*
- *Chapter 12, Leveraging the NVMEM Framework*
- *Chapter 13, Watchdog Device Drivers*
- *Chapter 14, Linux Kernel Debugging Tips and Best Practices*

11
Writing PCI Device Drivers

A PCI is more than just a bus. It is a standard with a complete set of specifications defining how different parts of a computer should interact. Over the years, a PCI bus has become the de facto bus standard for device inter-connections, such that almost every SoC has native support for such buses. The need for speed led to different versions and generations of this bus.

In the early days of the standard, the first bus that implemented the PCI standard was the PCI bus (the bus name is the same as the standard), as a replacement for the ISA bus. This improved (with 32-bit addressing and jumper-less autodetection and configuration) the address limitation encountered with ISA (limited to 24 bits, and which occasionally necessitated playing with jumpers in order to route IRQs and so on). Compared with the previous bus implementation of the PCI standard, the major factor that has been improved is speed.

PCI Express is the current family of PCI bus. It is a serial bus, while its ancestors were parallel. In addition to speed, PCIe extended the 32-bit addressing of its predecessors to 64 bits, with multiple improvements in the interrupt management system. This family is split into generations, GenX, which we will see in the following sections in this chapter. We will begin with an introduction to PCI buses and interfaces where we will learn about bus enumeration, and then we will look at the Linux kernel PCI APIs and core functionalities.

The good news with all of this is that, whatever the family, almost everything is transparent to the driver developer. The Linux kernel will abstract and hide most of the mechanisms behind a reduced set of APIs that can be used to write reliable PCI device drivers.

We will cover the following topics in this chapter:

- Introduction to PCI buses and interfaces
- The Linux kernel PCI subsystem and data structures
- PCI and **Direct Memory Access (DMA)**

Technical requirements

A good overview of Linux memory management and memory mapping is required, as is a familiarity with the concept of interrupts and locking, especially with the Linux kernel.

Linux kernel v4.19.X sources are available at `https://git.kernel.org/pub/scm/linux/kernel/git/stable/linux.git/refs/tags`.

Introduction to PCI buses and interfaces

Peripheral Component Interconnect (**PCI**) is a local bus standard used to attach peripheral hardware devices to the computer system. As a bus standard, it defines how different peripherals of a computer should interact. However, over the years, the PCI standard has evolved either in terms of features or in terms of speed. As of its creation until now, we have had several bus families implementing the PCI standard, such as PCI (yes, the bus with the same name as the standard), and **PCI Extended** (**PCI-X**), **PCI Express** (**PCIe** or **PCI-E**), which is the current generation of PCI. A bus that follows PCI standards is known as a PCI bus.

From a software point of view, all these technologies are compatible and can be handled by the same kernel drivers. This means the kernel doesn't need to know which exact bus variant is used. PCIe greatly *extends* PCI with a lot of similarities from a software point of view (especially Read/Write I/O or Memory transactions). While both are software compatible, PCIe is a serial bus instead of parallel (prior to PCIe, every PCI bus family was parallel), which also means you can't have a PCI card installed in a PCIe slot, or a PCIe card installed in a PCI slot.

PCI Express is the most popular bus standard on computers these days, so we are going to target PCIe in this chapter while mentioning similarities with or differences from PCI when necessary. Apart from the preceding, the following are some of the improvements in PCIe:

- PCIe is a serial bus technology, whereas PCI (or other implementations) is parallel, thereby reducing the number of I/O lanes required for connecting devices, thus reducing the design complexity.

- PCIe implements enhanced interrupt management features (providing Message-Based Interrupts, aka MSI, or its extended version, MSI-X), extending the number of interrupts a PCI device can deal with without increasing its latency.

- PCIe increases the transmission frequency and throughput: Gen1, Gen2, Gen3 ...

PCI devices are types of memory-mapped devices. A device that is connected to any PCI bus is assigned address ranges in the processor's address space. These address ranges have a different meaning in the PCI address domain, which contains three different types of memory according to what they contain (control, data, and status registers for the PCI-based device) or to the way they are accessed (I/O port or memory mapped). These memory regions will be accessed by the device driver/kernel to control the particular device connected over the PCI bus and share information with it.

The PCI address domain contains the three different memory types that have to be mapped in the processor's address space.

Terminology

Since the PCIe ecosystem is quite large, there are a number of terms we may need to be familiar with prior to going further. These are as follows:

- **Root complex** (**RC**): This refers to the PCIe host controller in the SoC. It can access the main memory without CPU intervening, which is a feature used by other devices to access the main memory. They are also known as Host-to-PCI bridges.

- **Endpoint** (**EP**): Endpoints are PCIe devices, and are represented by type `00h` configuration space headers. They never appear on a switch's internal bus and have no downstream port.

- **Lane**: This represents a set of differential signal pairs (one pair for Tx, one pair for Rx).

- **Link**: This represents a dual-simplex (a pair actually) communication channel between two components. To scale bandwidth, a link may aggregate multiple lanes denoted by xN (`x1`, `x2`, `x4`, `x8`, `x12`, `x16`, and `x32`), where N is the number of pairs.

Not all PCIe devices are endpoints. They may also be switches or bridges.

- **Bridges**: These provide an interface to other buses, such as PCI or PCI X, or even another PCIe bus. A bridge can also provide an interface to the same bus. For example, a PCI-to-PCI bridge facilitates the addition of more loads to a bus by creating a completely separate secondary bus (we will see what a secondary bus is in forthcoming sections). The concept of bridges aids in understanding and implementing the concept of switches.

- **Switches**: These provide an aggregation capability and allow more devices to be attached to a single root port. It goes without saying that switches have a single upstream port, but may have several downstream ports. They are smart enough to act as packet routers and recognize which path a given packet will need to take based on its address or other routing information, such as an ID. That being said, there is also implicit routing, which is used only for certain message transactions, such as broadcasts from the root complex and messages that always go to the root complex.

Switch downstream ports are (kinds of virtual) PCI-PCI bridges bridging from the internal bus to buses representing the downstream PCI Express links of this PCI Express switch. You should keep in mind that only the PCI-PCI bridges representing the switch downstream ports may appear on the internal bus.

> **Important note**
>
> A PCI-to-PCI bridge provides a connectivity path between two peripheral component interconnect (PCI) buses. You should keep in mind that **only the downstream ports of PCI-PCI bridges are taken into account during bus enumeration**. This is very important in understanding the enumeration process.

PCI bus enumeration, device configuration, and addressing

PCIe's most obvious improvement over PCI is its point-to-point bus topology. Each device sits on its own dedicated bus, which, in PCIe jargon, is known as a **link**. Understanding the enumeration process of PCIe devices requires some knowledge.

When you look at the register space of the devices (in the header-type register), they will say whether they are a type 0 or type 1 register space. Typically, type 0 means an endpoint device and type 1 means a bridge device. The software has to identify whether it talks to an endpoint device or to a bridge device. Bridge device configuration differs from endpoint device configuration. During bridge device (type 1) enumeration, the software has to assign the following elements to it:

- **Primary Bus Number**: This is the upstream bus number.
- **Secondary/Subordinate Bus Number**: This gives the range of specific PCI bridges' downstream bus numbers. The secondary bus number is the bus number immediately downstream of the PCI-PCI bridge, while the subordinate bus number represents the highest bus number of all of the buses that can be reached downstream of the bridge. During the first stage of enumeration, the subordinate bus number field is given a value of 0xFF, as 255 is the highest bus number. As and when enumeration continues, this field will be given the real value of how far downstream this bridge can go.

Device identification

Device identification consists of a few properties or parameters that make the device unique or addressable. In the PCI subsystem, these parameters are the following:

- **Vendor ID**: This identifies the manufacturer of the device.
- **Device ID**: This identifies the particular vendor device.

The preceding two elements may be enough, but you can also rely on the following elements:

- **Revision ID**: This specifies a device-specific revision identifier.
- **Class Code**: This identifies the generic function implemented by the device.
- **Header Type**: This defines the layout of the header.

All those parameters can be read from the device configuration registers. That's what the kernel does to identify devices when enumerating buses.

Bus enumeration

Prior to delving into the PCIe bus enumeration function, there are some basic limitations we need to take care of:

- There can be 256 buses on the system (0-255) as there are 8 bits to identify them.

- There can be 32 devices per bus (0-31) as there are 5 bits to identify them on each bus.

- A device can have up to 8 functions (0-7), hence 3 bits to identify them.

All external PCIe lanes, irrespective of whether they originate from the CPU or not, are behind PCIe bridges (and therefore get new PCIe bus numbers). Configuration software is able to enumerate up to 256 PCI buses on a given system. The number 0 is always assigned to the root complex. Remember that only the downstream ports (secondary sides) of PCI-PCI bridges are taken into account during bus enumeration.

The PCI enumeration process is based on the **Depth-first search** (DFS) algorithm, which normally starts at a random node (but in the case of PCI enumeration, this node is known in advance, and it is the RC in our case) and which explores as far as possible (actually looking for bridges) along each branch before backtracking.

Said like this, when a bridge is found, configuration software assigns a number to it, at least one larger than the bus number this bridge lives on. After this, the configuration software starts looking for new bridges on this new bus, and so on, before backtracking to this bridge's sibling (if the bridge was part of a multi-port switch) or neighbor bridge (in terms of topology).

Enumerated devices are identified with the BDF format, which stands for *Bus-Device-Function*, which uses triple bytes – in other words XX:YY:ZZ – in hexadecimal (without the 0x) notation for identification. For instance, 00:01:03 would literally mean Bus 0x00: Device 0x01: Function 0x03. We could interpret this as function 3 of device 1 on bus 0. This notation helps in quickly locating a device within a given topology. In case a double-byte notation is used, this would mean the function has been omitted or does not matter, in other words, XX:YY.

The following diagram shows the topology of a PCIe fabric:

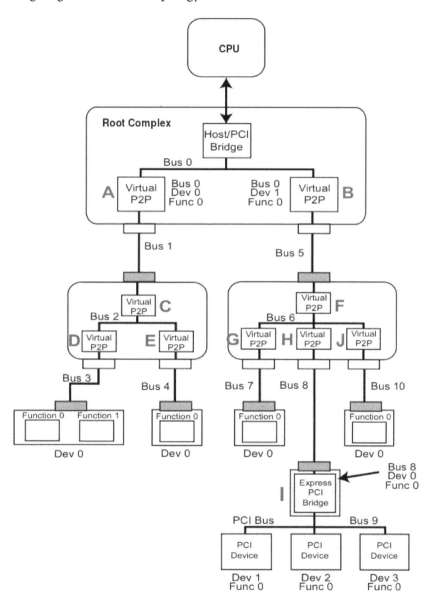

Figure 11.1 – PCI bus enumeration

Before we describe the preceding topology diagram, keep repeating the following four statements until you become familiar with them:

1. A PCI-to-PCI bridge facilitates the addition of more loads to a bus by creating a completely separate secondary bus. Thus, each bridge downstream port is a new bus, and must be given a bus number, at least +1 bigger than the bus number where it lives.

2. Switch downstream ports are (kinds of virtual) PCI-PCI (P2P) bridges bridging from the internal bus to buses representing the downstream PCI Express links of this PCI Express switch.

3. The CPU is connected to the root complex through the Host-to-PCI bridge, which represents the upstream bridge in the root complex.

4. Only the downstream ports of PCI-PCI bridges are taken into account during bus enumeration.

After applying the enumeration algorithm to the topology in the diagram, we can list 10 steps, from **A** to **J**. Steps **A** and **B** are internal to the root complex, which hosts the bus 0 along with two bridges (thus providing two buses), 00:00:00 and 00:01:00. The following are descriptions of the steps in the enumeration process in the preceding topology diagram, although step **C** is where standardized enumeration logic starts:

- 00:00 being a (virtual) bridge, undoubtedly, its downstream port is a bus. It is then assigned the number 1 (remember, it is always greater than the bus number where the bridge lives, which is 0 in this case). Bus 1 is then enumerated.

- Step **C**: There is a switch on bus 1 (one upstream virtual bridge that provides its internal bus and two downstream virtual bridges that expose its output buses). This switch's internal bus is given the number 2.

- We immediately fall into step **D**, where the first downstream bridge is found and its bus is enumerated after being assigned the bus number 3, behind which there is an endpoint (no downstream port). According to the principle of the DFS algorithm, we reached the leaf node of this branch, so we can start backtracking.

- Hence step **E**, where the virtual bridge, which is the sibling of the one found in step **D**, has found its turn. Its bus is then assigned the bus number 4, and there is a device behind it. Backtracking can happen again.

- We then reach step **F**, where the virtual bridge found in step B is assigned a bus number, which is 5. There is a switch behind this bus (one upstream virtual bridge implementing the internal bus, which is given the bus number 6, and 3 downstream virtual bridges representing its external buses, hence this is a 3-port switch).

- Step **G** is where the first downstream virtual bridge of the 3-port switch is found. Its bus is given the bus number 7, and there is an endpoint behind this bus. If we had to identify the function 0 of this endpoint using the BDF format, it would be 07:00:00 (function 0 of device 0 on bus 7). Back to the DFS algorithm, we have reached the bottom of the branch. We can then start backtracking, which leads us to step **H**.

- In step **H**, the second downstream virtual bridge in the 3-port switch is found. Its bus is assigned the bus number 8. There is a PCIe-to-PCI bridge behind this bus.

- In step **I**, this PCIe-to-PCI bridge is assigned the bus number 9 downstream, and there is a 3-function endpoint behind this bus. In BDF notation, these would be identified as 09:00:00, 09:00:01, and 09:00:02. Since endpoints mark the depth of a branch, it allows us to perform another backtrack, which leads us to step **J**.

- In the backtracking phase, we enter step **J**. The third and final downstream virtual bridge of the 3-port switch is found and its bus is given the bus number 10. There is an endpoint behind this bus, which would have been identified as 0a:00:00 in BDF format. This marks the end of the enumeration process.

PCI(e) bus enumeration may look complicated at first glance, but it is quite simple. Reading the preceding material twice can be enough to understand the whole process.

PCI address spaces

A PCI target can implement up to three different types of address spaces according to their content or by access method. These are **Configuration Address Space**, **Memory Address Space**, and **I/O Address Space**. Configuration and memory address spaces are memory mapped – they are assigned address ranges from the system address space, so that reads and writes to that range of addresses don't go to RAM, but are routed to the device directly from the CPU, while I/O address space is not. Without further ado, let's analyze the differences between them and their different use cases.

PCI configuration space

This is the address space from where the configuration of the device can be accessed and which stores basic information about the device, and which is also used by the OS to program the device with its operational settings. There are 256 bytes for the configuration space on PCI. The PCIe extends this to 4 KB of register space. Since configuration address space is memory mapped, any address that points to configuration space is allocated from the system memory map. Thus, these 4 KB spaces allocate memory addresses from the system memory map, but the actual values/bits/content are generally implemented in registers on the peripheral device. For instance, when you read the vendor ID or device ID, the target peripheral device will return the data even though the memory address being used is from the system memory map.

Part of this address space is standardized. Configuration address space is split as follows:

- The first 64 bytes (00h – 3Fh) represent the standard configuration header, which includes the PCI Bus ID, vendor ID, and device ID registers, to identify the device.

- The remaining 192 bytes (40h – FFh) make up the user-defined configuration space, such as the information specific to a PC card to be used by its accompanying software driver.

Generally speaking, the configuration space stores basic information about the device. It allows the central resource or OS to program a device with operational settings. There is no physical memory associated with configuration address space. It is a list of addresses used in the **TLP** (**Transaction Layer Packet**) in order to identify the target of the transaction.

Commands that are used to transfer data between each **Configuration Address Space** of PCI devices are known as either **Configuration Read** commands or **Configuration Write** commands.

PCI I/O address space

These days, I/O address space is used for compatibility with the x86 architecture's I/O port address space. The PCIe specification discourages the use of this address space. It will be no surprise if future revisions of the PCI Express specification deprecate the use of I/O address space. The only advantage of I/O mapped I/O is that, because of its separate address space, it does not steal address ranges from system memory space. As a result, computers can access the whole 4 GB of RAM on a 32-bit system.

I/O Read and **I/O Write** commands are used to transfer data in **I/O Address Space**.

PCI memory address space

In the early days of computers, Intel defined a way to access registers in I/O devices via the so-called I/O address space. It made sense in those days because of the memory address space of the processor, which was quite limited (think of 16-bit systems, for example) and it made little or even no sense to use some ranges of it for accessing devices. When the system memory space became less of a constraint (think of 32-bit systems, for example, where the CPU can address up to 4 GB), the separation between I/O address space and memory address space became less important, and even burdensome.

There were so many limitations and constraints on that address space that it resulted in registers in I/O devices being mapped directly to the system's memory address space, hence, memory-mapped I/O, or MMIO. Those limitations and constraints included the following:

- The need for a dedicated bus

- A separate instruction set

- And since it was implemented during the 16-bit systems era, the port address space was limited to 65536 ports (which corresponds to 2^{16}), although very old machines used 10 bits for I/O address space and had only 1024 unique port addresses

It has therefore become more practical to take advantage of the benefits of memory-mapped I/O.

Memory-mapped I/O allows hardware devices to be accessed by simply reading or writing to those "special" addresses using the normal memory access instructions, though it is more expensive to decode up to 4 GB of address (or more) as compared to 65536. That being said, PCI devices expose their memory regions through windows called BARs. A PCI device can have up to six BARs.

The concept of BAR

BAR stands for **Base Address Register** and is a PCI concept with which a device tells the host how much memory it needs, as well as its type. This is memory space (grabbed from the system memory map), not actual physical RAM (you can actually think of RAM itself as a "specialized memory-mapped I/O device" whose job is just to save and give back data, although with today's modern CPUs with caching and such, this is not physically straightforward). It is the responsibility of the BIOS or the OS to allocate the requested memory space to the target device.

Once assigned, the BARs are seen as memory windows by the host system (CPUs) to talk to the device. The device itself doesn't write into that window. This concept can be seen as an indirection mechanism to access the real physical memory, which is internal and local to the PCI device.

Actually, the real physical address of the memory and the address of the input/output registers are internal to the PCI device. The following is how a host deals with the storage space of peripherals:

1. The peripheral device tells the system by some means that it has several storage intervals and I/O address space, how big each interval is, and their respective local addresses. Obviously, these addresses are local and internal, all starting from 0.

2. After the system software knows how many peripherals there are, and what kind of storage intervals they have, they can assign "physical addresses" to these intervals, and establish the connection between these intervals and the bus. These addresses are accessible. Obviously, the so-called "physical address" here is somewhat different from the real physical address. It is actually a logical address, so it often becomes a "bus address" because this is the address that the CPU sees on the bus. As you can imagine, there must be some sort of address mapping mechanism on the peripheral. The so-called "allocation of addresses for peripherals" is to assign bus addresses to them and establish mappings.

Interrupt distribution

Here we will discuss the way in which interrupts are handled by a PCI device. There are three interrupt types in PCI Express. These are as follows:

- Legacy interrupts, also called INTx interrupts, the only one mechanism available in the old PCI implementations.

- **MSI** (**Message Based Interrupts**) extend the legacy mechanism, for example, by increasing the number of interrupts that are possible.

- MSI-X (eXtended MSI) extends and enhances MSI, for example, by allowing the targeting of individual interrupts to different processors (helpful in some high-speed networking applications).

The application logic in a PCI Express endpoint can implement one or more of the three methods enumerated above to signal an interrupt. Let's look at these in detail.

PCI legacy INT-X-based interrupts

Legacy interrupt management was based on PCI INT-X interrupt lines, made of up to four virtual interrupt wires, referred to as INTA, INTB, INTC, and INTD. These interrupt wires are shared by all the PCI devices in the system. The following are the steps that the legacy implementation had to go through in order to identify and handle the interrupt:

1. The device asserts one of its INT# pins in order to generate an interrupt.

2. The CPU acknowledges the interrupt and polls each device (actually its driver) connected to this INT# line (shared) by calling their interrupt handlers. The time needed to service the interrupt depends on the number of devices sharing the line. The device's interrupt service routine (ISR) may check whether the interrupt is originating from this device by reading the device's internal registers to identify the cause of the interrupt.

3. The ISR takes action to service the interrupt.

In the preceding method as well as the legacy method, interrupt lines are shared: Everyone answers the phone. Moreover, physical interrupt lines are limited. In the following section, we see how MSI addresses those issues and facilitates interrupt management.

> **Important note**
>
> The i.MX6 maps INTA/B/C/D to ARM GIC IRQ 155/154/153/152, respectively. This allows a PCIe-to-PCI bridge to function properly. Refer to IMX6DQRM.pdf, page 225.

Message-based interrupt type – MSI and MSI-X

There are two message-based interrupt mechanisms: MSI and MSI-X, the enhanced and extended version. MSI (or MSI-X) is simply a way of signaling interrupts using the PCI Express protocol layer, and the PCIe root complex (the host) takes care of interrupting the CPU.

Traditionally, a device is assigned pins as interrupt lines, which it has to assert when it wants to signal an interrupt to the CPU. This kind of signaling method is out-of-band since it uses yet another way (different from the main data path) to send such control information.

However, MSI allows the device to write a small amount of interrupt-describing data to a special memory-mapped I/O address, and the root complex then takes care of delivering the corresponding interrupt to the CPU. Once the endpoint device wants to generate the MSI interrupt, it issues a write request to (target) the address specified in the message address register with the data contents specified in the message data register. Since the data path is used for this, it is an in-band mechanism. Moreover, MSI increases the number of possible interrupts. This is described in the next section.

> **Important note**
>
> PCI Express does not have a separate interrupt pin at all. However, it is compatible with legacy interrupts on a software level. For this, it requires MSI or MSI-X since it uses special in-band messages to allow pin assertion or de-assertion to be emulated. In other words, PCI Express emulates this capability by providing `assert_INTx` and `deassert_INTx`. Message packets are sent through the PCI Express serial link.

In an implementation using MSI, the following are the usual steps:

1. The device generates an interrupt by sending an MSI memory write upstream.

2. The CPU acknowledges the interrupt and calls the appropriate device ISR since this is known in advance based on the MSI vector.

3. The ISR takes action to service the interrupt.

MSIs are not shared, so an MSI that is assigned to a device is guaranteed to be unique within the system. It goes without saying that MSI implementation significantly reduces the total servicing time needed for the interrupt.

> **Important note**
>
> Most people think that MSI allows the device to send data to a processor as part of the interrupt. This is a misconception. The truth is that the data that is sent as part of the memory write transaction is exclusively used by the chipset (the root complex actually) to determine which interrupt to trigger on which processor; that data is not available for the device to communicate additional information to the interrupt handler.

MSI mechanism

MSI was originally defined as part of the PCI 2.2 standard, allowing a device to allocate either 1, 2, 4, 8, 16, or up to 32 interrupts. The device is programmed with an address to write to in order to signal an interrupt (generally, a control register in an interrupt controller), and a 16-bit data word to identify the device. The interrupt number is added to the data word to identify the interrupt.

A PCI Express endpoint may signal an MSI by sending a standard PCI Express posted write packet to the root port. The packet is made of a specific address (allocated by the host) and one of up to 32 data values (thus, 32 interrupts) provided by the host to the endpoint. The varying data values and the address value provide more detailed identification of interrupt events than legacy interrupts. Interrupt masking capability is optional in MSI specifications.

This approach does have some limitations. Only one address is used for the 32 data values, which makes it difficult to target individual interrupts to different processors. This limitation is due to the fact that a memory-write operation associated with an MSI can only be distinguished from other memory-write operations by the address locations (not the data) they target, which are reserved by the system for interrupt delivery.

The following are the MSI configuration steps, performed by the PCI controller driver for a PCI Express device:

1. The bus enumeration process takes place during startup. It consists of kernel PCI core code scanning the PCI bus(es) in order to discover devices (in other words, it carries out configuration reads for valid vendor IDs). Upon discovering a PCI Express function, the PCI core code reads the capabilities list pointer to obtain the location of the first capability register within the chain of registers.

2. The PCI core code then searches the capability register sets. It keeps doing this until it discovers the MSI capability register set (capability ID of 05h).

3. After that, the PCI core code configures the device, assigning a memory address to the device's message address register. This is the destination address of the memory write used when delivering an interrupt request.

4. The PCI core code checks the Multiple Message Capable field in the device's message control register to determine how many event-specific messages the device would like assigned to it.

5. The core code then allocates a number of messages equal to or less than what the device requested. As a minimum, one message will be allocated to the device.

6. The core code writes the base message data pattern into the device's message data register.

7. Finally, the PCI core code sets the MSI enable bit in the device's message control register, thereby enabling it to generate interrupts using MSI memory writes.

MSI-X mechanism

MSI-X is just an extension of PCI MSI in PCIe – it serves the same function, but can carry more information and is more flexible. Note that PCIe supports both MSI and MSI-X. MSI-X was first defined in the PCI 3.0 (PCIe) standard. It allows a device to support from at least 64 (the minimum, but double the maximum MSI interrupts) to up to 2,048 interrupts. Actually, MSI-X allows a larger number of interrupts and gives each one a separate target address and data word. Since the single address used by the original MSI was found to be restrictive for some architectures, MSI-X-enabled devices use address and data pairs, allowing the device to use up to 2048 address and data pairs. Thanks to this large number of address values available to each endpoint, it's possible to route MSI-X messages to different interrupt consumers in a system, unlike the single address available to MSI packets. Moreover, endpoints with MSI-X capability also include application logic to mask and hold pending interrupts, as well as a memory table for the address and data pairs.

Apart from the above, an MSI-X interrupt is identical to an MSI. However, optional features in MSI (such as 64-bit addressing and interrupt masking) are made mandatory with MSI-X.

Legacy INTx emulation

Because PCIe claimed backward compatibility with legacy parallel PCI, it also needed to support INTx-based interrupt mechanisms. But how can this be implemented? Actually, there were four INTx (INTA, INTB, INTC, and INTD) physical IRQ lines in classical PCI systems, which were all level-triggered, active low actually (in other words, the interrupt request is active as long as the physical INTx wire is at a low voltage). Then how is each IRQ transported in the emulated version?

The answer is that PCIe virtualizes PCI physical interrupt signals by using an in-band signaling mechanism, the so-called MSI. Since there were two levels (asserted and de-asserted) per physical line, PCIe provides two messages per line, known as `assert_INTx` and `deassert_INTx` messages. There are eight message types in total: `assert_INTA, deassert_INTA, ... assert_INTD, deassert_INTD`. In fact, they are simply known as INTx messages. This way, INTx interrupts are propagated across the PCIe link just like MSI and MSI-X.

This backward compatibility exists mainly for PCI to PCIe bridge chips so that PCI devices will work properly in a PCIe system without modifying the drivers.

Now we are familiar with interrupt distribution in the PCI subsystem. We have covered both legacy INT-X-based mechanisms and message-based mechanisms. Now it's time to dive into the code, from data structures to APIs.

The Linux kernel PCI subsystem and data structures

The Linux kernel supports the PCI standard and provides APIs to deal with such devices. In Linux, the PCI implementation can be broadly divided into the following main components:

- **PCI BIOS**: This is an architecture-dependent part in charge of kicking off the PCI bus initialization. ARM-specific Linux implementation lies in `arch/arm/kernel/bios32.c`. The PCI BIOS code interfaces with PCI Host Controller code as well as the PCI core in order to perform bus enumeration and the allocation of resources, such as memory and interrupts.

  The successful completion of BIOS execution guarantees that all the PCI devices in the system are assigned parts of available PCI resources and their respective drivers (referred to as slave or endpoint drivers) can take control of them using the facilities provided by the PCI core.

  Here, the kernel invokes the services of architecture and board-specific PCI functionalities. Two important tasks of PCI configuration are completed here. The first task is to scan for all PCI devices on the bus, configure them, and allocate memory resources. The second task is to configure the device. Configured here means a resource (memory) was reserved and IRQ assigned. It does not mean initialized. Initialization is device-specific and should be done by the device driver. PCI BIOS may optionally skip resource allocation (if they were assigned before Linux was booted, for example, in a PC scenario).

- **Host Controller** (root complex): This part is SoC-specific (located in `drivers/pci/host/`, in other words, `drivers/pci/controller/pcie-rcar.c` for r-car SoCs). However, some SoCs may implement the same PCIe IP block from a given vendor, such as Synopsys DesignWare. Such controllers can be found in the same directory, such as `drivers/pci/controller/dwc/` in the kernel source. For instance, the i.MX6 whose PCIe IP block is from this vendor has its driver implemented in `drivers/pci/controller/dwc/pci-imx6.c`.
  This part handles SoC (and sometimes board)-specific initialization and configuration and may invoke the PCI BIOS. However, it should provide PCI bus access and facility callback functions for BIOS as well as the PCI core, which will be called during PCI system initialization and while accessing the PCI bus for configuration cycles. Moreover, it provides resource information for available memory/IO space, INTx interrupt lines, and MSI. It should facilitate IO space access (as supported) and may also need to provide indirect memory access (if supported by hardware).

- **Core** (`drivers/pci/probe.c`): This is in charge of creating and initializing the data structure tree for buses, devices, as well as bridges in the system. It handles bus/device numbering. It creates device entries and provides `proc/sysfs` information. It also provides services for PCI BIOS and slave (**End Point**) drivers and optionally hot plug support (if supported by h/w). It targets the (**EP**) driver interface query and initializes corresponding devices found during enumeration. It also provides an MSI interrupt handling framework and PCI Express port bus support. All of the above is enough to facilitate the development of device drivers in the Linux kernel.

PCI data structures

The Linux kernel PCI framework aids development on PCI device drivers, which are built on top of the two main data structures: `struct pci_dev`, which represents a PCI device from within the kernel, and `struct pci_driver`, which represents a PCI driver.

struct pci_dev

This is the structure with which the kernel instantiates each PCI device on the system. It describes the device and stores some of its state parameters. This structure is defined in `include/linux/pci.h` as follows:

```
struct pci_dev {
    struct pci_bus    *bus; /* Bus this device is on */
    struct pci_bus *subordinate; /* Bus this device bridges to */
```

```
    struct proc_dir_entry *procent;
    struct pci_slot         *slot;
    unsigned short     vendor;
    unsigned short     device;
    unsigned short     subsystem_vendor;
    unsigned short     subsystem_device;
    unsigned int       class;

    /* 3 bytes: (base,sub,prog-if) */
    u8 revision;       /* PCI revision, low byte of class word */
    u8 hdr_type; /* PCI header type (multi' flag masked out) */
    u8 pin;                   /* Interrupt pin this device uses */
    struct pci_driver *driver; /* Driver bound to this device */
    u64     dma_mask;
    struct device_dma_parameters dma_parms;

    struct device          dev;
    int   cfg_size;
    unsigned int irq;
[...]
    unsigned int       no_msi:1;   /* May not use MSI */
    unsigned int no_64bit_msi:1; /* May only use 32-bit MSIs */
    unsigned int msi_enabled:1;
    unsigned int msix_enabled:1;
    atomic_t enable_cnt;
[...]
};
```

In the preceding block, some elements have been removed for the sake of readability. For the remaining, the following elements have the following meanings:

- `procent` is the device entry in /proc/bus/pci/.

- `slot` is the physical slot this device is in.

- `vendor` is the vendor ID of the device manufacturer. The PCI Special Interest Group maintains a global registry of such numbers, and manufacturers must apply to have a unique number assigned to them. This ID is stored in a 16-bit register in the device configuration space.

- `device` is the ID that identifies this particular device once it is probed. This is vendor-dependent and there is no official registry as such. This is also stored in a 16-bit register.

- `subsystem_vendor` and `subsystem_device` specify the PCI subsystem vendor and subsystem device IDs. They can be used for further identification of the device, as we have seen earlier.

- `class` identifies the class this device belongs to. It is stored in a 16-bit register (in the device configuration space) whose top 8 bits identify the base class or group.

- `pin` is the interrupt pin this device uses, in the case of the legacy INTx-based interrupt.

- `driver` is the driver associated with this device.

- `dev` is the underlying device structure for this PCI device.

- `cfg_size` is the size of the configuration space.

- `irq` is the field that is worth spending time on. When the device boots, MSI(-X) mode is not enabled and it remains unchanged until it is explicitly enabled by means of the `pci_alloc_irq_vectors()` API (old drivers use `pci_enable_msi()`).

    Consequently, `irq` first corresponds to the default preassigned non-MSI IRQ. However, its value or usage may change according to one of the following situations:

    a) In MSI interrupt mode (upon a successful call to `pci_alloc_irq_vectors()` with the `PCI_IRQ_MSI` flag set), the (preassigned) value of this field is replaced by a new MSI vector. This vector corresponds to the base interrupt number of the allocated vectors, so that the IRQ number corresponding to vector X (index starting from 0) is equivalent to (the same as) `pci_dev->irq + X` (see the `pci_irq_vector()` function, which is intended to return the Linux IRQ number of a device vector).

b) In MSI-X interrupt mode (upon a successful call to `pci_alloc_irq_vectors()` with the `PCI_IRQ_MSIX` flag set), the (preassigned) value of this field is untouched (because of the fact that each MSI-X vector has its dedicated message address and message data pair, and this does not require the 1:1 vector-to-entry mapping). However, in this mode, `irq` is invalid. Using it in the driver to request a service interrupt may result in unpredictable behavior. Consequently, in case MSI(-X) is needed, the `pci_alloc_irq_vectors()` function (which enables MXI(-X) prior to allocating vectors) should be called before the driver calls `devm_equest_irq()`, because MSI(-X) is delivered via a vector that is different from the vector of a pin-based interrupt.

- `msi_enabled` holds the enabling state of the MSI IRQ mode.

- `msix_enabled` holds the enabling state of the MSI-X IRQ mode.

- `enable_cnt` holds the number of times `pci_enable_device()` has been called. This helps in really disabling the device only after all callers of `pci_enable_device()` have called `pci_disable_device()`.

struct pci_device_id

While `struct pci_dev` describes the device, `struct pci_device_id` is intended to identify the device. This structure is defined as follows:

```
struct pci_device_id {
    u32 vendor, device;
    u32 subvendor, subdevice;
    u32 class, class_mask;
    kernel_ulong_t driver_data;
};
```

To understand how this structure is important for the PCI driver, let's describe each of its elements:

- `vendor` and `device` represent the vendor ID and the device ID of the device, respectively. Both are paired to make a unique 32-bit identifier for a device. The driver relies on this 32-bit identifier to identify its devices.

- `subvendor` and `subdevice` represent the subsystem ID.

- `class`, `class_mask` are class-related PCI drivers that are intended to handle every device of a given class. For such drivers, `vendor` and `device` should be set to `PCI_ANY_ID`. The different classes of PCI devices are described in the PCI specification. These two values allow the driver to specify that it supports a type of PCI class device.

- `driver_data` is the data private to the driver. This field is not used to identify a device, but to pass different data to differentiate between devices.

There are three macros that allow you to create specific instances of `struct pci_device_id`:

- `PCI_DEVICE`: This macro is used to describe a specific PCI device by creating a `struct pci_device_id` that matches a specific PCI device with the vendor and device IDs given as parameters (`PCI_DEVICE(vend,dev)`), and with the sub-vendor, sub-device, and class-related fields set to `PCI_ANY_ID`.

- `PCI_DEVICE_CLASS`: This macro is used to describe a specific PCI device class by creating a `struct pci_device_id` that matches a specific PCI class with `class` and `class_mask` given as parameters (`PCI_DEVICE_CLASS(dev_class,dev_class_mask)`). The vendor, device, sub-vendor, and sub-device fields will be set to `PCI_ANY_ID`. A typical example is `PCI_DEVICE_CLASS(PCI_CLASS_STORAGE_EXPRESS, 0xffffff)`, which corresponds to the PCI class for NVMe devices, and which will match any of these whatever the vendor and device IDs are.

- `PCI_DEVICE_SUB`: This macro is used to describe a specific PCI device with a subsystem by creating a `struct pci_device_id` that matches a specific device with subsystem information given as parameters (`PCI_DEVICE_SUB(vend, dev, subvend, subdev)`).

Every device/class supported by the driver should be fed into the same array for later use (there are two places where we are going to use it), as in the following example:

```
static const struct pci_device_id bt8xxgpio_pci_tbl[] = {
    { PCI_DEVICE(PCI_VENDOR_ID_BROOKTREE, PCI_DEVICE_ID_BT848) },
    { PCI_DEVICE(PCI_VENDOR_ID_BROOKTREE, PCI_DEVICE_ID_BT849) },
    { PCI_DEVICE(PCI_VENDOR_ID_BROOKTREE, PCI_DEVICE_ID_BT878) },
    { PCI_DEVICE(PCI_VENDOR_ID_BROOKTREE, PCI_DEVICE_ID_BT879) },
    { 0, },
};
```

Each `pci_device_id` structure needs to be exported to user space in order to let the hotplug and device manager (`udev`, `mdev`, and so on ...) know what driver goes with what device. The first reason to feed them all in the same array is that they can be exported in a single shot. To achieve this, you should use the `MODULE_DEVICE_TABLE` macro, as in the following example:

```
MODULE_DEVICE_TABLE(pci, bt8xxgpio_pci_tbl);
```

This macro creates a custom section with the given information. At compilation time, the build process (`depmod` to be more precise) extracts this information out of the driver and builds a human-readable table called `modules.alias`, located in the `/lib/modules/<kernel_version>/` directory. When the kernel tells the hotplug system that a new device is available, the hotplug system will refer to the `modules.alias` file to find the proper driver to load.

struct pci_driver

This structure represents an instance of a PCI device driver, whatever it is and whatever subsystem it belongs to. It is the main structure that every PCI driver must create and fill in order to be able to get them registered with the kernel. `struct pci_driver` is defined as follows:

```
struct pci_driver {
    const char *name;
    const struct pci_device_id *id_table; int (*probe)(struct
                                              pci_dev *dev,
    const struct pci_device_id *id); void (*remove)(struct
                                          pci_dev *dev);
    int (*suspend)(struct pci_dev *dev, pm_message_t state);
    int (*resume) (struct pci_dev *dev);  /* Device woken up */
    void (*shutdown) (struct pci_dev *dev); [...]
};
```

Parts of the elements in this structure have been removed since they hold no interest for us. The following are the meanings of the remaining fields in the structure:

- `name`: This is the name of the driver. Since drivers are identified by their name, it must be unique among all PCI drivers in the kernel. It is common to set this field to the same name as the module name of the driver. If there is already a driver register with the same name in the same subsystem bus, the registration of your driver will fail. To see how it works under the hood, have a look at `driver_register()` at `https://elixir.bootlin.com/linux/v4.19/source/drivers/base/driver.c#L146`.

- `id_table`: This should point to the `struct pci_device_id` table described earlier. It is the second and last place this structure is used in the driver. It must be non-NULL for the probe to be called.

- `probe`: This is the pointer to the `probe` function of the driver. It is called by the PCI core when a PCI device matches (either by vendor/product IDs or class ID) an entry in `id_table` in the driver. This method should return `0` if it managed to initialize the device, or a negative error otherwise.

- `remove`: This is called by the PCI core when a device handled by this driver is removed from the system (disappears from the bus) or when the driver is being unloaded from the kernel.

- `suspend`, `resume`, and `shutdown`: These are optional but recommended power management functions. In those callbacks, you can use PCI-related power management helpers such as `pci_save_state()` or `pci_restore_state()`, `pci_disable_device()` or `pci_enable_device()`, `pci_set_power_state()`, and `pci_choose_state()`. These callbacks are invoked by the PCI core, respectively:

  - When the device is suspended, in which case the state is given as an argument to the callback.

  - When the device is being resumed. This may occur only after `suspend` has been called.

  - For a proper shutdown of the device.

The following is an example of a PCI driver structure being initialized:

```
static struct pci_driver bt8xxgpio_pci_driver = {
    .name       = "bt8xxgpio",
    .id_table   = bt8xxgpio_pci_tbl,
    .probe      = bt8xxgpio_probe,
```

```
    .remove      = bt8xxgpio_remove,
    .suspend     = bt8xxgpio_suspend,
    .resume      = bt8xxgpio_resume,
};
```

Registering a PCI driver

Registering a PCI driver with the PCI core consists of calling `pci_register_driver()`, given an argument as a pointer to the `struct pci_driver` structure set up earlier. This should be done in the `init` method of the module, as follows:

```
static int init pci_foo_init(void)
{
    return pci_register_driver(&bt8xxgpio_pci_driver);
}
```

`pci_register_driver()` returns 0 if everything went well while registering, or a negative error otherwise. This return value is handled by the kernel.

However, on the unloading path of the module, `struct pci_driver` needs to be unregistered so that the system does not try to use a driver whose corresponding module no longer exists. Thus, unloading a PCI driver entails calling `pci_unregister_driver()`, along with a pointer to the same structure as per the registration, shown as follows. This should be done in the module `exit` function:

```
static void exit pci_foo_exit(void)
{
    pci_unregister_driver(&bt8xxgpio_pci_driver);
}
```

That being said, with those operations often being repeated in PCI drivers, the PCI core exposes the `module_pci_macro()` macro in order to handle registering/unregistering automatically, as follows:

```
module_pci_driver(bt8xxgpio_pci_driver);
```

This macro is safer, as it takes care of both registering and unregistering, preventing some developers from providing one and forgetting the other.

Now we are familiar with the most important PCI data structures – `struct pci_dev`, `pci_device_id`, and `pci_driver`, as well as the Hyphenate helpers to deal with those data structures. The logical continuation is the driver structure, in which we learn where and how to use the previously enumerated data structures.

Overview of the PCI driver structure

While writing a PCI device driver, there are steps that need to be followed, some of which need to be done in a predefined order. Here, we try to discuss in detail each of these steps, explaining details where applicable.

Enabling the device

Before any operation can be performed on a PCI device (even for reading its configuration registers only), this PCI device must be enabled, and this has to be done explicitly by the code. The kernel provides `pci_enable_device()` for this purpose. This function initializes the device so that it can be used by a driver, asking low-level code to enable I/O and memory. It also handles PCI power management wake up, such that if the device was suspended, it will be woken up, too. The following is what `pci_enable_device()` looks like:

```
int pci_enable_device(struct pci_dev *dev)
```

Since `pci_enable_device()` can fail, the value it returns must be checked, as in the following example:

```
int err;
    err = pci_enable_device(pci_dev);
    if (err) {
    printk(KERN_ERR "foo_dev: Can't enable device.\n");
    return err;
}
```

Remember, `pci_enable_device()` will initialize both memory-mapped and I/O BARs. However, you may want to initialize one but not the other, either because your device does not support both, or because you'll not use both in the driver.

In order not to initialize I/O spaces, you can use another variant of the enabling method, `pci_enable_device_mem()`. On the other hand, if you need to deal with I/O space only, you can use the `pci_enable_device_io()` variant instead. The difference between both variants is that `pci_enable_device_mem()` will initialize only memory-mapped BARs, whereas `pci_enable_device_io()` will initialize I/O BARs. Note that if the device is enabled more than once, each operation will increment the `.enable_cnt` field in the `struct pci_dev` structure, but only the first operation will really act on the device.

When the PCI device is to be disabled, you should adopt the `pci_disable_device()` method, whatever the enabling variant you used. This method signals to the system that the PCI device is no longer in use by the system. The following is its prototype:

```
void pci_disable_device(struct pci_dev *dev)
```

`pci_disable_device()` also disables bus mastering on the device, if active. However, the device is not disabled until all the callers of `pci_enable_device()` (or one of its variants) have called `pci_disable_device()`.

Bus mastering capability

A PCI device can, by definition, initiate transactions on the bus, at the very moment at which it becomes the bus master. After the device is enabled, you may want to enable bus mastering.

This actually consists of enabling DMA in the device, by setting the bus master bit in the appropriate configuration register. The PCI core provides `pci_set_master()` for this purpose. This method also invokes `pci_bios` (`pcibios_set_master()` actually) in order to perform the necessary arch-specific settings. `pci_clear_master()` will disable DMA by clearing the bus master bit. It is the reverse operation:

```
void pci_set_master(struct pci_dev *dev)
void pci_clear_master(struct pci_dev *dev)
```

Note that `pci_set_master()` must be invoked if the device is intended to perform DMA operations.

Accessing configuration registers

Once the device is bound to the driver and after it has been enabled by the driver, it is common to access device memory spaces. The one often accessed first is the configuration space. Conventional PCI and PCI-X mode 1 devices have 256 bytes of configuration space. PCI-X mode 2 and PCIe devices have 4,096 bytes of configuration space. It is primordial for the driver to be able to access the device configuration space, either to read information mandatory for the proper operation of the driver, or to set up some vital parameters. The kernel exposes standard and dedicated APIs (read and write) for different sizes of data configuration space.

In order to read data from the device configuration space, you can use the following primitives:

```
int pci_read_config_byte(struct pci_dev *dev, int where,
                         u8 *val);
int pci_read_config_word(struct pci_dev *dev, int where,
                         u16 *val);
int pci_read_config_dword(struct pci_dev *dev, int where,
                         u32 *val);
```

The above reads, respectively, as one, two, or four bytes in the configuration space of the PCI device represented here by the dev argument. The read value is returned to the val argument. When it comes to writing data to the device configuration space, you can use the following primitives:

```
int pci_write_config_byte(struct pci_dev *dev, int where,
                          u8 val);
int pci_write_config_word(struct pci_dev *dev, int where,
                          u16 val);
int pci_write_config_dword(struct pci_dev *dev, int where,
                           u32 val);
```

The above primitives write, respectively, one, two, or four bytes into the device configuration space. The val argument represents the value to write.

In either read or write cases, the where argument is the byte offset from the beginning of the configuration space. However, there exist in the kernel some commonly accessed configuration offsets identified by symbolically named macros, defined in include/uapi/linux/pci_regs.h. The following is a short excerpt:

```
#define    PCI_VENDOR_ID    0x00  /*   16   bits  */
#define    PCI_DEVICE_ID    0x02  /*   16   bits  */
```

```
#define    PCI_STATUS           0x06   /*    16    bits  */
#define PCI_CLASS_REVISION    0x08   /* High 24 bits are class,
                                         low 8 revision */
#define    PCI_REVISION_ID    0x08   /* Revision ID */
#define    PCI_CLASS_PROG     0x09   /* Reg. Level Programming
                                         Interface */
#define    PCI_CLASS_DEVICE   0x0a   /* Device class */
[...]
```

Thus, to get the revision ID of a given PCI device, you could use the following example:

```
static unsigned char foo_get_revision(struct pci_dev *dev)
{
    u8 revision;
    pci_read_config_byte(dev, PCI_REVISION_ID, &revision);
    return revision;
}
```

In the above, we used `pci_read_config_byte()` because the revision is represented by one byte only.

> **Important note**
>
> Since data is stored in (and read from) PCI devices in little endian format, read primitives (actually `word` and `dword` variants) take care of converting read data into the native endianness of the CPU, and write primitives (`word` and `dword` variants) take care of converting data from the native CPU byte order to little endian prior to writing data to the device.

Accessing memory-mapped I/O resources

Memory registers are used for just about everything else, for example, for burst transactions. Those registers actually correspond to the device memory BARs. Each of them is then assigned a memory region from the system address space so that any access to those regions is redirected to the corresponding device, targeting the right local (in the device) memory corresponding to the BAR. This is memory-mapped I/O.

In the Linux kernel memory-mapped I/O world, it is common to request (to claim actually) a memory region before creating a mapping for it. You can use `request_mem_region()` and `ioremap()` primitives for both purposes. The following are their prototypes:

```
struct resource *request_mem_region (unsigned long start,
                                      unsigned long n,
                                      const char *name)
void iomem *ioremap(unsigned long phys_addr,
                    unsigned long size);
```

`request_mem_region()` is a pure reservation mechanism and does not perform any mapping. It relies on the fact that other drivers should be polite and should call `request_mem_region()` on their turns, which would prevent another driver from overlapping a memory region that has already been claimed. You should not map nor access the claimed region unless this call returns successfully. In its arguments, `name` represents the name to be given to the resource, `start` represents what address the mapping should be created for, and `n` indicates how large the mapping should be. To obtain this information for a given BAR, you can use `pci_resource_start()`, `pci_resource_len()`, or even `pci_resource_end()`, whose prototypes are the following:

- `unsigned long pci_resource_start (struct pci_dev *dev, int bar)`: This function returns the first address (memory address or I/O port number) associated with the BAR whose index is bar.

- `unsigned long pci_resource_len (struct pci_dev *dev, int bar)`: This function returns the size of the BAR bar.

- `unsigned long pci_resource_end (struct pci_dev *dev, int bar)`: This function returns the last address that is part of the I/O region number bar.

- `unsigned long pci_resource_flags (struct pci_dev *dev, int bar)`: This function is not only related to the memory resource BAR. It actually returns the flags associated with this resource. `IORESOURCE_IO` would mean the BAR bar is an I/O resource (thus suitable for I/O mapped I/O), while `IORESOURCE_MEM` would mean it is a memory resource (to be used for memory-mapped I/O).

On the other side, `ioremap()` does create real mapping, and returns a memory-mapped I/O cookie on the mapped region. As an example, the following code shows how to map the `bar0` of a given device:

```
unsigned long bar0_base; unsigned long bar0_size;
void iomem *bar0_map_membase;


/* Get the PCI Base Address Registers */
bar0_base = pci_resource_start(pdev, 0);
bar0_size = pci_resource_len(pdev, 0);


/*
 * think about managed version and use
 * devm_request_mem_regions()
 */
if (request_mem_region(bar0_base, bar0_size, "bar0-mapping")) {
    /* there is an error */
    goto err_disable;
}


/* Think about managed version and use devm_ioremap instead */
bar0_map_membase = ioremap(bar0_base, bar0_size);
if (!bar0_map_membase) {
    /* error */
    goto err_iomap;
}

/* Now we can use ioread32()/iowrite32() on bar0_map_membase*/
```

The preceding code works well, but it is tedious, since we would do this for each BAR. In fact, `request_mem_region()` and `ioremap()` are quite basic primitives. The PCI framework provides many more PCI-related functions to facilitate such common tasks:

```
int pci_request_region(struct pci_dev *pdev, int bar,
                    const char *res_name)
int pci_request_regions(struct pci_dev *pdev,
                    const char *res_name)
void iomem *pci_iomap(struct pci_dev *dev, int bar,
                    unsigned long maxlen)
```

```
void iomem *pci_iomap_range(struct pci_dev *dev, int bar,
                            unsigned long offset,
                            unsigned long maxlen)
void iomem *pci_ioremap_bar(struct pci_dev *pdev, int bar)

void pci_iounmap(struct pci_dev *dev, void iomem *addr)
void pci_release_regions(struct pci_dev *pdev)
```

The preceding helpers can be described as follows:

- `pci_request_regions()` marks all PCI regions associated with the `pdev` PCI device as being reserved by the owner `res_name`. In its arguments, `pdev` is the PCI device whose resources are to be reserved and `res_name` is the name to be associated with the resource. `pci_request_region()`, on the other hand, targets a single BAR, identified by the `bar` argument.

- `pci_iomap()` creates a mapping for a BAR. You can access it using `ioread*()` and `iowrite*()`. `maxlen` specifies the maximum length to map. If you want to get access to the complete BAR without checking for its length first, pass 0 here.

- `pci_iomap_range()` creates a mapping from starting from an offset in the BAR. The resulting mapping starts at `offset` and is `maxlen` wide. `maxlen` specifies the maximum length to map. If you want to get access to the complete BAR from `offset` to the end, pass 0 here.

- `pci_ioremap_bar()` provides an error-proof way (relative to `pci_ioremap()`) to carry out a PCI memory remapping.. It makes sure the BAR is actually a memory resource, not an I/O resource. However, it maps the whole BAR size.

- `pci_iounmap()` is the opposite of `pci_iomap()`, which undoes the mapping. Its `addr` argument corresponds to the cookie previously returned by `pci_iomap()`.

- `pci_release_regions()` is the opposite of `pci_request_regions()`. It releases reserved PCI I/O and memory resources previously claimed (reserved). `pci_release_region()` targets the single BAR variant.

Using these helpers, we can rewrite the same code as before, but for BAR1 this time. This would look as follows:

```
#define DRV_NAME "foo-drv"

void iomem *bar1_map_membase;
```

```
int err;
err = pci_request_regions(pci_dev, DRV_NAME);
if (err) {
    /* an error occured */ goto error;
}


bar1_map_membase = pci_iomap(pdev, 1, 0);
if (!bar1_map_membase) {
    /* an error occured */
    goto err_iomap;
}
```

After the memory areas are claimed and mapped, `ioread*()` and `iowrite*()` APIs, which provide platform abstraction, access the mapped registers.

Accessing I/O port resources

I/O port access requires going through the same steps as I/O memory, although the underlying mechanisms are different: requesting the I/O region, mapping the I/O region (this is not mandatory, it is just a matter of politeness), and accessing the I/O region.

The first two steps have already been addressed without you noticing. Actually, `pci_requestregion*()` primitives handle both I/O port and I/O memory. It relies on the resource flags (`pci_resource_flags()`) in order to call the appropriate low-level helper (`(request_region())` for I/O port or `request_mem_region()` for I/O memory:

```
unsigned long flags = pci_resource_flags(pci_dev, bar);
if (flags & IORESOURCE_IO)
    /* using request_region() */
else if (flag & IORESOURCE_MEM)
    /* using request_mem_region() */
```

Thus, whether the resource is I/O memory or I/O port, you can safely use either `pci_request_regions()` or its single bar variant, `pci_request_region()`.

The same applies to the I/O port mapping. pci_iomap*() primitives are able to deal with either I/O port or I/O memory. They rely on the resource flags too, and they invoke the appropriate helper to create the mapping. Based on the resource type, the underlying mapping functions are ioremap() for I/O memory, which are resources of the IORESOURCE_MEM type, and __pci_ioport_map() for I/O port, which corresponds to resources of the IORESOURCE_IO type. __pci_ioport_map() is an arch-dependent function (overridden by MIPS and SH architectures actually), which, most of the time, corresponds to ioport_map().

To confirm what we just said, we can have a look at the body of the pci_iomap_range() function, on which pci_iomap() relies:

```
void iomem *pci_iomap_range(struct pci_dev *dev, int bar,
                            unsigned long offset,
                            unsigned long maxlen)
{
    resource_size_t start = pci_resource_start(dev, bar);
    resource_size_t len = pci_resource_len(dev, bar);
    unsigned long flags = pci_resource_flags(dev, bar);

    if (len <= offset || !start)
        return NULL;
    len -= offset; start += offset;

    if (maxlen && len > maxlen)
        len = maxlen;
    if (flags & IORESOURCE_IO)
        return pci_ioport_map(dev, start, len);
    if (flags & IORESOURCE_MEM)
        return ioremap(start, len);

    /* What? */
    return NULL;
}
```

However, when it comes to accessing I/O ports, APIs completely change. The following are helpers for accessing I/O ports. These functions hide the details of the underlying mapping and of what type they are. The following lists the functions provided by the kernel to access I/O ports:

```
u8 inb(unsigned long port);
u16 inw(unsigned long port);
u32 inl(unsigned long port);
void outb(u8 value, unsigned long port);
void outw(u16 value, unsigned long port);
void outl(u32 value, unsigned long port);
```

In the preceding excerpt, the `in*()` family reads one, two, or four bytes, respectively, from the `port` location. The data fetched is returned by a value. On the other hand, the `out*()` family writes one, two, or four bytes, respectively, referred to as a `value` argument in the `port` location.

Dealing with interrupts

Drivers that need to service interrupts for a device need to request those interrupts first. It is common to request interrupts from within the `probe()` method. That being said, in order to deal with legacy and non-MSI IRQ, drivers can directly use the `pci_dev->irq` field, which is preassigned when the device is probed.

However, for a more generic approach, it is recommended to use the `pci_alloc_irq_vectors()` API. This function is defined as follows:

```
int pci_alloc_irq_vectors(struct pci_dev *dev,
                          unsigned int min_vecs,
                          unsigned int max_vecs,
                          unsigned int flags);
```

The preceding function returns the number of vectors allocated (which might be smaller than `max_vecs`) if successful, or a negative error code in the event of an error. The number of allocated vectors is always at least up to `min_vecs`. If less than `min_vecs` interrupt vectors are available for `dev`, the function will fail with `-ENOSPC`.

The advantage of this function is that it can deal with either legacy interrupts and MSI or MSI-X interrupts. Depending on the `flags` argument, the driver can instruct the PCI layer to set up the MSI or MSI-X capability for this device. This argument is used to specify the type of interrupt used by the device and the driver. Possible flags are defined in `include/linux/pci.h`:

- `PCI_IRQ_LEGACY`: A single legacy IRQ vector.

- `PCI_IRQ_MSI`: On the success path, `pci_dev->msi_enabled` is set to 1.

- `PCI_IRQ_MSIX`: On the success path, `pci_dev->msix_enabled` is set to 1.

- `PCI_IRQ_ALL_TYPES`: This allows trying to allocate any of the above kinds of interrupt, but in a fixed order. MSI-X mode is always tried first and the function returns immediately in case of success. If MSI-X fails, then MSI is tried. The legacy mode is used as a fallback in case both MSI-X and MSI fail. The driver can rely on `pci_dev->msi_enabled` and `pci_dev->msix_enabled` to determine which mode was successful.

- `PCI_IRQ_AFFINITY`: This allows affinity auto-assign. If set, `pci_alloc_irq_vectors()` will spread the interrupts around the available CPUs.

To get the Linux IRQ numbers to be passed to `request_irq()` and `free_irq()`, which corresponds to a vector, use the following function:

```
int pci_irq_vector(struct pci_dev *dev, unsigned int nr);
```

In the preceding, `dev` is the PCI device to operate on, and `nr` is the device-relative interrupt vector index (0-based). Let's now look at how this function works more closely:

```
int pci_irq_vector(struct pci_dev *dev, unsigned int nr)
{
    if (dev->msix_enabled) {
        struct msi_desc *entry;
        int i = 0;
        for_each_pci_msi_entry(entry, dev) {
            if (i == nr)
                return entry->irq;
            i++;
        }
        WARN_ON_ONCE(1);
        return -EINVAL;
    }
```

```
    if (dev->msi_enabled) {
        struct msi_desc *entry = first_pci_msi_entry(dev);
        if (WARN_ON_ONCE(nr >= entry->nvec_used))
            return -EINVAL;
    } else {
        if (WARN_ON_ONCE(nr > 0))
            return -EINVAL;
    }


    return dev->irq + nr;
}
```

In the preceding excerpt, we can see that MSI-X is the first attempt (`if  (dev->msix_ enabled)`). Additionally, the returned IRQ has nothing to do with the original `pci_dev->irq` preassigned at device probe time. But if MSI is enabled (`dev->msi_ enabled` is true) instead, then this function will perform some sanity check and will return `dev->irq + nr`. This confirms the fact that `pci_dev->irq` is replaced with a new value when we operate in MSI mode, and that this new value corresponds to the base interrupt number of the allocated MSI vectors. Finally, you'll notice that there are no special checks for legacy mode.

Actually, in legacy mode, the preassigned `pci_dev->irq` remains untouched, and it is only a single allocated vector. Thus, `nr` should be `0` when operating in legacy mode. In this case, the vector returned is nothing but `dev->irq`.

Some devices might not support using legacy line interrupts, in which case the driver can specify that only MSI or MSI-X is acceptable:

```
nvec =
    pci_alloc_irq_vectors(pdev, 1, nvec,
                        PCI_IRQ_MSI | PCI_IRQ_MSIX);
if (nvec < 0)
    goto out_err;
```

> **Important note**
>
> Note that MSI/MSI-X and legacy interrupts are mutually exclusive and the reference design supports legacy interrupts by default. Once MSI or MSI-X interrupts are enabled on a device, it stays in this mode until they are disabled again.

Legacy INTx IRQ assignment

The probe method of the PCI bus type (`struct bus_type pci_bus_type`) is `pci_device_probe()`, implemented in `drivers/pci/pci-driver.c`. This method is invoked each time a new PCI device is added to the bus or when a new PCI driver is registered with the system. This function calls `pci_assign_irq(pci_dev)` and then `pcibios_alloc_irq(pci_dev)` in order to assign an IRQ to the PCI device, the famous `pci_dev->irq`. The trick starts happening in `pci_assign_irq()`. `pci_assign_irq()` reads the pin to which the PCI device is connected, as follows:

```
u8 pin;
pci_read_config_byte(dev, PCI_INTERRUPT_PIN, &pin);
/* (1=INTA, 2=INTB, 3=INTD, 4=INTD) */
```

The next steps rely on the PCI host bridge, whose driver should expose a number of callbacks, including a special one, `.map_irq`, whose purpose is to create IRQ mappings for devices according to their slot and the previously read pin:

```
void pci_assign_irq(struct pci_dev *dev)
{
    int irq = 0; u8 pin;
    struct pci_host_bridge *hbrg =
                pci_find_host_bridge(dev->bus);

    if (!(hbrg->map_irq)) {
    pci_dbg(dev, "runtime IRQ mapping not provided by arch\n");
        return;
    }

    pci_read_config_byte(dev, PCI_INTERRUPT_PIN, &pin);
    if (pin) {
        [...]
        irq = (*(hbrg->map_irq))(dev, slot, pin);
        if (irq == -1)
            irq = 0;
    }

    dev->irq = irq;
    pci_dbg(dev, "assign IRQ: got %d\n", dev->irq);
```

```
        /* Always tell the device, so the driver knows what is the
         * real IRQ to use; the device does not use it.
         */
        pci_write_config_byte(dev, PCI_INTERRUPT_LINE, irq);
}
```

This is the first assignment of the IRQ during the device probing. Going back to the `pci_device_probe()` function, the next method invoked after `pci_assign_irq()` is `pcibios_alloc_irq()`. However, `pcibios_alloc_irq()` is defined as a weak and empty function, overridden only by AArch64 architecture, in `arch/arm64/kernel/pci.c`, and which relies on ACPI (if enabled) to mangle the assigned IRQ. Perhaps in the feature other architecture will want to override this function as well.

The final code of `pci_device_probe()` is the following:

```
static int pci_device_probe(struct device *dev)
{
    int error;
    struct pci_dev *pci_dev = to_pci_dev(dev);
    struct pci_driver *drv = to_pci_driver(dev->driver);
    pci_assign_irq(pci_dev);
    error = pcibios_alloc_irq(pci_dev);
    if (error < 0)
        return error;

    pci_dev_get(pci_dev);
    if (pci_device_can_probe(pci_dev)) {
        error = pci_device_probe(drv, pci_dev);
        if (error) {
            pcibios_free_irq(pci_dev);
            pci_dev_put(pci_dev);
        }
    }
    return error;
}
```

> **Important note**
>
> The IRQ value contained in `PCI_INTERRUPT_LINE` is BAD until after `pci_enable_device()` is called. However, a peripheral driver should never alter `PCI_INTERRUPT_LINE` because it reflects how a PCI interrupt is connected to the interrupt controller, which is not changeable.

Emulated INTx IRQ swizzling

Note that most PCIe devices in legacy INTx mode will default to the local INTA "virtual wire output," and the same holds true for many physical PCI devices connected by PCIe/PCI bridges. OSes would end up sharing the INTA input among all the peripherals in the system; all devices sharing the same IRQ line – I will let you picture the disaster.

The solution to this is "virtual wire INTx IRQ swizzling." Back to the code of the `pci_device_probe()` function, it invokes `pci_assign_irq()`. If you look at the body of this function (in `drivers/pci/setup-irq.c`), you'll notice some swizzling operations, which are intended to solve this.

Locking considerations

It is common for many device drivers to have a per-device spinlock that is taken in the interrupt handler. Since interrupts are guaranteed to be non-reentrant on a Linux-based system, it is not necessary to disable interrupts when you are working with pin-based interrupts or a single MSI. However, if a device uses multiple interrupts, the driver must disable interrupts while the lock is held. This would prevent deadlock if the device sends a different interrupt, whose handler will try to acquire the spinlock that is already locked by the interrupt being serviced. Thus, the locking primitives to use in such situations are `spin_lock_irqsave()` or `spin_lock_irq()`, which disable local interrupts and acquire the lock. You can refer to *Chapter 1, Linux Kernel concepts for Embedded Developers,* for more details on locking primitive and interrupt management.

A word on legacy APIs

There are many drivers that still use the old and now deprecated MSI or MSI-X APIs, which are `pci_enable_msi()`, `pci_disable_msi()`, `pci_enable_msix_range()`, `pci_enable_msix_exact()`, and `pci_disable_msix()`.

The previously listed APIs should not be used in new code at all. However, the following is an example of a code excerpt trying to use MSI and falling back to legacy interrupt mode in case MSI is not available:

```
    int err;

    /* Try using MSI interrupts */
    err = pci_enable_msi(pci_dev);
    if (err)
        goto intx;

    err = devm_request_irq(&pci_dev->dev, pci_dev->irq,
                        my_msi_handler, 0, "foo-msi", priv);
    if (err) {
        pci_disable_msi(pci_dev);
        goto intx;
    }
    return 0;

    /* Try using legacy interrupts */
intx:
    dev_warn(&pci_dev->dev,
    "Unable to use MSI interrupts, falling back to legacy\n");
    err = devm_request_irq(&pci_dev->dev, pci_dev->irq,
            my_shared_handler, IRQF_SHARED, "foo-intx", priv);
    if (err) {
        dev_err(pci_dev->dev, "no usable interrupts\n");
        return err;
    }
    return 0;
```

Since the preceding code contains deprecated APIs, it may be a good exercise to convert it to new APIs.

Now that we are done with the generic PCI device driver structure and have addressed interrupt management in such drivers, we can move a step forward and leverage the direct memory access capabilities of the device.

PCI and Direct Memory Access (DMA)

In order to speed up data transfer and offload the CPU by allowing it not to perform heavy memory copy operations, both the controller and the device can be configured to perform Direct Memory Access (DMA), which is a means by which data is exchanged between device and host without the CPU being involved. Depending on the root complex, the PCI address space can be either 32 or 64 bits.

System memory regions that are the source or destination of DMA transfers are called DMA buffers. However, DMA buffer memory ranges depend on the size of the bus address. This originated from the ISA bus, which was 24-bits wide. In such a bus, DMA buffers could live only in the bottom 16 MB of system memory. This bottom memory is also referred to as `ZONE_DMA`. However, PCI buses do not have such limitations. While the classic PCI bus supports 32-bit addressing, PCIe extended this to 64 bits. Thus, two different address formats can be used: the 32-bit address format and the 64-bit address format. In order to pull the DMA API, the driver should contain `#include <linux/dma-mapping.h>`.

To inform the kernel of any special needs of DMA-able buffers (which consist of specifying the width of the bus), you can use `dma_set_mask()`, defined as follows:

```
dma_set_mask(struct device *dev, u64 mask);
```

This would help the system in terms of efficient memory allocation, especially if the device can directly address "consistent memory" in system RAM above 4 GB physical RAM. In the above helper, `dev` is the underlying device for the PCI device, and `mask` is the actual mask to use, which you can specify by using the `DMA_BIT_MASK` macro along with the actual bus width. `dma_set_mask()` returns 0 on success. Any other value means an error occurred.

The following is an example for a 32-bit (or 64-bit) bit system:

```
int err = 0;
err = pci_set_dma_mask(pci_dev, DMA_BIT_MASK(32));

/*
 * OR the below on a 64 bits system:
 * err = pci_set_dma_mask(dev, DMA_BIT_MASK(64));
 */
if (err) {
    dev_err(&pci_dev->dev,
            "Required dma mask not supported, \
```

```
            failed to initialize device\n");
    goto err_disable_pci_dev;
}
```

That being said, DMA transfer requires suitable memory mapping. This mapping consists of allocating DMA buffers and generating a bus address for each, which are of the type dma_addr_t. Since I/O devices view DMA buffers through the lens of the bus controller and any intervening I/O memory management unit (IOMMU), the resulting bus addresses will be given to the device in order for it to be notified of the location of the DMA buffers. Since each memory mapping also produces a virtual address, not only bus addresses but also virtual addresses will be generated for the mapping. In order for the CPU to be able to access the buffers, DMA service routines also map the kernel virtual address of DMA buffers to bus addresses.

There are two types of (PCI) DMA mapping: coherent ones and streaming ones. For either, the kernel provides a healthy API that masks many of the internal details of dealing with the DMA controller.

PCI coherent (aka consistent) mapping

Such mapping is called consistent because it allocates uncached (coherent) and unbuffered memory for the device to perform DMA operation. Since a write by either the device or the CPU can be immediately read by either without worrying about cache coherency, such mappings are also synchronous. All this makes consistent mapping too expensive for the system, although most devices require it. However, in terms of code, it is easier to implement.

The following function sets up a coherent mapping:

```
void * pci_alloc_consistent(struct pci_dev *hwdev, size_t size,
                             dma_addr_t *dma_handle)
```

With the above, memory allocated for the mapping is guaranteed to be physically contiguous. size is the length of the region you need to allocate. This function returns two values: the virtual address that you can use to access it from the CPU and dma_handle, the third argument, which is an output parameter and which corresponds to the bus address the function call generated for the allocated region. The bus address is actually the one you pass to the PCI device.

Do note that `pci_alloc_consistent()` is actually a dumb wrapper of `dma_alloc_coherent()` with the `GFP_ATOMIC` flag set, meaning allocation does not sleep and it is safe to call it from within an atomic context. You may want to use `dma_alloc_coherent()` (which is strongly encouraged) instead if you wish to change the allocation flags, for example, using `GFP_KERNEL` instead of `GFP_ATOMIC`.

Keep in mind that mapping is expensive, and the minimum it can allocate is a page. Under the hood, it only allocates the number of pages in the power of 2. The order of pages is obtained with `int order = get_order(size)`. Such a mapping is to be used for buffers that last the lifetime of the device.

To unmap and free such a DMA region, you can call `pci_free_consistent()`:

```
pci_free_consistent(dev, size, cpu_addr, dma_handle);
```

Here, `cpu_addr` and `dma_handle` correspond to the kernel virtual address and to the bus address returned by `pci_alloc_consistent()`. Though the mapping function can be called from an atomic context, this one may not be called in such a context.

Do also note that `pci_free_consistent()` is a simple wrapper of `dma_free_coherent()`, which can be used if the mapping has been done using `dma_alloc_coherent()`:

```
#define DMA_ADDR_OFFSET        0x14
#define DMA_REG_SIZE_OFFSET         0x32
[...]


int do_pci_dma (struct pci_dev *pci_dev, int direction,
                size_t count)
{
    dma_addr_t dma_pa;
    char *dma_va;
    void iomem *dma_io;

    /* should check errors */
    dma_io = pci_iomap(dev, 2, 0);

    dma_va = pci_alloc_consistent(&pci_dev->dev, count,
                              &dma_pa);
    if (!dma_va)
        return -ENOMEM;
```

```
        /* may need to clear allocated region */
        memset(dma_va, 0, count);

        /* set up the device */
        iowrite8(CMD_DISABLE_DMA, dma_io + REG_CMD_OFFSET);
        iowrite8(direction ? CMD_WR : CMD_RD);
        /* Send bus address to the device */
        iowrite32(dma_pa, dma_io + DMA_ADDR_OFFSET);
        /* Send size to the device */
        iowrite32(count, dma_io + DMA_REG_SIZE_OFFSET);

        /* Start the operation */
        iowrite8(CMD_ENABLE_DMA, dma_io + REG_CMD_OFFSET);

        return 0;
}
```

The preceding code shows how to perform a DMA mapping and send the resulting bus address to the device. In the real world, an interrupt may be raised. You should then handle it from within the driver.

Streaming DMA mapping

Streaming mapping, on the other hand, has more constraints in terms of code. First of all, such mappings need to work with a buffer that has already been allocated. Additionally, a buffer that has been mapped belongs to the device and not to the CPU anymore. Thus, before the CPU can use the buffer, it should be unmapped first, in order to address possible caching issues.

If you need to initiate a write transaction (CPU to device), the driver should place data in the buffer prior to the mapping. Moreover, the direction the data should move into has to be specified, and the data should only be used based on this direction.

The reason the buffers must be unmapped before they can be accessed by the CPU is because of the cache. It goes without saying that CPU mapping is cacheable. The dma_map_*() family functions (actually wrapped by pci_map_*() functions), which are used for streaming mappings, will first clean/invalidate the caches related to the buffers and will rely on the CPU not to access those buffers until the corresponding dma_unmap_*() (wrapped by pci_unmap_*() functions). Those unmappings will then invalidate (if necessary) the caches again, in case of any speculative fetches in the meantime, before the CPU may read any data written to memory by the device. Only at this time can the CPU access the buffers.

There are streaming mappings that can accept several non-contiguous and scattered buffers. We can then enumerate two forms of streaming mapping:

- Single buffer mapping, which allows only single-page mapping
- Scatter/gather mapping, which allows the passing of several buffers (scattered over memory)

Each of them is introduced in the following sections.

Single buffer mapping

This consists of mapping a single buffer. It is for occasional mapping. That being said, you can set up a single buffer with this:

```
dma_addr_t pci_map_single(struct pci_dev *hwdev, void *ptr,
                          size_t size, int direction)
```

direction should be either PCI_DMA_BIDIRECTION, PCI_DMA_TODEVICE, PCI_DMA_FROMDEVICE, or PCI_DMA_NONE. ptr is the kernel virtual address of the buffer, and dma_addr_t is the returned bus address that can be sent to the device. You should make sure to use the direction that really matches the way data is intended to move, not just always DMA_BIDIRECTIONAL. pci_map_single() is a dumb wrapper of dma_map_single(), with the directions mapping to DMA_TO_DEVICE, DMA_FROM_DEVICE, or DMA_BIDIRECTIONAL.

You should free the mapping with the following:

```
Void pci_unmap_single(struct pci_dev *hwdev,
                      dma_addr_t dma_addr,
                      size_t size, int direction)
```

This is a wrapper around dma_unmap_single(). dma_addr should be the same as the one returned by pci_map_single() (or the one returned by dma_map_single() in case you used it). direction and size should match what you have specified in the mapping.

The following shows a simplified example of streaming mapping (actually a single buffer):

```
int do_pci_dma (struct pci_dev *pci_dev, int direction,
                void *buffer, size_t count)
{
    dma_addr_t dma_pa;
    /* bus address */
    void iomem *dma_io;

    /* should check errors */
    dma_io = pci_iomap(dev, 2, 0);
    dma_dir = (write ? DMA_TO_DEVICE : DMA_FROM_DEVICE);

    dma_pa = pci_map_single(pci_dev, buffer, count, dma_dir);
    if (!dma_va)
        return -ENOMEM;
    /* may need to clear allocated region */
    memset(dma_va, 0, count);

    /* set up the device */
    iowrite8(CMD_DISABLE_DMA, dma_io + REG_CMD_OFFSET);
    iowrite8(direction ? CMD_WR : CMD_RD);
    /* Send bus address to the device */
    iowrite32(dma_pa, dma_io + DMA_ADDR_OFFSET);
    /* Send size to the device */
    iowrite32(count, dma_io + DMA_REG_SIZE_OFFSET);

    /* Start the operation */
    iowrite8(CMD_ENABLE_DMA, dma_io + REG_CMD_OFFSET);

    return 0;
}
```

In the preceding example, `buffer` is supposed to be already allocated and to contain the data. It is then mapped, its bus address is sent to the device, and the DMA operation is started. The next code sample (implemented as an interrupt handler for the DMA transaction) demonstrates how to deal with the buffer from the CPU side:

```
void pci_dma_interrupt(int irq, void *dev_id)
{
    struct private_struct *priv =
    (struct private_struct *) dev_id;

    /* Unmap the DMA buffer */
    pci_unmap_single(priv->pci_dev, priv->dma_addr,
                        priv->dma_size, priv->dma_dir);

    /* now it is safe to access the buffer */
    [...]
}
```

In the preceding, the mapping is released before the CPU can play with the buffer.

Scatter/gather mapping

Scatter/gather mapping is the second family of streaming DMA mapping with which you can transfer several (not necessarily physically contiguous) buffer regions in a single shot instead of mapping each buffer individually and transferring them one by one. In order to set up a `scatterlist` mapping, you should allocate your scattered buffers first, which must be of page size, except the last one, which may have a different size. After this, you should allocate an array of `scatterlist` and fill it with the previously allocated buffers using `sg_set_buf()`. Finally, you must call `dma_map_sg()` on the `scatterlist` array. Once done with DMA, call `dma_unmap_sg()` on the array to unmap the `scatterlist` entries.

While you can send contents of several buffers over DMA one by one by mapping each one of them, scatter/gather can send them all at once by sending the pointer to `scatterlist` to the device, along with a length, which is the number of entries in the list:

```
u32 *wbuf1, *wbuf2, *wbuf3;
struct scatterlist sgl[3];
int num_mapped;
```

```
wbuf1 = kzalloc(PAGE_SIZE, GFP_DMA);
wbuf2 = kzalloc(PAGE_SIZE, GFP_DMA);
/* size may be different for the last entry */
wbuf3 = kzalloc(CUSTOM_SIZE, GFP_DMA);

sg_init_table(sg, 3);
sg_set_buf(&sgl[0], wbuf1, PAGE_SIZE);
sg_set_buf(&sgl[1], wbuf2, PAGE_SIZE);
sg_set_buf(&sgl[2], wbuf3, CUSTOM_SIZE);
num_mapped = pci_map_sg(NULL, sgl, 3, PCI_DMA_BIDIRECTIONAL);
```

First of all, do note that `pci_map_sg()` is a dumb wrapper of `dma_map_sg()`. In
the preceding code, we used `sg_init_table()`, which results in a statically allocated
table. We could have used `sg_alloc_table()` for a dynamic allocation. Moreover, we
could have used the `for_each_sg()` macro, in order to loop over each `sg` (**scatterlist**)
element, along with the `sg_set_page()` helper in order to set the page to which
this scatterlist is bound (you should never assign the page directly). The following is an
example involving such helpers:

```
static int pci_map_memory(struct page **pages,
                          unsigned int num_entries,
                          struct sg_table *st)
{
    struct scatterlist *sg;
    int i;

    if (sg_alloc_table(st, num_entries, GFP_KERNEL))
        goto err;
    for_each_sg(st->sgl, sg, num_entries, i)
        sg_set_page(sg, pages[i], PAGE_SIZE, 0);

    if (!pci_map_sg(priv.pcidev, st->sgl, st->nents,
                    PCI_DMA_BIDIRECTIONAL))
        goto err;
    return 0;
err:
    sg_free_table(st);
```

```
    return -ENOMEM;
}
```

In the preceding block, pages should have been allocated and should obviously be of `PAGE_SIZE` size. `st` is an output parameter that will be set up appropriately on the success path of this function.

Again, note that scatterlist entries must be of page size (except the last entry, which may have a different size). For each buffer in the input scatterlist, `dma_map_sg()` determines the proper bus address to give to the device. The bus address and length of each buffer are stored in the struct scatterlist entries, but their location in the structure varies from one architecture to another. Thus, there are two macros that you can use to make your code portable:

- `dma_addr_t sg_dma_address(struct scatterlist *sg)`: This returns the bus (DMA) address from this scatterlist entry.

- `unsigned int sg_dma_len(struct scatterlist *sg)`: This returns the length of this buffer.

`dma_map_sg()` and `dma_unmap_sg()` take care of cache coherency. However, if you have to access (read/write) the data between the DMA transfers, the buffers must be synchronized between each transfer in an appropriate manner, by means of either `dma_sync_sg_for_cpu()` if the CPU needs to access the buffers, or `dma_sync_sg_for_device()` if it is the device that needs access. Similar functions for single region mapping are `dma_sync_single_for_cpu()` and `dma_sync_single_for_device()`.

Given all of the above, we can conclude that coherent mappings are simple to code but expensive to use, whereas streaming mappings have the reverse characteristics. Streaming mapping is to be used when the I/O device owns the buffer for long durations. Streamed DMA is common for asynchronous operations when each DMA operates on a different buffer, such as network drivers, where each `skbuf` data is mapped and unmapped on the fly. However, the device may have the last word on what method you should use. That being said, if you had the choice, you should use streaming mapping when you can and coherent mapping when you must.

Summary

In this chapter, we have dealt with PCI specification buses and implementations, as well as its support in the Linux kernel. We went through the enumeration process and how the Linux kernel allows different address spaces to be accessed. We then followed a detailed step-by-step guide on how to write a PCI device driver, from the device table population to the module's `exit` method. We took a deeper look at the interrupt mechanisms and their underlying behaviors as well as the differences between them. Now you are able to write a PCI device driver on your own, and you are familiar with their enumeration process. Moreover, you understand their interrupt mechanisms and are aware of the differences between them (MSI or not). Finally, you learned how to access their respective memory regions.

In the next chapter, we will deal with the NVMEM framework, which helps to develop drivers for non-volatile storage devices such as EEPROM. This will serve to end the complexity that we have experienced so far while learning about PCI device drivers.

12
Leveraging the NVMEM Framework

The **NVMEM** (**Non-Volatile MEMory**) framework is the kernel layer to handle non-volatile storage, such as EEPROM, eFuse, and so on. The drivers for these devices used to be stored in `drivers/misc/`, where most of the time each one had to implement its own API to handle identical functionalities, either for kernel users or to expose its content to user space. It turned out that these drivers seriously lacked abstraction code. Moreover, the increasing support for the number of these devices in the kernel led to a lot of code duplication.

The introduction of this framework in the kernel aims at solving these previously mentioned issues. It also introduces DT representation for consumer devices to get the data they require (MAC addresses, SoC/revision ID, part numbers, and so on) from the NVMEM. We will begin this chapter by introducing NVMEM data structures, which are mandatory to walk through the framework, and then we will look at the NVMEM provider drivers, where we will learn how to expose the NVMEM memory region to consumers. Finally, we will learn about NVMEM consumer drivers, to leverage the content exposed by the providers.

In this chapter, we will cover the following topics:

- Introducing the NVMEM data structures and APIs
- Writing the NVMEM provider driver
- NVMEM consumer driver APIs

Technical requirements

The following are prerequisites for this chapter:

- C programming skills
- Kernel programming and device driver development skills
- Linux kernel v4.19.X sources, available at `https://git.kernel.org/pub/scm/linux/kernel/git/stable/linux.git/refs/tags`

Introducing NVMEM data structures and APIs

NVMEM is a small framework with a reduced set of APIs and data structures. In this section, we will introduce those APIs and data structures, as well as the concept of a **cell**, which is the base of this framework.

NVMEM is based on the producer/consumer pattern, just like the clock framework described in *Chapter 4*, *Storming the Common Clock Framework*. There is a single driver for the NVMEM device, exposing the device cells so that they can be accessed and manipulated by consumer drivers. While the NVMEM device driver must include `<linux/nvmem-provider.h>`, consumers have to include `<linux/nvmem-consumer.h>`. This framework has only a few data structures, among which is `struct nvmem_device`, which looks as follows:

```
struct nvmem_device {
    const char  *name;
    struct module *owner;
    struct device dev;
    int stride;
    int word_size;
    int id;
    int users;
    size_t size;
    bool read_only;
```

```
    int flags;
    nvmem_reg_read_t reg_read;
    nvmem_reg_write_t reg_write; void *priv;
    [...]
};
```

This structure actually abstracts the real NVMEM hardware. It is created and populated by the framework upon device registration. That said, its fields are actually set with a complete copy of the fields in struct nvmem_config, which is described as follows:

```
struct nvmem_config {
    struct device *dev;
    const char *name;
    int id;
    struct module *owner;
    const struct nvmem_cell_info *cells;
    int ncells;
    bool read_only;
    bool root_only;
    nvmem_reg_read_t reg_read;
    nvmem_reg_write_t reg_write;
    int size;
    int word_size;
    int stride;
    void *priv;
    [...]
};
```

This structure is the runtime configuration of the NVMEM device, providing either information on it or the helper functions to access its data cells. Upon device registration, most of its fields are used to populate the newly created nvmem_device structure.

The meanings of the fields in the structure are described as follows (knowing these are used to build the underlying struct nvmem_device):

- dev is the parent device.

- name is an optional name for this NVMEM device. It is used with id filled to build the full device name. The final NVMEM device name will be <name><id>. It is better to append - in the name so that the full name can have this pattern: <name>-<id>. This is what is used in the PCF85363 driver. If omitted, nvmem<id> will be used as the default name.

- id is an optional ID for this NVMEM device. It is ignored if name is NULL. If set to -1, the kernel will take care of providing a unique ID to the device.

- owner is the module that owns this NVMEM device.

- cells is an array of predefined NVMEM cells. It is optional.

- ncells is the number of elements in cells.

- read_only marks this device as read-only.

- root_only tells whether this device is accessible only to the root.

- reg_read and reg_write are the underlying callbacks used by the framework to read and write data, respectively. They are defined as follows:

```
typedef int (*nvmem_reg_read_t)(void *priv,
                                unsigned int offset,
                                void *val, size_t bytes);
typedef int (*nvmem_reg_write_t)(void *priv,
                                unsigned int offset,
                                void *val,
                                size_t bytes);
```

- size represents the size of the device.

- word_size is the minimum read/write access granularity for this device. stride is the minimum read/write access stride. Its principle has already been explained in previous chapters.

- priv is context data passed to read/write callbacks. It could, for example, be a bigger structure wrapping this NVMEM device.

Previously, we used the term **data cell**. A data cell represents a memory region (or data region) in the NVMEM device. This may also be the whole memory of the device. Actually, data cells are to be assigned to consumer drivers. These memory regions are maintained by the framework using two different data structures, depending on whether we are on the consumer side or on the provider side. These are the `struct nvmem_cell_info` structure for the provider, and `struct nvmem_cell` for the consumer. From within the NVMEM core code, the kernel uses `nvmem_cell_info_to_nvmem_cell()` to switch from the former structure to the second one.

These structures are introduced as follows:

```
struct nvmem_cell {
        const char *name;
        int offset;
        int bytes;
        int bit_offset;
        int nbits;
        struct nvmem_device *nvmem;
        struct list_head node;
};
```

The other data structure, that is, `struct nvmem_cell`, looks like the following:

```
struct nvmem_cell_info {
        const char *name;
        unsigned int offset;
        unsigned int bytes;
        unsigned int bit_offset;
        unsigned int nbits;
};
```

As you can see, the preceding two data structures share almost the same properties. Let's look at their meanings, as follows:

- `name` is the name of the cell.
- `offset` is the offset (where it starts) of the cell from within the whole hardware data registers.
- `bytes` is the size (in bytes) of the data cells, starting from `offset`.

- A cell may have bit-level granularity. For these cells, `bit_offset` should be set in order to specify the bit offset from within the cell, and `nbits` should be defined according to the size (in bits) of the region of interest.

- `nvmem` is the NVMEM device to which this cell belongs.

- `node` is used to track the cell system-wide. This field ends up in the `nvmem_cells` list, which holds all the cells available on the system, regardless of the NVMEM device they belong to. This global list is actually protected by a mutex, `nvmem_cells_mutex`, both statically defined in `drivers/nvmem/core.c`.

To clarify the preceding explanation, let's take as an example a cell with the following config:

```
static struct nvmem_cellinfo mycell = {
        .offset = 0xc,
        .bytes = 0x1,
        [...],
}
```

In the preceding example, if we consider `.nbits` and `.bit_offset` as both equal to `0`, it means we are interested in the whole data region of the cell, which is 1 byte-sized in our case. But what if we are interested only in bits 2 to 4 (3 bits, actually)? The structure would be as follows:

```
staic struct nvmem_cellinfo mycell = {
        .offset = 0xc,
        .bytes = 0x1,
        .bit_offset = 2,
        .nbits = 2 [...]
}
```

> **Important note**
>
> The preceding examples are only for pedagogical purposes. Even though you can have predefined cells in the driver code, it is recommended that you rely on the device tree to declare the cells, as we will see later in the chapter, in the *Device tree bindings for NVMEM providers* section, to be precise.

Neither the consumer nor the provider driver should create instances of `struct nvmem_cell`. The NVMEM core internally handles this, either when the producer provides an array of cell info, or when the consumer requests a cell.

So far, we have gone through the data structures and APIs provided by this framework. However, NVMEM devices can be accessed either from the kernel or user space. Moreover, in the kernel, there must be a driver exposing the device storage in order to have other drivers accessing it. This is the producer/consumer design, where the provider driver is the producer, and the other driver is the consumer. Right now, let's start with the provider (aka the producer) part of this framework.

Writing the NVMEM provider driver

The provider is the one exposing the device memory so that other drivers (the consumers) can access it. The main tasks of these drivers are as follows:

- Providing suitable NVMEM configuration with respect to the device's datasheet, along with the routines allowing you to access the memory

- Registering the device with the system

- Providing device tree binding documentation

That is all the provider has to do. Most (the rest) of the mechanism/logic is handled by the NVMEM framework's code.

NVMEM device (un)registration

Registering/unregistering the NVMEM device is actually part of the provider-side driver, which can use the nvmem_register()/nvmem_unregister() functions, or their managed versions, devm_nvmem_register()/devm_nvmem_unregister():

```
struct nvmem_device *nvmem_register(const
                                    struct nvmem_config *config)
struct nvmem_device *devm_nvmem_register(struct device *dev,
                                const struct nvmem_config *config)
int nvmem_unregister(struct nvmem_device *nvmem)
int devm_nvmem_unregister(struct device *dev,
                          struct nvmem_device *nvmem)
```

Upon registration, the `/sys/bus/nvmem/devices/dev-name/nvmem` binary entry will be created. In these interfaces, the `*config` parameter is the NVMEM config describing the NVMEM device that has to be created. The `*dev` parameter is only for the managed version and represents the device using the NVMEM device. On the success path, these functions return a pointer to `nvmem_device`, or return `ERR_PTR()` on error otherwise.

On the other hand, unregistration functions accept the pointer to the NVMEM device created on the success path of the registration function. They return 0 upon successful unregistration and a negative error otherwise.

NVMEM storage in RTC devices

There are many **Real-Time Clock** (**RTC**) devices that embed non-volatile storage. This embedded storage can be either EEPROM or battery-backed RAM. Looking at the RTC device data structure in `include/linux/rtc.h`, you will notice that there are NVMEM-related fields, as follows:

```
struct rtc_device {
    [...]
    struct nvmem_device *nvmem;
    /* Old ABI support */
    bool nvram_old_abi;
    struct bin_attribute *nvram;
    [...]
}
```

Note the following in the preceding structure excerpt:

- `nvmem` abstracts the underlying hardware memory.

- `nvram_old_abi` is a Boolean that tells whether the NVMEM of this RTC is to be registered using the old (and now deprecated) NVRAM ABI, which uses `/sys/class/rtc/rtcx/device/nvram` to expose the memory. This field should be set to `true` only if you have existing applications (that you do not want to break) using this old ABI interface. New drivers should not set this.

- `nvram` is actually the binary attribute for the underlying memory, used by the RTC framework only for old ABI support; that is, if `nvram_old_abi` is `true`.

The RTC-related NVMEM framework API can be enabled through the RTC_NVMEM kernel config option. This API is defined in drivers/rtc/nvmem.c, and exposes both rtc_nvmem_register() and rtc_nvmem_unregister(), respectively, for RTC-NVMEM registration and unregistration. These are described as follows:

```
int rtc_nvmem_register(struct rtc_device *rtc,
                        struct nvmem_config *nvmem_config)
void rtc_nvmem_unregister(struct rtc_device *rtc)
```

rtc_nvmem_register() returns 0 on success. It accepts a valid RTC device as its first parameter. This has an impact on the code. It means the RTC's NVMEM should be registered only after the actual RTC device has been successfully registered. In other words, rtc_nvmem_register() is to be called only after rtc_register_device() has succeeded. The second argument should be a pointer to a valid nvmem_config object. Moreover, as we have already seen, this config can be declared in the stack since all its fields are entirely copied for building the nvmem_device structure. The opposite is rtc_nvmem_unregister(), which unregisters the NVMEM.

Let's summarize this with an excerpt of the probe function of the DS1307 RTC driver, drivers/rtc/rtc-ds1307.c:

```
static int ds1307_probe(struct i2c_client *client,
                        const struct i2c_device_id *id)
{
    struct ds1307 *ds1307;
    int err = -ENODEV;
    int tmp;
    const struct chip_desc *chip;
    [...]

    ds1307->rtc->ops = chip->rtc_ops ?: &ds13xx_rtc_ops;
    err = rtc_register_device(ds1307->rtc);
    if (err)
        return err;
    if (chip->nvram_size) {
        struct nvmem_config nvmem_cfg = {
            .name = "ds1307_nvram",
            .word_size = 1,
            .stride = 1,
```

```
            .size = chip->nvram_size,
            .reg_read = ds1307_nvram_read,
            .reg_write = ds1307_nvram_write,
            .priv = ds1307,
        };
        ds1307->rtc->nvram_old_abi = true;
        rtc_nvmem_register(ds1307->rtc, &nvmem_cfg);
    }
    [...]
}
```

The preceding code first registers the RTC with the kernel prior to registering the NVMEM device, giving an NVMEM config that corresponds to the RTC's storage space. The preceding is RTC-related and not generic. Other NVMEM devices must have their driver expose callbacks to which the NVMEM framework will forward any read/write requests, either from user space or internally from within the kernel itself. The next section explains how this is done.

Implementing NVMEM read/write callbacks

In order for the kernel and other frameworks to be able to read/write data from/to the NVMEM device and its cells, each NVMEM provider must expose a couple of callbacks allowing those read/write operations. This mechanism allows hardware-independent consumer code, so any reading/writing request from the consumer side is redirected to the underlying provider's read/write callback. The following are the read/write prototypes that every provider must conform to:

```
typedef int (*nvmem_reg_read_t)(void *priv,
                                unsigned int offset,
                                void *val, size_t bytes);
typedef int (*nvmem_reg_write_t)(void *priv,
                                 unsigned int offset,
                                 void *val, size_t bytes);
```

These are independent of the underlying bus that the NVMEM device is behind. nvmem_reg_read_t is for reading data from the NVMEM device. priv is the user context provided in the NVMEM config, offset is where reading should start, val is an output buffer where the read data has to be stored, and bytes is the size of the data to be read (the number of bytes, actually). This function should return the number of successful bytes read on success, and a negative error code on error.

On the other hand, `nvmem_reg_write_t` is for writing purposes. `priv` has the same meaning as for reading, `offset` is where writing should start at, `val` is a buffer containing the data to be written, and `bytes` is the number of bytes in data in `val`, which should be written. `bytes` is not necessarily the size of `val`. This function should return the number of bytes written successfully on success, and a negative error code on error.

Now that we have seen how to implement provider read/write callbacks, let's see how we can extend the provider capabilities with the device tree.

Device tree bindings for NVMEM providers

The NVMEM data provider does not have any bindings particularly. It should be described with respect to its parent bus DT binding. This means, for example, that if it is an I2C device, it should be described (in respect to the I2C binding) as a child of the node that represents the I2C bus that it sits behind. However, there is an optional `read-only` property that makes the device read-only. Moreover, each child node will be considered as a data cell (a memory region in the NVMEM device).

Let's consider the following MMIO NVMEM device along with its child nodes for explanation:

```
ocotp: ocotp@21bc000 {
    #address-cells = <1>;
    #size-cells = <1>;
    compatible = "fsl,imx6sx-ocotp", "syscon";
    reg = <0x021bc000 0x4000>;
    [...]

    tempmon_calib: calib@38 {
        reg = <0x38 4>;
    };
    tempmon_temp_grade: temp-grade@20 {
        reg = <0x20 4>;
    };
    foo: foo@6 {
        reg = <0x6 0x2> bits = <7 2>
    };
    [...]
};
```

According to the properties defined in the child nodes, the NVMEM framework builds the appropriate `nvmem_cell` structures and inserts them into the system-wide `nvmem_cells` list. The following are the possible properties for data cell bindings:

- `reg`: This property is mandatory. It is a two-cell property, describing the offset in bytes (the first cell in the property) and the size in bytes (the second cell of the property) of the data region within the NVMEM device.

- `bits`: This is an optional two-cell property that specifies the offset (possible values from 0-7) in bits and the number of bits within the address range specified by the `reg` property.

Having defined the data cells from within the provider node, these can be assigned to consumers using the `nvmem-cells` property, which is a list of phandles to NVMEM providers. Moreover, there should be an `nvmem-cell-names` property too, whose main purpose is to name each data cell. This assigned name can therefore be used to look for the appropriate data cell using the consumer APIs. The following is an example assignment:

```
tempmon: tempmon {
    compatible = "fsl,imx6sx-tempmon", "fsl,imx6q-tempmon";
    interrupt-parent = <&gpc>;
    interrupts = <GIC_SPI 49 IRQ_TYPE_LEVEL_HIGH>;
    fsl,tempmon = <&anatop>;
    clocks = <&clks IMX6SX_CLK_PLL3_USB_OTG>;
    nvmem-cells = <&tempmon_calib>, <&tempmon_temp_grade>;
    nvmem-cell-names = "calib", "temp_grade";
};
```

The full NVMEM device tree binding is available in `Documentation/devicetree/bindings/nvmem/nvmem.txt`.

We just came across the implementation of drivers (the so-called producers) that expose the storage of the NVMEM device. Though it is not always the case, there may be other drivers in the kernel that would need access to the storage exposed by the producer (aka the provider). The next section will describe these drivers in detail.

NVMEM consumer driver APIs

NVMEM consumers are drivers who access the storage exposed by the producer. These drivers can pull the NVMEM consumer API by including `<linux/nvmem-consumer.h>`, which will bring the following cell-based APIs in:

```
struct nvmem_cell *nvmem_cell_get(struct device *dev,
                                  const char *name);
struct nvmem_cell *devm_nvmem_cell_get(struct device *dev,
                                       const char *name);
void nvmem_cell_put(struct nvmem_cell *cell);
void devm_nvmem_cell_put(struct device *dev,
                         struct nvmem_cell *cell);
void *nvmem_cell_read(struct nvmem_cell *cell, size_t *len);
int nvmem_cell_write(struct nvmem_cell *cell,
                     void *buf, size_t len);
int nvmem_cell_read_u32(struct device *dev,
                        const char *cell_id,
                        u32 *val);
```

The `devm_`-prefixed APIs are resource-managed versions, which are to be used whenever possible.

That being said, the consumer interface entirely depends on the ability of the producer to expose (part of) its cells so that they can be accessed by others. As discussed previously, this capability of providing/exposing cells should be done via the device tree. `devm_nvmem_cell_get()` serves to grab a given cell with respect to the name assigned through the `nvmem-cell-names` property. The `nvmem_cell_read` API always reads the whole cell size (that is, `nvmem_cell->bytes`) if possible. Its third parameter, `len`, is an output parameter holding the actual number of `nvmem_config.word_size` (actually, it holds `1` most of the time, which means a single byte) being read.

On successful read, the content pointed to by `len` will be equal to the number of bytes in the cell: `*len = nvmem_cell->bytes`. `nvmem_cell_read_u32()`, on the other side, reads a cell value as `u32`.

The following is the code that grabs the cells allocated to the `tempmon` node described in the previous section, and reads their content as well:

```
static int imx_init_from_nvmem_cells(struct
                                     platform_device *pdev)
{
```

```
    int ret; u32 val;
    ret = nvmem_cell_read_u32(&pdev->dev, "calib", &val);
    if (ret)
        return ret;

    ret = imx_init_calib(pdev, val);
    if (ret)
        return ret;

    ret = nvmem_cell_read_u32(&pdev->dev, "temp_grade", &val);
    if (ret)
        return ret;
    imx_init_temp_grade(pdev, val);
    return 0;
}
```

Here, we have gone through both the consumer and producer aspects of this framework. Often, drivers need to expose their services to user space. The NVMEM framework (just like other Linux kernel frameworks) can transparently handle exposing NVMEM services to user space. The next section explains this in detail.

NVMEM in user space

The NVMEM user space interface relies on sysfs, as most of the kernel frameworks do. Each NVMEM device registered with the system has a directory entry created in /sys/bus/nvmem/devices, along with an nvmem binary file (on which you can use hexdump or even echo) created in that directory, which represents the device's memory. The full path has the following pattern: /sys/bus/nvmem/devices/<dev-name>X/nvmem. In this path pattern, <dev-name> is the nvmem_config.name name provided by the producer driver. The following code excerpt shows how the NVMEM core constructs the <dev-name>X pattern:

```
int rval;
rval = ida_simple_get(&nvmem_ida, 0, 0, GFP_KERNEL);
nvmem->id = rval;
if (config->id == -1 && config->name) {
    dev_set_name(&nvmem->dev, "%s", config->name);
} else {
    dev_set_name(&nvmem->dev, "%s%d", config->name ? : "nvmem",
```

```
        config->name ? config->id : nvmem->id);
}
```

The preceding code says if nvmem_config->id == -1, then X in the pattern is omitted and only nvmem_config->name is used to name the sysfs directory entry. If nvmem_config->id != -1 and nvmem_config->name is set, it will be used along with the nvmem_config->id field set by the driver (which is X in the pattern). However, if nvmem_config->name is not set by the driver, the core will use the nvmem string along with an ID that has been generated (which is X in the pattern).

> **Important note**
> Whatever cells are defined, the NVMEM framework exposes the full register space via the NVMEM binary, not the cells. Accessing the cells from user space requires knowing their offsets and size in advance.

NVMEM content can then be read in user space, thanks to the sysfs interface, using either hexdump or the simple cat command. For example, assuming we have an I2C EEPROM sitting on I2C number 2 at address 0x55 registered on the system as an NVMEM device, its sysfs path would be /sys/bus/nvmem/devices/2-00550/nvmem. The following is how you can write/read some content:

```
cat /sys/bus/nvmem/devices/2-00550/nvmem
echo "foo" > /sys/bus/nvmem/devices/2-00550/nvmem
cat /sys/bus/nvmem/devices/2-00550/nvmem
```

Now we have seen how the NVMEM registers are exposed to user space. Though this section is short, we have covered enough to leverage this framework from user space.

Summary

In this chapter, we went through the NVMEM framework implementation in the Linux kernel. We introduced its APIs from the producer side as well as from the consumer side, and also discussed how to use it from user space. I have no doubt that these devices have their place in the embedded world.

In the next chapter, we will address the issue of reliability by means of watchdog devices, discussing how to set up these devices and writing their Linux kernel drivers.

13
Watchdog Device Drivers

A watchdog is a hardware (sometimes emulated by software) device intended to ensure the availability of a given system. It helps make sure that the system always reboots upon a critical hang, thus allowing to monitor the "normal" behavior of the system.

Whether it is hardware-based or emulated by software, the watchdog is, most of the time, nothing but a timer initialized with a reasonable timeout that should be periodically refreshed by software running on the monitored system. If for any reason the software stops/fails at refreshing the timer (and has not explicitly shut it down) before it expires (it runs to timeout), this will trigger a (hardware) reset of the whole system (the computer, in our case). Such a mechanism can even help with recovering from a kernel panic. By the end of this chapter, you will be able to do the following:

- Read/understand an existing watchdog kernel driver and use what it exposes in user space.
- Write new watchdog device drivers.
- Master some not-so-well-known concepts, such as *watchdog governor* and *pretimeout*.

In this chapter, we will also address the concepts behind the Linux kernel watchdog subsystem with the following topics:

- Watchdog data structures and APIs
- The watchdog user space interface

Technical requirements

Before we start walking through this chapter, the following elements are required:

- C programming skills
- Basic electronics knowledge
- Linux kernel v4.19.X sources, available at `https://git.kernel.org/pub/scm/linux/kernel/git/stable/linux.git/refs/tags`

Watchdog data structures and APIs

In this section, we will walk through the watchdog framework and learn how it works under the hood. The watchdog subsystem has a few data structures. The main one is `struct watchdog_device`, which is the Linux kernel representation of a watchdog device, containing all the information about it. It is defined in `include/linux/watchdog.h`, as follows:

```
struct watchdog_device {
    int id;
    struct device *parent;
    const struct watchdog_info *info;
    const struct watchdog_ops *ops;
    const struct watchdog_governor *gov;
    unsigned int bootstatus;
    unsigned int timeout;
    unsigned int pretimeout;
    unsigned int min_timeout;
    struct watchdog_core_data *wd_data;
    unsigned long status;
    [...]
};
```

The following are descriptions of the fields in this data structure:

- `id`: The watchdog's ID allocated by the kernel during the device registration.

- `parent`: Represents the parent for this device.

- `info`: This `struct watchdog_info` structure pointer provides some additional information about the watchdog timer itself. This is the structure that is returned to the user when calling the `WDIOC_GETSUPPORT` ioctl on the watchdog char device in order to retrieve its capabilities. We will introduce this structure later in detail.

- `ops`: A pointer to the list of watchdog operations. Once again, we will introduce this data structure later.

- `gov`: A pointer to the watchdog pretimeout governor. A governor is nothing but a policy manager that reacts according to certain events or system parameters.

- `bootstatus`: The status of the watchdog device at boot. This is a bitmask of reasons that triggered the system reset. Possible values will be enumerated later when describing the `struct watchdog_info` structure.

- `timeout`: This is the watchdog device's timeout value in seconds.

- `pretimeout`: The concept of *pretimeout* can be explained as an event that occurs sometime before the real timeout occurs, so if the system is in an unhealthy state, it triggers an interrupt before the real timeout reset. These interrupts are usually non-maskable ones (**NMI**, which stands for **Non-Maskable Interrupt**), and this can be used to secure important data and shut down specific applications or panic the system (which allows us to gather useful information prior to the reset, instead of a blind and sudden reboot).

  In this context, the `pretimeout` field is actually the time interval (the number of seconds) before triggering the real timeout's interrupt. This is not the number of seconds until the pretimeout. As an example, if you set the timeout to `60` seconds and the pretimeout to `10`, you'll have the pretimeout event triggered in `50` seconds. Setting a pretimeout to `0` disables it.

- `min_timeout` and `max_timeout` are, respectively, the watchdog device's minimum and maximum timeout values (in seconds). These are actually the lower and upper bounds for a valid timeout range. If the values are 0, then the framework will leave a check for the watchdog driver itself.

- `wd_data`: A pointer to the watchdog core internal data. This field must be accessed via the `watchdog_set_drvdata()` and `watchdog_get_drvdata()` helpers.

- `status` is a field that contains the device's internal status bits. Possible values are listed here:

  --`WDOG_ACTIVE`: Tells whether the watchdog is running/active.

  --`WDOG_NO_WAY_OUT`: Informs whether the `nowayout` feature is set. You can use `watchdog_set_nowayout()` to set the `nowayout` feature; its signature is `void watchdog_set_nowayout(struct watchdog_device *wdd, bool nowayout)`.

  --`WDOG_STOP_ON_REBOOT`: Should be stopped on reboot.

  --`WDOG_HW_RUNNING`: Informs that the hardware watchdog is running. You can use the `watchdog_hw_running()` helper to check whether this flag is set or not. However, you should set this flag on the success path of the watchdog's start function (or in the probe function if for any reason you start it there or you discover that the watchdog is already started). You can use the `set_bit()` helper for this.

  --`WDOG_STOP_ON_UNREGISTER`: Specifies that the watchdog should be stopped on unregistration. You can use the `watchdog_stop_on_unregister()` helper to set this flag.

As we introduced it previously, let's delve in detail into the `struct watchdog_info` structure, defined in `include/uapi/linux/watchdog.h`, actually, because it is part of the user space API:

```
struct watchdog_info {
    u32 options;
    u32 firmware_version;
    u8 identity[32];
};
```

This structure is also the one returned to the user space on the success path of the `WDIOC_GETSUPPORT` ioctl. In this structure, the fields have the following meanings:

- `options` represents the supported capabilities of the card/driver. It is a bitmask of the capabilities supported by the watchdog device/driver since some watchdog cards offer more than just a countdown. Some of these flags may also be set in the `watchdog_device.bootstatus` field in response to the `GET_BOOT_STATUS` ioctl. These flags are listed as follows, with dual explanations given where necessary:

  --`WDIOF_SETTIMEOUT` means the watchdog device can have its timeout set. If this flag is set, then a `set_timeout` callback has to be defined.

`--WDIOF_MAGICCLOSE` means the driver supports the magic close char feature. As the closing of the watchdog char device file does not stop the watchdog, this feature means writing a *V* character (also called a magic character or magic *V*) sequence in this watchdog file will allow the next close to turn off the watchdog (if `nowayout` is not set).

`--WDIOF_POWERUNDER` means the device can monitor/detect bad powers or power faults.. When set in `watchdog_device.bootstatus`, this flag means that it is the fact that the machine showed an under-voltage that triggered the reset.

`--WDIOF_POWEROVER`, on the other hand, means the device can monitor the operating voltage. When set in `watchdog_device.bootstatus`, it means the system reset may be due to an over-voltage status. Note that if one level is under and one over, both bits will be set.

`--WDIOF_OVERHEAT` means the watchdog device can monitor the chip/SoC temperature. When set in `watchdog_device.bootstatus`, it means the reason for the last machine reboot via the watchdog was due to exceeding the thermal limit.

`--WDIOF_FANFAULT` informs us that this watchdog device can monitor the fan. When set, it means a system fan monitored by the watchdog card has failed.

Some devices even have separate event inputs. If defined, electrical signals are present on these inputs, which also leads to a reset. This is the aim of `WDIOF_EXTERN1` and `WDIOF_EXTERN2`. When set in `watchdog_device.bootstatus`, it means the machine was last rebooted because of an external relay/source 1 or 2.

`--WDIOF_PRETIMEOUT` means this watchdog device supports a pretimeout feature.

`--WDIOF_KEEPALIVEPING` means this driver supports the `WDIOC_KEEPALIVE` ioctl (it can be pinged via an ioctl); otherwise, the ioctl will return `-EOPNOTSUPP`. When set in `watchdog_device.bootstatus`, this flag means the watchdog saw a keep-alive ping since it was last queried.

`--WDIOF_CARDRESET`: This is a special flag that may appear in `watchdog_device.bootstatus` only. It means the last reboot was caused by the watchdog itself (its timeout, actually).

- `firmware_version` is the firmware version of the card.
- `identity` should be a string describing the device.

The other data structure without which nothing is possible is struct watchdog_ops, defined as follows:

```
struct watchdog_ops { struct module *owner;
    /* mandatory operations */
    int (*start)(struct watchdog_device *);
    int (*stop)(struct watchdog_device *);
    /* optional operations */
    int (*ping)(struct watchdog_device *);
    unsigned int (*status)(struct watchdog_device *);
    int (*set_timeout)(struct watchdog_device *, unsigned int);
    int (*set_pretimeout)(struct watchdog_device *,
                        unsigned int);
    unsigned int (*get_timeleft)(struct watchdog_device *);
    int (*restart)(struct watchdog_device *,
                unsigned long, void *);
    long (*ioctl)(struct watchdog_device *, unsigned int,
                unsigned long);
};
```

The preceding structure contains the list of operations allowed on the watchdog device. Each operation's meaning is presented in the following descriptions:

- start and stop: These are mandatory operations that, respectively, start and stop the watchdog.

- A ping callback is used to send a keep-alive ping to the watchdog. This method is optional. If not defined, then the watchdog will be restarted via the .start operation, as it would mean that the watchdog does not have its own ping method.

- status is an optional routine that returns the status of the watchdog device. If defined, its return value will be sent in response to a WDIOC_GETBOOTSTATUS ioctl.

- set_timeout is the callback to set the watchdog timeout value (in seconds). If defined, you should also set the X option flag; otherwise, any attempt to set the timeout will result in an -EOPNOTSUPP error.

- set_pretimeout is the callback to set the pretimeout. If defined, you should also set the WDIOF_PRETIMEOUT option flag; otherwise, any attempt to set the pretimeout will result in an -EOPNOTSUPP error.

- `get_timeleft` is an optional operation that returns the number of seconds left before a reset.

- `restart`: This is actually the routine to restart the machine (not the watchdog device). If set, you may want to call `watchdog_set_restart_priority()` on the watchdog device in order to set the priority of this restart handler prior to registering the watchdog with the system.

- `ioctl`: You should not implement this callback unless you have to – for example, if you need to handle extra/non-standard ioctl commands. If defined, this method will override the watchdog core default ioctl, unless it returns -`ENOIOCTLCMD`.

This structure contains the callback functions supported by the device according to its capabilities.

Now that we are familiar with the data structures, we can switch to watchdog APIs and particularly see how to register and unregister such a device with the system.

Registering/unregistering a watchdog device

The watchdog framework provides two elementary functions to register/unregister watchdog devices with the system. These are `watchdog_register_device()` and `watchdog_unregister_device()`, and their respective prototypes are the following:

```
int watchdog_register_device(struct watchdog_device *wdd)
void watchdog_unregister_device(struct watchdog_device *wdd)
```

The preceding registration method returns zero on a success path or a negative *errno* code on failure. On the other hand, `watchdog_unregister_device()` performs the reverse operation. In order to no longer bother with unregistration, you can use the managed version of this function, `devm_watchdog_register_device`, whose prototype is as follows:

```
int devm_watchdog_register_device(struct device *dev,
                                   struct watchdog_device *wdd)
```

The preceding managed version will automatically handle unregistration on driver detach.

The registration method (whatever it is, managed or not) will check whether the
`wdd->ops->restart` function is provided and will register this method as a restart
handler. Thus, prior to registering the watchdog device with the system, the driver
should set the restart priority using the `watchdog_set_restart_priority()`
helper, knowing that the priority value of the restart handler should follow the following
guidelines:

- `0`: This is the lowest priority, which means using the watchdog's restart function as
  a last resort; that is, when there is no other restart handler provided in the system.

- `128`: This is the default priority, and means using this restart handler by default if
  no other handler is expected to be available and/or if a restart is sufficient to restart
  the entire system.

- `255`: This is the highest priority, preempting all other handlers.

The device registration should be done only after you have dealt with all the elements
we have discussed; that is, after providing the valid `.info`, `.ops`, and timeout-related
fields of the watchdog device. Prior to all this, memory space should be allocated for the
`watchdog_device` structure. Wrapping this structure in a bigger and per-driver data
structure is good practice, as shown in the following example, which is an excerpt from
`drivers/watchdog/imx2_wdt.c`:

```
[...]
struct imx2_wdt_device {
    struct clk *clk;
    struct regmap *regmap;
    struct watchdog_device wdog;
    bool ext_reset;
};
```

You can see how the watchdog device data structure is embedded in a bigger structure,
`struct imx2_wdt_device`. Now comes the `probe` method, which initializes
everything and sets the watchdog device in the bigger structure:

```
static int init imx2_wdt_probe(struct platform_device *pdev)
{
    struct imx2_wdt_device *wdev;
    struct watchdog_device *wdog; int ret;
    [...]
    wdev = devm_kzalloc(&pdev->dev, sizeof(*wdev), GFP_KERNEL);
```

```
    if (!wdev)
        return -ENOMEM;

    [...]

    Wdog = &wdev->wdog;
    if (imx2_wdt_is_running(wdev)) {
        imx2_wdt_set_timeout(wdog, wdog->timeout);
        set_bit(WDOG_HW_RUNNING, &wdog->status);
    }

    ret = watchdog_register_device(wdog);
    if (ret) {
        dev_err(&pdev->dev, "cannot register watchdog
device\n");
        [...]
    }
    return 0;
}


static int exit imx2_wdt_remove(struct platform_device *pdev)
{
    struct watchdog_device *wdog = platform_get_drvdata(pdev);
    struct imx2_wdt_device *wdev = watchdog_get_drvdata(wdog);
    watchdog_unregister_device(wdog);
    if (imx2_wdt_is_running(wdev)) {
      imx2_wdt_ping(wdog);
      dev_crit(&pdev->dev, "Device removed: Expect reboot!\n");
    }
    return 0;
}
[...]
```

Additionally, the bigger structure can be used in the move method to track the device state, and particularly the watchdog data structure embedded inside. This is what the preceding code excerpt highlights.

So far, we have dealt with watchdog basics, walked through the base data structures, and described the main APIs. Now, we can learn about fancy features such as pretimeouts and governors in order to define the behavior of the system upon the watchdog event.

Handling pretimeouts and governors

The concept of *governor* appears in several subsystems in the Linux kernel (thermal governors, CPUFreq governors, and now watchdog governors). It is nothing but a driver that implements policy management (sometimes in the form of an algorithm) that reacts to some states/events of the system.

The way each subsystem implements its governor drivers may be different from other subsystems, but the main idea remains the same. Moreover, governors are identified by a unique name and the governor (policy manager) in use. They may be changed on the fly, most often from within the sysfs interface.

Now, back to watchdog pretimeouts and governors. Support for them can be added to the Linux kernel by enabling the `CONFIG_WATCHDOG_PRETIMEOUT_GOV` kernel config option. There are actually two watchdog governor drivers in the kernel: `drivers/watchdog/pretimeout_noop.c` and `drivers/watchdog/pretimeout_panic.c`. Their unique names are, respectively, `noop` and `panic`. Either can be used by default by enabling `CONFIG_WATCHDOG_PRETIMEOUT_DEFAULT_GOV_NOOP` or `CONFIG_WATCHDOG_PRETIMEOUT_DEFAULT_GOV_PANIC`.

The main goal of this section is to deliver the pretimeout event to the watchdog governor that is currently active. This can be achieved by means of the `watchdog_notify_pretimeout()` interface, which has the following prototype:

```
void watchdog_notify_pretimeout(struct watchdog_device *wdd)
```

As we have discussed, some watchdog devices generate an IRQ in response to a pretimeout event. The main idea is to call `watchdog_notify_pretimeout()` from within this IRQ handler. Under the hood, this interface will internally find the watchdog governor (by looking for its name in the global list of watchdog governors registered with the system) and call its `.pretimeout` callback.

Just for your information, the following is what a watchdog governor structure looks like (you can find more information on watchdog governor drivers by looking at the source in `drivers/watchdog/pretimeout_noop.c` or `drivers/watchdog/pretimeout_panic.c`):

```
struct watchdog_governor {
    const char name[WATCHDOG_GOV_NAME_MAXLEN];
```

```
    void (*pretimeout)(struct watchdog_device *wdd);
};
```

Obviously, its fields have to be filled in by the underlying watchdog governor driver. For the real usage of a pretimeout notification, you can refer to the IRQ handler of the i.MX6 watchdog driver, defined in `drivers/watchdog/imx2_wdt.c`. An excerpt of this was shown earlier in the previous section. There, you will notice that `watchdog_notify_pretimeout()` gets called from within the watchdog (the pretimeout, actually) IRQ handler. Moreover, you will notice that the driver uses a different `watchdog_info` structure depending on whether there is a valid IRQ for the watchdog. If there is a valid one, the structure with the `WDIOF_PRETIMEOUT` flag set in `.options` is used, meaning that the device has a pretimeout feature. Otherwise, it uses the structure without the `WDIOF_PRETIMEOUT` flag set.

Now that we are familiar with the concept of governor and pretimeout, we can think about learning an alternative way of implementing watchdogs, such as GPIO-based ones.

GPIO-based watchdogs

Sometimes, it may be better to use an external watchdog device instead of the one provided by the SoC itself, such as for power efficiency reasons, for example, as there are SoCs whose internal watchdog requires much more power than external ones. Most of the time, if not always, this kind of external watchdog device is controlled through a GPIO line and has the possibility to reset the system. It is pinged by toggling the GPIO line to which it is connected. This kind of configuration is used in UDOO QUAD (not checked on other UDOO variants).

The Linux kernel is able to handle this device by enabling the `CONFIG_GPIO_WATCHDOG` config option, which will pull the underlying driver, `drivers/watchdog/gpio_wdt.c`. If enabled, it will periodically *ping* the hardware connected to the GPIO line by toggling it from `1-to-0-to-1`. If that hardware does not receive its ping periodically, it will reset the system. You should use this instead of talking directly to the GPIOs using sysfs; it offers a better sysfs user space interface than the GPIO and it integrates with kernel frameworks better than your user space code could.

The support for this comes from the device tree only, and better documentation on its binding can be found in `Documentation/devicetree/bindings/watchdog/gpio-wdt.txt`, obviously from within the kernel sources.

The following is a binding example:

```
watchdog: watchdog {
    compatible = "linux,wdt-gpio";
```

```
    gpios = <&gpio3 9 GPIO_ACTIVE_LOW>;
    hw_algo = "toggle";
    hw_margin_ms = <1600>;
};
```

The `compatible` property must always be `linux,wdt-gpio`. `gpios` is a GPIO specifier that controls the watchdog device. `hw_algo` should be either `toggle` or `level`. The former means that either low-to-high or high-to-low transitions should be used to ping the external watchdog device, and that the watchdog is disabled when the GPIO line is left floating or connected to a three-state buffer. To achieve this, configuring the GPIO as input is sufficient. The second `algo` means that applying a signal level (high or low) is enough to ping the watchdog.

The way it works is the following: when user space code pings the watchdog through the `/dev/watchdog` device file, the underlying driver (`gpio_wdt.c`, actually) will either toggle the GPIO line (`1-0-1` if `hw_algo` is `toggle`) or assign a specific level (high or low if `hw_algo` is `level`) on that GPIO line. For example, the UDOO QUAD uses `APX823-31W5`, a GPIO-controlled watchdog, whose event output is connected to the i.MX6 PORB line (reset line, actually). Its schematic is available here: `http://udoo.` `org/download/files/schematics/UDOO_REV_D_schematics.pdf`.

Now, we are done with the watchdog on the kernel side. We went through the underlying data structure, dealt with its APIs, introduced the concept of pretimeouts, and even dealt with the GPIO-based watchdog alternative. In the next section, we will look into user space implementation, which is a kind of consumer of the watchdog services.

The watchdog user space interface

On Linux-based systems, the standard user space interface to the watchdog is the `/dev/` `watchdog` file, through which a daemon will notify the kernel watchdog driver that the user space is still alive. The watchdog starts right after the file is opened, and gets pinged by periodically writing into this file.

When the notification occurs, the underlying driver will notify the watchdog device, which will result in resetting its timeout; the watchdog will then wait for yet another `timeout` duration prior to resetting the system. However, if for any reason the user space does not perform the notification before the timeout is elapsed, the watchdog will reset the system (causing a reboot). This mechanism provides a way to enforce the system availability. Let's start with the basics, learning how to start and stop the watchdog.

Starting and stopping the watchdog

The watchdog is automatically started once you open the /dev/watchdog device file, as in the following example:

```
int fd;
fd = open("/dev/watchdog", O_WRONLY);

if (fd == -1) {
    if (errno == ENOENT)
        printf("Watchdog device not enabled.\n");
    else if (errno == EACCES)
        printf("Run watchdog as root.\n");
    else
        printf("Watchdog device open failed %s\n",
strerror(errno));

    exit(-1);
}
```

Only, closing the watchdog device file does not stop it. You will be surprised to face a system reset after closing the file. To properly stop the watchdog, you will first need to write the magic character *V* into the watchdog device file. This instructs the kernel to turn off the watchdog next time the device file is closed, as shown:

```
const char v = 'V';
printf("Send magic character: V\n"); ret = write(fd, &v, 1);
if (ret < 0)
    printf("Stopping watchdog ticks failed (%d)...\n", errno);
```

Then, you'll need to close the watchdog device file in order to stop it:

```
printf("Close for stopping..\n");
close(fd);
```

> **Important note**
>
> There is an exception when stopping the watchdog by closing the file device: it is when the kernel's `CONFIG_WATCHDOG_NOWAYOUT` config option is enabled. When this option is enabled, the watchdog cannot be stopped at all. Hence, you will need to service it all the time or it will reset the system. Moreover, the watchdog driver should have set the `WDIOF_MAGICCLOSE` flag in its option; otherwise, the magic close feature won't work.

Now that we have seen how to start and stop the watchdog, it is time to learn how to refresh the device in order to prevent the system from suddenly rebooting.

Pinging/kicking the watchdog – sending keep-alive pings

There are two ways to kick or feed the watchdog:

1. Writing any character into `/dev/watchdog`: A write to the watchdog device file is defined as a keep-alive ping. It is recommended not to write a `V` character at all (as it has a particular meaning), even if it is in a string.

2. Using the `WDIOC_KEEPALIVE` ioctl, `ioctl(fd, WDIOC_KEEPALIVE, 0);`: The argument to the ioctl is ignored. The watchdog driver should have set the `WDIOF_KEEPALIVEPING` flag in its options prior to this ioctl so that it works.

It is good practice to feed the watchdog every half of its timeout value. This means if its timeout is `30s`, you should feed it every `15s`. Now, let's learn about gathering some information on how the watchdog is ruling our system.

Getting watchdog capabilities and identity

Getting the watchdog capabilities and/or identity consists of grabbing the underlying `struct watchdog_info` structure associated with the watchdog. If you remember, this info structure is mandatory and is provided by the watchdog driver.

To achieve this, you need to use the `WDIOC_GETSUPPORT` ioctl. The following is an example:

```
struct watchdog_info ident;
ioctl(fd, WDIOC_GETSUPPORT, &ident);
printf("WDIOC_GETSUPPORT:\n");

/* Printing the watchdog's identity, its unique name actually
```

```
*/
printf("\tident.identity = %s\n",ident.identity);

/* Printing the firmware version */
printf("\tident.firmware_version = %d\n",
        ident.firmware_version);

/* Printing supported options (capabilities) in hex format */
printf("WDIOC_GETSUPPORT: ident.options = 0x%x\n",
        ident.options);
```

We can go further by testing some fields in the capabilities, as follows:

```
if (ident.options & WDIOF_KEEPALIVEPING)
    printf("\tKeep alive ping reply.\n");

if (ident.options & WDIOF_SETTIMEOUT)
    printf("\tCan set/get the timeout.\n");
```

You can (or should I say "must") use this in order to check the watchdog features prior to performing certain actions on it. Now, we can go further and learn how to get and set more fancy watchdog properties.

Setting and getting the timeout and pretimeout

Prior to setting/getting the timeout, the watchdog info should have the WDIOF_ SETTIMEOUT flag set. There are drivers with which it is possible to modify the watchdog timeout on the fly using the WDIOC_SETTIMEOUT ioctl. These drivers must have the WDIOF_SETTIMEOUT flag set in their watchdog info structure and provide a .set_timeout callback.

While the argument here is an integer representing the timeout value in seconds, the return value is the real timeout applied to the hardware device, as it may differ from the one requested in the ioctl due to hardware limitations:

```
int timeout = 45;
ioctl(fd, WDIOC_SETTIMEOUT, &timeout);
printf("The timeout was set to %d seconds\n", timeout);
```

When it comes to querying the current timeout, you should use the WDIOC_
GETTIMEOUT ioctl, as in the following example:

```
int timeout;
ioctl(fd, WDIOC_GETTIMEOUT, &timeout);
printf("The timeout is %d seconds\n", timeout);
```

Finally, when it comes to the pretimeout, the watchdog driver should have set
WDIOF_PRETIMEOUT in the options and provided a .set_pretimeout callback in its
ops. You should then use WDIOC_SETPRETIMEOUT with the pretimeout value as
a parameter:

```
pretimeout = 10;
ioctl(fd, WDIOC_SETPRETIMEOUT, &pretimeout);
```

If the desired pretimeout value is either 0 or bigger than the current timeout, you will get
an -EINVAL error.

Now that we have seen how to get and set the timeout/pretimeout on the watchdog device,
we can learn how to get the time left before the watchdog fires.

Getting the time left

The WDIOC_GETTIMELEFT ioctl allows checking how much time is left on the watchdog
counter before a reset occurs. Moreover, the watchdog driver should support this feature
by providing a .get_timeleft() callback; otherwise, you'll have an EOPNOTSUPP
error. The following is an example showing how to use this ioctl:

```
int timeleft;
ioctl(fd, WDIOC_GETTIMELEFT, &timeleft);
printf("The remaining timeout is %d seconds\n", timeleft);
```

The timeleft variable is filled on the return path of the ioctl.

Once the watchdog fires, it triggers a reboot when it is configured to do so. In the next
section, we will learn how to get the last reboot reason, in order to see whether the reboot
was caused by the watchdog.

Getting the (boot/reboot) status

There are two ioctl commands to play with in this section. These are WDIOC_GETSTATUS and WDIOC_GETBOOTSTATUS. The way that these are handled depends on the driver implementation, and there are two types of driver implementations:

- Old drivers that provide watchdog features through a miscellaneous device. These drivers do not use the generic watchdog framework interface and provide their own file_ops along with their own .ioctl ops. Moreover, these drivers only support WDIOC_GETSTATUS, while others may support both WDIOC_GETSTATUS and WDIOC_GETBOOTSTATUS. The difference between the two is that the former will return the raw content of the device's status register, while the latter is supposed to be a bit smarter as it parses the raw content and only returns the boot status flag. These drivers need to be migrated to the new generic watchdog framework. Note that some of those drivers supporting both commands may return the same value (the same case statement) for both ioctls, while others may return a different one (each command has its own case statement).

- New drivers use the generic watchdog framework. These drivers rely on the framework and do not care about file_ops anymore. Everything is done from within the drivers/watchdog/watchdog_dev.c file (you can have a look, especially at how the ioctl command is implemented). With these kinds of drivers, WDIOC_GETSTATUS and WDIOC_GETBOOTSTATUS are handled by the watchdog core separately. This section will deal with these drivers.

Now, let's focus on the generic implementation. For these drivers, WDIOC_GETBOOTSTATUS will return the value of the underlying watchdog_device.bootstatus field. For WDIOC_GETSTATUS, if the watchdog .status ops is provided, it will be called and its return value will be copied to the user; otherwise, the content of watchdog_device.bootstatus will be tweaked with an AND operation in order to clear (or flag) the bits that are not meaningful. The following code snippet shows how it is done in kernel space:

```
static unsigned int watchdog_get_status(struct
                                        watchdog_device *wdd)
{
    struct watchdog_core_data *wd_data = wdd->wd_data;
    unsigned int status;

    if (wdd->ops->status)
        status = wdd->ops->status(wdd);
    else
```

```
          status = wdd->bootstatus &
                         (WDIOF_CARDRESET | WDIOF_OVERHEAT |
                         WDIOF_FANFAULT | WDIOF_EXTERN1 |
                         WDIOF_EXTERN2 | WDIOF_POWERUNDER |
                         WDIOF_POWEROVER);

      if (test_bit(_WDOG_ALLOW_RELEASE, &wd_data->status))
          status |= WDIOF_MAGICCLOSE;

      if (test_and_clear_bit(_WDOG_KEEPALIVE, &wd_data->status))
          status |= WDIOF_KEEPALIVEPING;

      return status;
}
```

The preceding code is a generic watchdog core function to get the watchdog status. It is actually a wrapper that is responsible for calling the underlying `ops.status` callback. Now, back to our user space usage. We can do the following:

```
int flags = 0;
int flags;
ioctl(fd, WDIOC_GETSTATUS, &flags);
/* or ioctl(fd, WDIOC_GETBOOTSTATUS, &flags); */
```

Obviously, we can proceed to individual flag checking as we did earlier in the *Getting watchdog capabilities and identity* section.

So far, we have written code to play with the watchdog device. The next section will show us how to deal with the watchdog from user space without writing code, essentially using the sysfs interface.

The watchdog sysfs interface

The watchdog framework offers the possibility of managing watchdog devices from user space through the sysfs interface. This is possible if the CONFIG_WATCHDOG_SYSFS config option is enabled in the kernel, and the root directory is /sys/class/watchdogX/. X is the index of the watchdog device in the system. Each watchdog directory in sysfs has the following:

- nowayout: Gives 1 if the device supports the nowayout feature, and 0 otherwise.
- status: This is the sysfs equivalent of the WDIOC_GETSTATUS ioctl. This sysfs file reports the watchdog's internal status bits.
- timeleft: This is the sysfs equivalent of the WDIOC_GETTIMELEFT ioctl. This sysfs entry returns the time (the number of seconds, actually) left before the watchdog resets the system.
- timeout: Gives the current value of the timeout programmed.
- identity: Contains an identity string of the watchdog device.
- bootstatus: This is the sysfs equivalent of the WDIOC_GETBOOTSTATUS ioctl. This entry informs whether the system reset was caused by the watchdog device or not.
- state: Gives the active/inactive status of the watchdog device.

Now that the preceding watchdog properties have been described, we can focus on pretimeout management from user space.

Handling a pretimeout event

Setting the governor is done via sysfs. A governor is nothing but a policy manager that takes certain actions depending on some external (but input) parameters. There are thermal governors, CPUFreq governors, and now watchdog governors. Each governor is implemented in its own driver.

You can check the available governors for a watchdog (let's say watchdog0) using the following command:

```
# cat /sys/class/watchdog/watchdog0/pretimeout_available_
governors
noop panic
```

Now, we can check whether pretimeout governors can be selected:

```
# cat /sys/class/watchdog/watchdog0/pretimeout_governor
panic
# echo -n noop > /sys/class/watchdog/watchdog0/pretimeout_
governor
# cat /sys/class/watchdog/watchdog0/pretimeout_governor
noop
```

To check the pretimeout value, you can simply do the following:

```
# cat /sys/class/watchdog/watchdog0/pretimeout
10
```

Now we are familiar with using the watchdog sysfs interface from the user space. Though we are not in the kernel, we can leverage the whole framework, particularly playing with the watchdog parameters.

Summary

In this chapter, we discussed all aspects of watchdog devices: their APIs, the GPIO alternative, and how they help keep the system reliable. We saw how to start, how (when it is possible) to stop, and how to service the watchdog devices. Moreover, we introduced the concept of pretimeout and watchdog-dedicated governors.

In the next chapter, we will discuss some Linux kernel development and debugging tips, such as analyzing kernel panic messages and kernel tracing.

14
Linux Kernel Debugging Tips and Best Practices

Most of the time, as part of development, writing code is not the hardest part. Things are rendered difficult by the fact that the Linux kernel is a standalone software that is at the lowest layer of the operating system. This makes it challenging to debug the Linux kernel. However, this is compensated by the fact that the majority of the time, we don't need additional tools to debug kernel code because most of the kernel debugging tools are part of the kernel itself. We will begin by familiarizing ourselves with the Linux kernel release model and you will learn the Linux kernel release process and steps. Then, we will look at the Linux kernel debugging-related development tips (especially debugging by printing) and finally, we will focus on tracing the Linux kernel, ending with off-target debugging and learning to leverage kernel oops.

This chapter will cover the following topics:

- Understanding the Linux kernel release process
- Linux kernel development tips
- Linux kernel tracing and performance analysis
- Linux kernel debugging tips

Technical requirements

The following are prerequisites for this chapter:

- Advanced computer architecture knowledge and C programming skills
- Linux kernel v4.19.X sources, available at `https://git.kernel.org/pub/scm/linux/kernel/git/stable/linux.git/refs/tags`

Understanding the Linux kernel release process

According to the Linux kernel release model, there are always three types of active kernel release: mainline, the stable release, and the **Long-Term Support** (**LTS**) release. First, bug fixes and new features are gathered and prepared by subsystem maintainers and then submitted to Linus Torvalds in order for him to include them in his own Linux tree, which is called the *mainline Linux tree*, also known as the *master* Git repository. This is where every stable release originates from.

Before each new kernel version is released, it is submitted to the community through *release candidate* tags, so that developers can test and polish all the new features and, most importantly, share feedback. During this cycle, Linus will rely on the feedback in order to decide whether the final version is ready to be released. When he is convinced that the new kernel is ready to go, he makes (tags it actually) the final release, and we call this release *stable* to indicate that it's no longer a *release candidate*: those releases are *vX.Y* versions.

There is no strict timeline for making releases. However, new mainline kernels are generally released every 2–3 months. Stable kernel releases are based on Linus's releases, that is, the mainline tree releases.

Once a mainline kernel is released by Linus, it also appears in the *linux-stable* tree (available at `https://git.kernel.org/pub/scm/linux/kernel/git/stable/linux.git/`), where it becomes a branch and from where it can receive bug fixes for a stable release. *Greg Kroah-Hartman* is responsible for maintaining this tree, which is also referred to as the stable tree because it is used to track previously released stable kernels. That said, in order for a fix to be applied to this tree, this fix must first be incorporated in the Linus tree. Because the fix must go forth before coming back, it is said that this fix is back-ported. Once the bug is fixed in the mainline repository, it can then be applied to previously released kernels that are still maintained by the kernel development community. All fixes back-ported to stable releases must meet a set of mandatory acceptance criteria — and one of these criteria is that they **must already exist in Linus's tree**.

> **Important note**
> Bugfix kernel releases are considered stable.

For example, the `4.9` kernel is released by Linus, and then the stable kernel releases based on this kernel are numbered `4.9.1`, `4.9.2`, `4.9.3`, and so on. Such releases are known as *bugfix kernel releases*, and the sequence is usually shortened with the number *4.9.y* when referring to their branch in the stable kernel release tree. Each stable kernel release tree is maintained by a single kernel developer, who is responsible for picking the requisite patches for the release, and for performing the review/release process. There are usually only a few bugfix kernel releases until the next mainline kernel becomes available, unless it is designated a *long-term maintenance kernel*.

Every subsystem and kernel maintainer repository is hosted here: `https://git.kernel.org/pub/scm/linux/kernel/git/`. There, we can also find either Linus or stable trees. In the Linus tree (`https://git.kernel.org/pub/scm/linux/kernel/git/torvalds/linux.git/`), there is only one branch in Linus's tree, that is, the master branch. Tags in there are either stable releases or release candidates. In the stable tree (`https://git.kernel.org/pub/scm/linux/kernel/git/stable/linux.git/`), there is one branch per stable kernel release (named *<A.B>.y*, where *<A.B>* is the release version in the Linus tree) and each branch contains its bugfix kernel releases.

> **Important note**
>
> There are a few links that you can keep to hand in order to follow the Linux kernel release. The first one is `https://www.kernel.org/`, from where you can download kernel archives, and then there is `https://www.kernel.org/category/releases.html`, from where you can access the latest LTS kernel releases and their support timelines. You can also refer to this link, `https://patchwork.kernel.org/`, from where you can follow kernel patch submissions on a subsystem basis.

Now that we are familiar with the Linux kernel release model, we can delve into some development tips and best practices, which helps to consolidate and leverage other kernel developer experiences.

Linux kernel development tips

The best Linux kernel development practices are inspired by existing kernel code. This way, you could certainly learn good practices. That said, we will not reinvent the wheel. We will focus on what is necessary for this chapter, that is, debugging. The most frequently used debugging method involves logging and printing. In order to leverage this time-tested debugging technique, the Linux kernel provides suitable logging APIs and exposes a kernel message buffer to store the logs. Though it may seem obvious, we will focus on the kernel logging APIs and learn how to manage the message buffer, either from within the kernel code or from user space.

Message printing

Message printing and logging are inherent to development, irrespective of whether we are in kernel space or user space. In a kernel, the `printk()` function has long since been the de facto kernel message printing function. It is similar to `printf()` in the C library, but with the concept of log levels.

If you look at an example of actual driver code, you'll notice it is used as follows:

```
printk(<LOG_LEVEL> "printf like formatted message\n");
```

Here, `<LOG_LEVEL>` is one of the eight different log levels defined in `include/linux/kern_levels.h` and specifies the severity of the error message. You should also note that there is no comma between the log level and the format string (as the preprocessor concatenates both strings).

Kernel log levels

The Linux kernel uses the concept of levels to determine how critical the message is. There are eight of them, each defined as a string, and they are described as follows:

- KERN_EMERG, defined as "0". It is to be used for emergency messages, meaning the system is about to crash or is unstable (unusable).

- KERN_ALERT, defined as "1", meaning that something bad happened and action must be taken immediately.

- KERN_CRIT, defined as "2", meaning that a critical condition occurred, such as a serious hardware/software failure.

- KERN_ERR, defined as "3" and used during an error condition, often used by drivers to indicate difficulties with the hardware or a failure to interact with a subsystem.

- KERN_WARNING, defined as "4" and used as a warning, meaning nothing serious by itself, but may indicate problems.

- KERN_NOTICE, defined as "5", meaning nothing serious, but notable nevertheless. This is often used to report security events.

- KERN_INFO, defined as "6", used for informational messages, for example, startup information at driver initialization.

- KERN_DEBUG, defined as "7", used for debugging purposes, and active only if the DEBUG kernel option is enabled. Otherwise, its content is simply ignored.

If you don't specify a log level in your message, it defaults to DEFAULT_MESSAGE_LOGLEVEL (usually "4" = KERN_WARNING), which can be set via the CONFIG_DEFAULT_MESSAGE_LOGLEVEL kernel configuration option.

That said, for new drivers, you are encouraged to use more convenient printing APIs, which embed the log level in their names. Those printing helpers are pr_emerg, pr_alert, pr_crit, pr_err, pr_warning, pr_warn, pr_notice, pr_info, pr_debug, or pr_dbg. Besides being more concise than the equivalent printk() calls, they can use a common definition for the format string through the pr_fmt() macro; for instance, defining this at the top of a source file (before any #include directive):

```
#define pr_fmt(fmt) "%s:%s: " fmt, KBUILD_MODNAME, __func__
```

This would prefix every pr_*() message in that file with the module and function name that originated the message. pr_devel and pr_debug are replaced with printk(KERN_DEBUG ...) if the kernel was compiled with DEBUG, otherwise they are replaced with an empty statement.

The pr_*() family macros are to be used in core code. For device drivers, you should use the device-related helpers, which also accept the concerned device structure as a parameter. They also print the name of the relevant device in standard form, ensuring that it's always possible to associate a message with the device that generated it:

```
dev_emerg(const struct device *dev, const char *fmt, ...);
dev_alert(const struct device *dev, const char *fmt, ...);
dev_crit(const struct device *dev, const char *fmt, ...);
dev_err(const struct device *dev, const char *fmt, ...);
dev_warn(const struct device *dev, const char *fmt, ...);
dev_notice(const struct device *dev, const char *fmt, ...);
dev_info(const struct device *dev, const char *fmt, ...);
dev_dbg(const struct device *dev, const char *fmt, ...);
```

While the concept of log levels is used by the kernel to determine the importance of a message, it is also used to decide whether this message should be presented to the user immediately, by printing it to the current console (where the console could also be a serial line or even a printer, not an xterm).

In order to decide, the kernel compares the log level of the message with the console_loglevel kernel variable, and if the message log level importance is higher (that is, a lower value) than console_loglevel, the message will be printed to the current console. Since the default kernel log level is usually "4", this is the reason why you don't see pr_info() or pr_notice() or even pr_warn() messages on the console, as they have higher or equal values (which means lower priority) than the default one.

To determine the current console_loglevel on your system, you can simply type the following:

```
$ cat /proc/sys/kernel/printk
4       4    1    7
```

The first integer (4) is the current console log level, the second number (4) is the default one, the third number (1) is the minimum console log level that can be set, and the fourth number (7) is the boot-time default console log level.

To change your current console_loglevel, simply write to the same file, that is, /proc/sys/kernel/printk. Hence, in order to get all messages printed to the console, perform the following simple command:

```
# echo 8 > /proc/sys/kernel/printk
```

Every kernel message will appear on your console. You'll then have the following content:

```
# cat /proc/sys/kernel/printk
8    4    1    7
```

Another way to change the console log level is to use dmesg with the -n parameter:

```
# dmesg -n 5
```

With the preceding command, console_loglevel is set to print KERN_WARNING (4) or more severe messages. You can also specify the console_loglevel at boot time using the loglevel boot parameter (refer to Documentation/kernel-parameters.txt for more details).

> **Important note**
>
> There are also KERN_CONT and pr_cont, which are sort of special since they do not specify a level of urgency, but rather indicate a continued message. They should only be used by core/arch code during early bootup (a continued line is not SMP-safe otherwise). This can be useful when part of a message line to be printed depends on the result of a computation, as in the following example:

```
[...]
pr_warn("your last operation was ");
if (success)
    pr_cont("successful\n");
else
    pr_cont("NOT successful\n");
```

You should keep in mind that only the final print statement has the trailing \n character.

Kernel log buffer

Whether they are immediately printed on the console or not, each kernel message is logged in a buffer. This kernel message buffer is a fixed-size circular buffer, which means that if the buffer fills up, it wraps around and you may lose a message. Thus, increasing the buffer size could be helpful. In order to change the kernel message buffer size, you can play with the LOG_BUF_SHIFT option, the value of which is used to left-shift by 1 in order to obtain the final size, the kernel log buffer size (for example, 16 => 1<<16 => 64KB, 17 => 1 << 17 => 128KB). That said, it is a static size defined at compile time. This size can also be defined through kernel boot parameters, by using the log_buf_len parameter, in other words, log_buf_len=1M (accept only power of 2 values).

Adding timing information

Sometimes, it is useful to add timing information to the printed messages, so you can see when a particular event occurred. The kernel includes a feature for doing this, called printk times, enabled through the CONFIG_PRINTK_TIME option. This option is found on the **Kernel Hacking** menu when configuring the kernel. Once enabled, this timing information prefixes each log message as follows:

```
$ dmesg
[…]
[    1.260037] loop: module loaded
[    1.260194] libphy: Fixed MDIO Bus: probed
[    1.260195] tun: Universal TUN/TAP device driver, 1.6
[    1.260224] PPP generic driver version 2.4.2
[    1.260260] ehci_hcd: USB 2.0 'Enhanced' Host Controller
(EHCI) Driver
[    1.260262] ehci-pci: EHCI PCI platform driver
[    1.260775] ehci-pci 0000:00:1a.7: EHCI Host Controller
[    1.260780] ehci-pci 0000:00:1a.7: new USB bus registered,
assigned bus number 1
[    1.260790] ehci-pci 0000:00:1a.7: debug port 1
[    1.264680] ehci-pci 0000:00:1a.7: cache line size of 64 is
not supported
[    1.264695] ehci-pci 0000:00:1a.7: irq 22, io mem 0xf7ffa000
[    1.280103] ehci-pci 0000:00:1a.7: USB 2.0 started, EHCI
1.00
[    1.280146] usb usb1: New USB device found, idVendor=1d6b,
idProduct=0002
[    1.280147] usb usb1: New USB device strings: Mfr=3,
```

```
Product=2, SerialNumber=1
[…]
```

The timestamps that are inserted into the kernel message output consist of seconds and microseconds (`seconds.microseconds` actually) as absolute values from the start of machine operation (or from the start of kernel timekeeping), which corresponds to the time when the bootloader passes control to the kernel (when you see something like [0.000000] Booting Linux on physical CPU 0x0 on the console).

Printk times can be controlled at runtime by writing to `/sys/module/printk/parameters/time` in order to enable and disable `printk` timestamps. The following are examples:

```
# echo 1 >/sys/module/printk/parameters/time
# cat /sys/module/printk/parameters/time
N
# echo 1 >/sys/module/printk/parameters/time
# cat /sys/module/printk/parameters/time
Y
```

It does not control whether the timestamp is logged. It only controls whether it is printed while the kernel message buffer is being dumped, at boot time, or while using `dmesg`. This may be an area for boot-time optimization. If disabled, it would take less time for logs to be printed.

We are now familiar with kernel printing APIs and their log buffer. We have seen how to tweak the message buffer, and add or remove information according to requirements. Those skills can be used for debugging by printing. However, other debugging and tracing tools are shipped in the Linux kernel, and the following section will introduce some of them.

Linux kernel tracing and performance analysis

Though debugging by printing covers most of the debugging needs, there are situations where we need to monitor the Linux kernel at runtime to track strange behavior, including latencies, CPU hogging, scheduling issues, and so on. In the Linux world, the most useful tool for achieving this is part of the kernel itself. The most important is `ftrace`, which is a Linux kernel internal tracing tool, and is the main topic of this section.

Using Ftrace to instrument the code

Function Trace, in short **Ftrace**, does much more than what its name says. For example, it can be used to measure the time it takes to process interrupts, to track time-consuming functions, calculate the time to activate high-priority tasks, to track context switches, and much more.

Developed by *Steven Rostedt*, Ftrace has been included in the kernel since version 2.6.27 in 2008. This is the framework that provides a debugging ring buffer for recording data. This data is gathered by the kernel's integrated tracing programs. Ftrace works on top of the `debugfs` filesystem and is, most of the time, mounted in its own directory called `tracing` when it is enabled. In most modern Linux distributions, it is mounted by default in the `/sys/kernel/debug/` directory (this is only available to the root user), meaning that you can leverage Ftrace from within `/sys/kernel/debug/tracing/`.

The following are the kernel options to be enabled in order to support Ftrace on your system:

```
CONFIG_FUNCTION_TRACER
CONFIG_FUNCTION_GRAPH_TRACER
CONFIG_STACK_TRACER
CONFIG_DYNAMIC_FTRACE
```

The preceding options depend on the architecture supporting tracing features by having the `CONFIG_HAVE_FUNCTION_TRACER`, `CONFIG_HAVE_DYNAMIC_FTRACE`, and `CONFIG_HAVE_FUNCTION_GRAPH_TRACER` options enabled.

To mount the `tracefs` directory, you can add the following line to your `/etc/fstab` file:

```
tracefs    /sys/kernel/debug/tracing    tracefs defaults    0    0
```

Or you can mount it at runtime with the help of the following command:

```
mount -t tracefs nodev /sys/kernel/debug/tracing
```

The contents of the directory should look like this:

```
# ls /sys/kernel/debug/tracing/
README                          set_event_pid
available_events                set_ftrace_filter
available_filter_functions      set_ftrace_notrace
available_tracers               set_ftrace_pid
```

`buffer_size_kb`	`set_graph_function`
`buffer_total_size_kb`	`set_graph_notrace`
`current_tracer`	`snapshot`
`dyn_ftrace_total_info`	`stack_max_size`
`enabled_functions`	`stack_trace`
`events`	`stack_trace_filter`
`free_buffer`	`trace`
`function_profile_enabled`	`trace_clock`
`instances`	`trace_marker`
`max_graph_depth`	`trace_options`
`options`	`trace_pipe`
`per_cpu`	`trace_stat`
`printk_formats`	`tracing_cpumask`
`saved_cmdlines`	`tracing_max_latency`
`saved_cmdlines_size`	`tracing_on`
`set_event`	`tracing_thresh`

We won't describe all of these files and subdirectories, as this has already been covered in the official documentation. Instead, we'll just briefly describe the files relevant to our context:

- `available_tracers`: Available tracing programs.
- `tracing_cpumask`: This allows selected CPUs to be traced. The mask should be specified in a hex string format. For example, to trace only core 0, you should include a 1 in this file. To trace core 1, you should include a 2 in there. For core 3, the number 8 should be included.
- `current_tracer`: The tracing program that is currently running.
- `tracing_on`: The system file responsible for enabling or disabling data writing to the ring buffer (to enable this, the number 1 has to be added to the file; to disable it, the number 0 is added).
- `trace`: The file where tracing data is saved in a human-readable format.

Now that we have introduced Ftrace and described its functions, we can delve into its usage and learn how useful it can be for tracing and debugging purposes.

Available tracers

We can view the list of available tracers with the following command:

```
# cat /sys/kernel/debug/tracing/available_tracers
blk function_graph wakeup_dl wakeup_rt wakeup irqsoff function
nop
```

Let's take a quick look at the features of each tracer:

- `function`: A function call tracer without arguments.

- `function_graph`: A function call tracer with subcalls.

- `blk`: A call and event tracer related to block device I/O operations (this is what `blktrace` uses).

- `mmiotrace`: A memory-mapped I/O operation tracer. It traces all the calls that a module makes to the hardware. It is enabled with `CONFIG_ MMIOTRACE`, which depends on `CONFIG_HAVE_MMIOTRACE_SUPPORT`.

- `irqsoff`: Traces the areas that disable interrupts and saves the trace with the longest maximum latency. This tracer depends on `CONFIG_IRQSOFF_TRACER`.

- `preemptoff`: Depends on `CONFIG_PREEMPT_TRACER`. It is similar to `irqsoff`, but traces and records the amount of time for which preemption is disabled.

- `preemtirqsoff`: Similar to `irqsoff` and `preemptoff`, but it traces and records the largest time for which irqs and/or preemption is disabled.

- `wakeup` and `wakeup_rt`, enabled by `CONFIG_SCHED_TRACER`: The former traces and records the maximum latency that it takes for the highest priority task to get scheduled after it has been woken up, while the latter traces and records the maximum latency that it takes for just **real-time** (**RT**) tasks (in the same way as the current `wakeup` tracer does).

- `nop`: The simplest tracer, which, as the name suggests, doesn't do anything. The `nop` tracer simply displays the output of `trace_printk()` calls.

`irqsoff`, `preemptoff`, and `preemtirqsoff` are the so-called latency tracers. They measure how long interrupts are disabled for, how long preemption is disabled for, and how long interrupts and/or preemption are disabled for. Wakeup latency tracers measure how long it takes a process to run after it has been awoken for either all tasks or just RT tasks.

The function tracer

We'll begin our introduction to Ftrace with the function tracer. Let's look at a test script:

```
# cd /sys/kernel/debug/tracing
# echo function > current_tracer
# echo 1 > tracing_on
# sleep 1
# echo 0 > tracing_on
# less trace
```

This script is fairly straightforward, but there are a few things worth noting. We enable the current tracer by writing its name to the `current_tracer` file. Next, we write a `1` to `tracing_on`, which enables the ring buffer. The syntax requires a space between `1` and the `>` symbol; `echo1> tracing_on` will not work. One line later, we disable it (if `0` is written to `tracing_on`, the buffer won't clear and Ftrace won't be disabled).

Why would we do this? Between the two `echo` commands, we see the `sleep 1` command. We enable the buffer, run this command, and then disable it. This lets the tracer include information relating to all of the system calls that occur while the command runs. In the last line of the script, we give the command to display tracing data in the console. Once the script has run, we'll see the following printout (this is just a small fragment):

```
# entries-in-buffer/entries-written: 72097/184701    #P:1
#
#                                _-----=> irqs-off
#                               / _----=> need-resched
#                              | / _---=> hardirq/softirq
#                              || / _--=> preempt-depth
#                              ||| /     delay
#           TASK-PID   CPU#   ||||    TIMESTAMP  FUNCTION
#              | |       |    ||||       |          |
     mmcqd/0-917    [000] d.h5   413.431967: irq_may_run <-handle_fasteoi_irq
     mmcqd/0-917    [000] d.h5   413.431967: handle_irq_event <-handle_fasteoi_irq
     mmcqd/0-917    [000] d.h5   413.431967: preempt_count_sub <-handle_irq_event
     mmcqd/0-917    [000] d.h4   413.431967: handle_irq_event_percpu <-handle_irq_event
     mmcqd/0-917    [000] d.h4   413.431967: dw_mci_interrupt <-handle_irq_event_percpu
     mmcqd/0-917    [000] d.h4   413.431967: dw_mci_cmd_interrupt <-dw_mci_interrupt
     mmcqd/0-917    [000] d.h4   413.431967: __tasklet_schedule <-dw_mci_cmd_interrupt
     mmcqd/0-917    [000] d.h4   413.431967: __raise_softirq_irqoff <-__tasklet_schedule
     mmcqd/0-917    [000] d.h4   413.431967: add_interrupt_randomness <-handle_irq_event_percpu
     mmcqd/0-917    [000] d.h4   413.431967: read_current_timer <-add_interrupt_randomness
     mmcqd/0-917    [000] d.h4   413.431967: note_interrupt <-handle_irq_event_percpu
     mmcqd/0-917    [000] d.h4   413.431967: preempt_count_add <-handle_irq_event
     mmcqd/0-917    [000] d.h5   413.431967: gic_eoi_irq <-handle_fasteoi_irq
     mmcqd/0-917    [000] d.h5   413.431967: preempt_count_sub <-handle_fasteoi_irq
     mmcqd/0-917    [000] d.h4   413.431967: irq_exit <-__handle_domain_irq
     mmcqd/0-917    [000] d.h4   413.431967: preempt_count_sub <-irq_exit
     mmcqd/0-917    [000] ..s4   413.431967: tasklet_action <-__do_softirq
```

Figure 14.1 – Ftrace function tracer snapshot

The printout starts with information pertaining to the number of entries in the buffer and the total number of entries written. The difference between these two numbers is the number of events lost while filling the buffer. Then, there's a list of functions that includes the following information:

- The process name (TASK).

- The process identifier (PID).

- The CPU the process runs on (CPU#).

- The function start time (TIMESTAMP). This timestamp is the time since boot.

- The name of the function being traced (FUNCTION) and the parent function that was called following the < - symbol. For example, in the first line of our output, the irq_may_run function was called by handle_fasteoi_irq.

Now that we are familiar with the function tracer and its specificities, we can learn about the next tracer, which is more feature-rich and provides much more tracing information, such as the call graph.

The function_graph tracer

The function_graph tracer works just like a function, but in a more detailed manner: the entry and exit point is shown for each function. With this tracer, we can trace functions with subcalls and measure the execution time of each function.

Let's edit the script from our previous example:

```
# cd /sys/kernel/debug/tracing
# echo function_graph > current_tracer
# echo 1 > tracing_on
# sleep 1
# echo 0 > tracing_on
# less trace
```

After running this script, we get the following printout:

```
# tracer: function_graph
#
# CPU  DURATION                  FUNCTION CALLS
# |     |   |                      |   |   |   |
  5)   0.400 us    |                } /* set_next_buddy */
  5)   0.305 us    |                __update_load_avg_se();
```

```
5)    0.340 us   |                    __update_load_avg_cfs_rq();
5)               |                    update_cfs_group() {
5)               |                      reweight_entity() {
5)               |                        update_curr() {
5)    0.376 us   |                          __calc_delta();
5)    0.308 us   |                          update_min_vruntime();
5)    1.754 us   |                        }
5)    0.317 us   |                        account_entity_dequeue();
5)    0.260 us   |                        account_entity_enqueue();
5)    3.537 us   |                      }
5)    4.221 us   |                    }
5)    0.261 us   |                    hrtick_update();
5) + 16.852 us   |                  } /* dequeue_task_fair */
5) + 23.353 us   |                } /* deactivate_task */
5)               |                pick_next_task_fair() {
5)    0.286 us   |                  update_curr();
5)    0.271 us   |                  check_cfs_rq_runtime();
5)               |                  pick_next_entity() {
5)    0.441 us   |                    wakeup_preempt_entity.isra.77();
5)    0.306 us   |                    clear_buddies();
5)    1.645 us   |                  }
 ------------------------------------------------
5) SCTP ti-27174  =>  Composi-2089
 ------------------------------------------------
5)    0.632 us   |                    __switch_to_xtra();
5)    0.350 us   |                    finish_task_switch();
5) ! 271.440 us   |                } /* schedule */
5)               |                _cond_resched() {
5)    0.267 us   |                  rcu_all_qs();
5)    0.834 us   |                }
5) ! 273.311 us   |              } /* futex_wait_queue_me */
```

In this graph, DURATION shows the time spent running a function. Pay careful attention to the points marked by the + and ! symbols. The plus sign (+) means the function took more than 10 microseconds, while the exclamation point (!) means it took more than 100 microseconds. Under FUNCTION_CALLS, we find information pertaining to each function call. The symbols used to show the initiation and completion of each function are the same as in the C programming language: bracers ({ }) demarcate functions, one at the start and one at the end; leaf functions that don't call any other function are marked with a semicolon (;).

Ftrace also allows tracing to be restricted just to functions that exceed a certain amount of time, using the tracing_thresh option. The time threshold at which the functions should be recorded must be written in that file in microsecond units. This can be used to find routines that are taking a long time in the kernel. It may be interesting to use this at kernel startup, to help optimize boot-up time. To set the threshold at startup, you can set it in the kernel command line as follows:

```
tracing_thresh=200 ftrace=function_graph
```

This traces all functions taking longer than 200 microseconds (0.2 ms). You can use any duration threshold you want.

At runtime, you can simply execute echo 200 > tracing_thresh.

Function filters

Pick and choose what functions to trace. It goes without saying that fewer functions to trace equals less overhead. The Ftrace printout can be big, and finding exactly what you're looking for can be extremely difficult. However, we can use filters to simplify our search: the printout will only display information about the functions we're interested in. To do this, we just have to write the name of our function in the set_ftrace_filter file, as follows:

```
# echo kfree > set_ftrace_filter
```

To disable the filter, we add an empty line to this file:

```
# echo  > set_ftrace_filter
```

We run the following command:

```
# echo kfree > set_ftrace_notrace
```

The result is the opposite: the printout will give us information about every function except `kfree()`. Another useful option is `set_ftrace_pid`. This tool is for tracing functions that can be called on behalf of a particular process.

Ftrace has many more filtering options. For a more detailed look at these, you can read the official documentation available at `https://www.kernel.org/doc/Documentation/trace/ftrace.txt`.

Tracing events

Before introducing trace events, let's talk about **tracepoints**. Tracepoints are special code inserts that trigger system events. Tracepoints may be dynamic (meaning they have several checks attached to them) or static (no checks attached).

Static tracepoints do not affect the system in any way; they just add a few bytes for the function call at the end of the instrumented function and add a data structure in a separate section. Dynamic tracepoints call a trace function when the relevant code fragment is executed. Tracing data is written to the ring buffer. Tracepoints can be included anywhere in code. In fact, they can already be found in a lot of kernel functions. Let's look at the `kmem_cache_free` function excerpt from `mm/slab.c`:

```
void kmem_cache_free(struct kmem_cache *cachep, void *objp)
{
    [...]
    trace_kmem_cache_free(_RET_IP_, objp);
}
```

`kmem_cache_free` is then itself a tracepoint. We can find countless more examples just by looking at the source code of other kernel functions.

The Linux kernel has a special API for working with tracepoints from the user space. In the `/sys/kernel/debug/tracing` directory, there is an `events` directory where system events are saved. These are available for tracing. System events in this context can be understood as the tracepoints included in the kernel.

A list of these can be viewed by running the following command:

```
# cat /sys/kernel/debug/tracing/available_events
mac80211:drv_return_void
mac80211:drv_return_int
mac80211:drv_return_bool
mac80211:drv_return_u32
```

```
mac80211:drv_return_u64
mac80211:drv_start
mac80211:drv_get_et_strings
mac80211:drv_get_et_sset_count
mac80211:drv_get_et_stats
mac80211:drv_suspend
[...]
```

A long list will be printed out in the console with the `<subsystem>:<tracepoint>` pattern. This is slightly inconvenient. We can print out a more structured list by using the following command:

```
# ls /sys/kernel/debug/tracing/events
block           gpio            napi            regmap          syscalls
cfg80211        header_event    net             regulator       task
clk             header_page     oom             rpm             timer
compaction      i2c             pagemap         sched           udp
enable          irq             power           signal          vmscan
fib             kmem            printk          skb             workqueue
filelock        mac80211        random          sock            writeback
filemap         migrate         raw_syscalls    spi
ftrace          module          rcu             swiotlb
```

All possible events are combined in the subdirectory by subsystem. Before we can start tracing events, we will make sure we've enabled writing to the ring buffer.

In *Chapter 1*, *Linux Kernel Concepts for Embedded Developers*, we introduced *hrtimers*. By listing the content of `/sys/kernel/debug/tracing/events/timer`, we will have timer-related tracepoints, including `hrtimer`-related ones, as follows:

```
# ls /sys/kernel/debug/tracing/events/timer
enable                  hrtimer_init        timer_cancel
filter                  hrtimer_start       timer_expire_entry
hrtimer_cancel          itimer_expire       timer_expire_exit
hrtimer_expire_entry    itimer_state        timer_init
hrtimer_expire_exit     tick_stop           timer_start
#
```

Let's now trace the access to `hrtimer`-related kernel functions. For our tracer, we'll use `nop` because `function` and `function_graph` record too much information, including event information that we're just not interested in. The following is the script we will use:

```
# cd /sys/kernel/debug/tracing/
# echo 0 > tracing_on
# echo > trace
# echo nop > current_tracer
# echo 1 > events/timer/enable
# echo 1 > tracing_on;
# sleep 1;
# echo 0 > tracing_on;
# echo 0 > events/timer/enable
# less trace
```

We first disable tracing in case it was already running. Then we clear the ring buffer data before setting the current tracer to `nop`. Next, we enable timer-related tracepoints, or should we say, we enable timer event tracing. Finally, we enable tracing and dump the ring buffer content, which looks like the following:

```
# tracer: nop
#
# entries-in-buffer/entries-written: 35988/35988   #P:8
#
#                                _-----=> irqs-off
#                               / _----=> need-resched
#                              | / _---=> hardirq/softirq
#                              || / _--=> preempt-depth
#                              ||| /     delay
#           TASK-PID   CPU#    ||||    TIMESTAMP  FUNCTION
#              | |       |     ||||       |          |
         bash-16561 [002] ....  639537.102581: hrtimer_init: hrtimer=000000002ba8a2be clockid=CLOCK_MONOTONIC mode=0x9
         bash-16561 [002] ....  639537.102582: hrtimer_init: hrtimer=000000000ded79d7 clockid=CLOCK_MONOTONIC mode=0x9
         bash-16561 [002] ....  639537.102590: hrtimer_init: hrtimer=000000003d041aad clockid=CLOCK_MONOTONIC mode=REL
        <idle>-0      [004] d.h.  639537.102680: hrtimer_cancel: hrtimer=000000007df5b21a
```

Figure 14.2 – Ftrace event tracing with the nop tracer snapshot

At the end of the printout, we'll find information about `hrtimer` function calls (here is a small section). More detailed information about configuring event tracing can be found here: `https://www.kernel.org/doc/Documentation/trace/events.txt`.

Tracing a specific process with the Ftrace interface

Using Ftrace as is lets you have tracing-enabled kernel tracepoints/functions irrespective of the process those functions run on behalf of. To trace just the kernel functions executed on behalf of a particular function, you should set the pseudo set_ftrace_pid variable to the **Process ID (PID)** of the process, which can be obtained using pgrep, for example. If the process is not already running, you can use a wrapper shell script and the exec command to execute a command as a known PID, as follows:

```
#!/bin/sh
echo $$ > /debug/tracing/set_ftrace_pid
# [can set other filtering here]
echo function_graph > /debug/tracing/current_tracer
exec $*
```

In the preceding example, $$ is the PID of the currently executing process (the shell script itself). This is set in the set_ftrace_pid variable, and then the function_graph tracer is enabled, after which this script executes the command (specified by the first argument to the script).

Assuming the script name is trace_process.sh, an example of usage could be the following:

```
sudo ./trace_command ls
```

Now we are familiar with tracing events and tracepoints. We are able to track and trace specific kernel events or subsystems. While tracing is a must in terms of kernel development, there are situations, which, sadly, affect the stability of the kernel. Such cases may require off-target analysis, which is addressed in debugging, and is discussed in the next section.

Linux kernel debugging tips

Writing the code is not always the hardest aspect of kernel development. Debugging is the real bottleneck, even for experienced kernel developers. That said, most kernel debugging tools are part of the kernel itself. Sometimes, finding where the fault originated is assisted by the kernel via messages called **Oops**. Debugging then comes down to analyzing the message.

Oops and panic analysis

Oops are messages printed by the Linux kernel when an error or an unhandled exception occurs. It tries its best to describe the exception and dumps the callstack just before the error or the exception occurs.

Take the following kernel module, for example:

```
#include <linux/kernel.h>
#include <linux/module.h>
#include <linux/init.h>

static void __attribute__ ((__noinline__)) create_oops(void) {
        *(int *)0 = 0;
}

static int __init my_oops_init(void) {
        printk("oops from the module\n");
        create_oops();
        return 0;
}
static void __exit my_oops_exit(void) {
        printk("Goodbye world\n");
}

module_init(my_oops_init);
module_exit(my_oops_exit);
MODULE_LICENSE("GPL");
```

In the preceding module code, we try to dereference a null pointer in order to panic the kernel. Moreover, we use the __noinline__ attribute in order for create_oops() not to be inlined, allowing it to appear as a separate function during disassembly and in the callstack. This module has been built and tested on both the ARM and x86 platforms. Oops messages and content will vary from machine to machine:

```
# insmod /oops.ko
[29934.977983] Unable to handle kernel NULL pointer dereference
at virtual address 00000000
[29935.010853] pgd = cc59c000
[29935.013809] [00000000] *pgd=00000000
```

```
[29935.017425] Internal error: Oops - BUG: 805 [#1] PREEMPT ARM
[...]
[29935.193185] systime: 1602070584s
[29935.196435] CPU: 0 PID: 20021 Comm: insmod Tainted: P
O    4.4.106-ts-armv7l #1
[29935.204629] Hardware name: Columbus Platform
[29935.208916] task: cc731a40 ti: cc66c000 task.ti: cc66c000
[29935.214354] PC is at create_oops+0x18/0x20 [oops]
[29935.219082] LR is at my_oops_init+0x18/0x1000 [oops]
[29935.224068] pc : [<bf2a8018>]    lr : [<bf045018>]    psr:
60000013
[29935.224068] sp : cc66dda8  ip : cc66ddb8  fp : cc66ddb4
[29935.235572] r10: cc68c9a4  r9 : c08058d0  r8 : c08058d0
[29935.240813] r7 : 00000000  r6 : c0802048  r5 : bf045000  r4
: cd4eca40
[29935.247359] r3 : 00000000  r2 : a6af642b  r1 : c05f3a6a  r0
: 00000014
[29935.253906] Flags: nZCv  IRQs on  FIQs on  Mode SVC_32  ISA
ARM  Segment none
[29935.261059] Control: 10c5387d  Table: 4c59c059  DAC:
00000051
[29935.266822] Process insmod (pid: 20021, stack limit =
0xcc66c208)
[29935.272932] Stack: (0xcc66dda8 to 0xcc66e000)
[29935.277311] dda0:                    cc66ddc4 cc66ddb8
bf045018 bf2a800c cc66de44 cc66ddc8
[29935.285518] ddc0: c01018b4 bf04500c cc66de0c cc66ddd8
c01efdbc a6af642b cff76eec cff6d28c
[29935.293725] dde0: cf001e40 cc24b600 c01e80b8 c01ee628
cf001e40 c01ee638 cc66de44 cc66de08
[...]
[29935.425018] dfe0: befdcc10 befdcc00 004fda50 b6eda3e0
a0000010 00000003 00000000 00000000
[29935.433257] Code: e24cb004 e52de004 e8bd4000 e3a03000
(e5833000)
[29935.462814] ---[ end trace ebc2c98aeef9342e ]---
[29935.552962] Kernel panic - not syncing: Fatal exception
```

Let's have a closer look at the preceding dump to understand some of the important bits of information:

```
[29934.977983] Unable to handle kernel NULL pointer dereference
at virtual address 00000000
```

The first line describes the bug and its nature, which in this case states that the code tried to dereference a NULL pointer:

```
[29935.214354] PC is at create_oops+0x18/0x20 [oops]
```

PC stands for **program counter**, which denotes the currently executed instruction address in memory. Here, we see that we were in the create_oops function, which is located in the oops module (which is listed in square brackets). The hex numbers indicate that the instruction pointer was 24 (0x18 in hex) bytes into the function, which appears to be 32 (0x20 in hex) bytes long:

```
[29935.219082] LR is at my_oops_init+0x18/0x1000 [oops]
```

LR is the link register, which contains the address to which the program counter should be set when it reaches a "return from subroutine" instruction. In other words, LR holds the address of the function that called the currently executing function (the one where PC is located). First, this means my_oops_init is the function that called the executing code. It also means that if the function in PC had returned, the next line to be executed would be my_oops_init+0x18, which means the CPU would branch at the 0x18 offset from the start address of my_oops_init:

```
[29935.224068] pc : [<bf2a8018>]     lr : [<bf045018>]     psr:
60000013
```

In the preceding line of code, pc and lr are the real hexadecimal content of PC and LR, with no symbol name shown. Those addresses can be used with the addr2line program, which is another tool we can use to find a faulty line. This is what we would see in the printout if the kernel was built with the CONFIG_KALLSYMS option disabled. We can then deduce that the addresses of create_oops and my_oops_init are 0xbf2a8000 and 0xbf045000, respectively:

```
[29935.224068] sp : cc66dda8  ip : cc66ddb8  fp : cc66ddb4
```

sp stands for **stack pointer** and holds the current position in the stack, while **fp** stands for **frame pointer** and points to the currently active frame in the stack. When a function returns, the stack pointer is restored to the frame pointer, which is the value of the stack pointer just before the function was called. The following example from Wikipedia explains it quite well:

> *For example, the stack frame of* DrawLine *would have a memory location holding the frame pointer value that* DrawSquare *uses. The value is saved upon entry to the subroutine and restored upon return:*

```
[29935.235572]  r10: cc68c9a4   r9 : c08058d0   r8 : c08058d0
[29935.240813]  r7 : 00000000   r6 : c0802048   r5 : bf045000   r4
: cd4eca40
[29935.247359]  r3 : 00000000   r2 : a6af642b   r1 : c05f3a6a   r0
: 00000014
```

The preceding is the dump of a number of CPU registers:

```
[29935.266822] Process insmod (pid: 20021, stack limit =
0xcc66c208)
```

The preceding line shows the process on behalf of which the panic occurred, which is insmod in this case, and its PID was 20021.

There are also oops where the backtrace is present, a bit like the following, which is an excerpt from the oops generated by typing echo c > /proc/sysrq-trigger:

```
[29255.091518] [<c0301780>] (sysrq_handle_crash) from [<c0302128>] (__handle_sysrq+0x98/0x134)
[29255.099903] [<c0302128>] (__handle_sysrq) from [<c030259c>] (write_sysrq_trigger+0x68/0x78)
[29255.108296] [<c030259c>] (write_sysrq_trigger) from [<c0250a40>] (proc_reg_write+0x78/0x8c)
[29255.116691] [<c0250a40>] (proc_reg_write) from [<c01fcc0c>] (__vfs_write+0x48/0xf4)
[29255.124382] [<c01fcc0c>] (__vfs_write) from [<c01fd3fc>] (vfs_write+0xbc/0x144)
[29255.131724] [<c01fd3fc>] (vfs_write) from [<c01fdbdc>] (SyS_write+0x68/0xc0)
[29255.138811] [<c01fdbdc>] (SyS_write) from [<c0107780>] (ret_fast_syscall+0x0/0x1c)
```

Figure 14.3 – Backtrace excerpt in a kernel oops

The backtrace traces the function call history before the one that generated the oops:

```
[29935.433257] Code: e24cb004 e52de004 e8bd4000 e3a03000
(e5833000)
```

Code is a hex-dump of the section of machine code that was being run at the time the oops occurred.

Trace dump on oops

When the kernel crashes, it is possible to use kdump/kexec with the crash utility in order to examine the state of the system at the point of the crash. However, this technique does not let you see what has happened prior to the event that caused the crash, which may be a good input for understanding or fixing the bug.

Ftrace is shipped with a feature that tries to address this issue. In order to enable it, you can either echo a 1 into /proc/sys/kernel/ftrace_dump_on_oops or enable ftrace_dump_on_oops in the kernel boot parameters. Having Ftrace configured along with this feature enabled will instruct Ftrace to dump the entire trace buffer to the console in ASCII format on oops or panic. Having the console output to a serial line makes debugging crashes much easier. This way, you can set everything up and just wait for the crash. Once it occurs, you'll see the trace buffer on the console. You'll then be able to trace back the events that led up to the crash. How far back you can go in tracing events depends on the size of the trace buffer, since this is what stores the event history data.

That said, dumping to the console may take a long time and it is common to shrink the trace buffer before putting everything in place, since the default Ftrace ring buffer is in excess of 1 megabyte per CPU. You can use /sys/kernel/debug/tracing/ buffer_size_kb in order to reduce the trace buffer size by writing in that file the number of kilobytes you want the ring buffer to be. Note that the value is per CPU, and not the total size of the ring buffer.

The following is an example of modifying the trace buffer size:

```
# echo 3 > /sys/kernel/debug/tracing/buffer_size_kb
```

The preceding command will shrink the Ftrace ring buffer down to 3 kilobytes per CPU (1 kb might be enough; it depends on how far you need to go back prior to the crash).

Using objdump to identify the faulty code line in the kernel module

We can use objdump to disassemble the object file and identify the line that generated the oops. We use the disassembled code to play with the symbol name and offset in order to point to the exact faulty line.

The following line will disassemble the kernel module in the oops.as file:

```
arm-XXXX-objdump -fS   oops.ko > oops.as
```

The generated output file will have content similar to the following:

```
[...]
architecture: arm, flags 0x00000011:
HAS_RELOC, HAS_SYMS
start address 0x00000000


Disassembly of section .text.unlikely:


00000000 <create_oops>:
   0:    e1a0c00d    mov    ip, sp
   4:    e92dd800    push   {fp, ip, lr, pc}
   8:    e24cb004    sub    fp, ip, #4
   c:    e52de004    push   {lr}            ; (str lr, [sp, #-4]!)
  10:    ebfffffe    bl     0 <__gnu_mcount_nc>
  14:    e3a03000    mov    r3, #0
  18:    e5833000    str    r3, [r3]
  1c:    e89da800    ldm    sp, {fp, sp, pc}


Disassembly of section .init.text:


00000000 <init_module>:
   0:    e1a0c00d    mov    ip, sp
   4:    e92dd800    push   {fp, ip, lr, pc}
   8:    e24cb004    sub    fp, ip, #4
   c:    e59f000c    ldr    r0, [pc, #12]    ; 20
 <init_module+0x20>
  10:    ebfffffe    bl     0 <printk>
  14:    ebfffffe    bl     0 <init_module>
  18:    e3a00000    mov    r0, #0
  1c:    e89da800    ldm    sp, {fp, sp, pc}
  20:    00000000    .word 0x00000000


Disassembly of section .exit.text:


00000000 <cleanup_module>:
   0:    e1a0c00d    mov    ip, sp
```

```
    4:        e92dd800      push   {fp, ip, lr, pc}
    8:        e24cb004      sub    fp, ip, #4
    c:        e59f0004      ldr    r0, [pc, #4]        ; 18
<cleanup_module+0x18>
   10:        ebfffffe      bl     0 <printk>
   14:        e89da800      ldm    sp, {fp, sp, pc}
   18:        00000016      .word 0x00000016
```

> **Important note**
>
> Enabling the debug option while compiling the module would make the debug
> info available in the .ko object. In this case, objdump -S would interpose
> the source code and assembly for a better view.

From the oops, we have seen that the PC is at create_oops+0x18, which is at the
0x18 offset from the address of create_oops. This leads us to the 18: e5833000
str   r3, [r3] line. In order to understand the line of interest to us, let's describe the
line before it, mov   r3, #0. After this line, we have r3 = 0. Back to our line of
interest, for people familiar with ARM assembly language, it means writing r3 to the
original address pointed to by r3 (the C equivalent of [r3] is *r3). Remember, this
corresponds to *(int *)0 = 0 in our code.

Summary

This chapter introduced a number of kernel debugging tips, and explained how to use
Ftrace to trace the code in order to identify strange behavior, such as time-consuming
functions and irq latencies. We covered the printing of APIs, either for core- or device
driver-related code. Finally, we learned how to analyze and debug kernel oops.

This chapter marks the end of this book and I hope you have enjoyed the journey through
this book while reading it as much as I did while writing it. I also hope that my best efforts
in imparting my knowledge throughout this book will prove useful to you.

Other Books You May Enjoy

If you enjoyed this book, you may be interested in these other books by Packt:

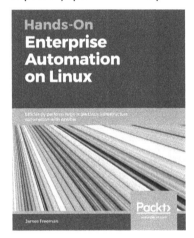

Hands-On Enterprise Automation on Linux

James Freeman

ISBN: 978-1-78913-161-1

- Perform large-scale automation of Linux environments in an enterprise
- Overcome the common challenges and pitfalls of extensive automation
- Define the business processes needed to support a large-scale Linux environment
- Get well-versed with the most effective and reliable patch management strategies
- Automate a range of tasks from simple user account changes to complex security policy enforcement
- Learn best practices and procedures to make your Linux environment automatable

Hands-On Linux for Architects

Denis Salamanca, Esteban Flores

ISBN: 978-1-78953-410-8

- Study the basics of infrastructure design and the steps involved

- Expand your current design portfolio with Linux-based solutions

- Discover open source software-based solutions to optimize your architecture

- Understand the role of high availability and fault tolerance in a resilient design

- Identify the role of containers and how they improve your continuous integration and continuous deployment pipelines

- Gain insights into optimizing and making resilient and highly available designs by applying industry best practices

Leave a review - let other readers know what you think

Please share your thoughts on this book with others by leaving a review on the site that you bought it from. If you purchased the book from Amazon, please leave us an honest review on this book's Amazon page. This is vital so that other potential readers can see and use your unbiased opinion to make purchasing decisions, we can understand what our customers think about our products, and our authors can see your feedback on the title that they have worked with Packt to create. It will only take a few minutes of your time, but is valuable to other potential customers, our authors, and Packt. Thank you!

Index

Y

www.ingramcontent.com/pod-product-compliance
Lightning Source LLC
Chambersburg PA
CBHW060920060326

40690CB00041B/2726